T5-CQR-873

# GAP JUNCTION-MEDIATED INTERCELLULAR SIGNALLING IN HEALTH AND DISEASE

The Novartis Foundation is an international scientific and educational charity (UK Registered Charity No. 313574). Known until September 1997 as the Ciba Foundation, it was established in 1947 by the CIBA company of Basle, which merged with Sandoz in 1996, to form Novartis. The Foundation operates independently in London under English trust law. It was formally opened on 22 June 1949.

The Foundation promotes the study and general knowledge of science and in particular encourages international co-operation in scientific research. To this end, it organizes internationally acclaimed meetings (typically eight symposia and allied open meetings, 15–20 discussion meetings, a public lecture and a public debate each year) and publishes eight books per year featuring the presented papers and discussions from the symposia. Although primarily an operational rather than a grant-making foundation, it awards bursaries to young scientists to attend the symposia and afterwards work for up to three months with one of the other participants.

The Foundation's headquarters at 41 Portland Place, London W1N 4BN, provide library facilities, open every weekday, to graduates in science and allied disciplines. The library is home to the Media Resource Service which offers journalists access to expertise on any scientific topic. Media relations are also strengthened by regular press conferences and book launches, and by articles prepared by the Foundation's Science Writer in Residence. The Foundation offers accommodation and meeting facilities to visiting scientists and their societies.

Information on all Foundation activities can be found at http://www.novartisfound.demon.co.uk

Novartis Foundation Symposium 219

# GAP JUNCTION-MEDIATED INTERCELLULAR SIGNALLING IN HEALTH AND DISEASE

1999

JOHN WILEY & SONS

Chichester · New York · Weinheim · Brisbane · Toronto · Singapore

Published in 1999 by John Wiley & Sons Ltd,
Baffins Lane, Chichester,
West Sussex PO19 1UD, England

National 01243 779777
International (+44) 1243 779777
e-mail (for orders and customer service enquiries): cs-books@wiley.co.uk
Visit our Home Page on http://www.wiley.co.uk
or http://www.wiley.com



*Other Wiley Editorial Offices*

John Wiley & Sons, Inc., 605 Third Avenue,
New York, NY 10158-0012, USA

WILEY-VCH Verlag GmbH, Pappelallee 3,
D-69469 Weinheim, Germany

Jacaranda Wiley Ltd, 33 Park Road, Milton,
Queensland 4064, Australia

John Wiley & Sons (Asia) Pte Ltd, 2 Clementi Loop #02-01,
Jin Xing Distripark, Singapore 129809

John Wiley & Sons (Canada) Ltd, 22 Worcester Road,
Rexdale, Ontario M9W 1L1, Canada

Novartis Foundation Symposium 219
ix+284 pages, 49 figures, 9 tables

*Library of Congress Cataloging-in-Publication Data*

Gap junction-mediated intercellular signalling in health and disease.
p. cm. – (Novartis Foundation symposium ; 219)
Organizer and editor : Gail Cardew.
Includes bibliographical references and index.
ISBN 0-471-98259-8 (hbk. : alk. paper)
1. Gap junctions (Cell biology)–Congresses. 2. Connexins–Congresses. 3. Cellular signal transduction–Congresses.
I. Cardew, Gail. II. Series.
[DNLM: 1. Cell Communication–physiology congresses. 2. Gap Junctions–physiology congresses. W1 N09285 v.204 1999 / QH 604.2 G211 1999]
QH603.C4G355 1999
611'.018–dc21
DNLM/DLC
for Library of Congress 98-46843
CIP

*British Library Cataloguing in Publication Data*

A catalogue record for this book is available from the British Library

ISBN 0 471 98259 8

Typeset in 10½ on 12½ pt Garamond by Dobbie Typesetting Limited, Tavistock, Devon.
Printed and bound in Great Britain by Biddles Ltd, Guildford and King's Lynn.
This book is printed on acid-free paper responsibly manufactured from sustainable forestry, in which at least two trees are planted for each one used for paper production.

# Contents

*Symposium on Gap junction-mediated intercellular signalling in health and disease, held at the Novartis Foundation, London 2–5 March 1998*

*This symposium is based on a proposal made by David Becker and Gail Cardew*

*Organizer and editor: Gail Cardew*


**N. B. Gilula** Introduction 1

**N. M. Kumar** Molecular biology of the interactions between connexins 6
*Discussion* 16

**V. M. Unger, N. M. Kumar, N. B. Gilula** and **M. Yeager** Electron cryo-crystallography of a recombinant cardiac gap junction channel 22
*Discussion* 30

**General discussion I** Channel rectification 38
The structural diversity of connexins 41

**W. H. Evans, S. Ahmad, J. Diez, C. H. George, J. M. Kendall** and **P. E. M. Martin** Trafficking pathways leading to the formation of gap junctions 44
*Discussion* 54

**A. Warner** Interactions between growth factors and gap junctional communication in developing systems 60
*Discussion* 72

**K. Willecke, S. Kirchhoff, A. Plum, A. Temme, E. Thönnissen** and **T. Ott** Biological functions of connexin genes revealed by human genetic defects, dominant negative approaches and targeted deletions in the mouse 76
*Discussion* 88

**J. Kistler, J. S. Lin, J. Bond, C. Green, R. Eckert, R. Merriman, M. Tunstall** and **P. Donaldson** Connexins in the lens: are they to blame in diabetic cataractogenesis? 97
*Discussion* 108

**D. I. Vaney** Neuronal coupling in the central nervous system: lessons from the retina 113
*Discussion* 125

**A. Forge, D. Becker, S. Casalotti, J. Edwards, W. H. Evans, N. Lench** and **M. Souter** Gap junctions and connexin expression in the inner ear 134
*Discussion* 151

**B. Nadarajah** and **J. G. Parnavelas** Gap junction-mediated communication in the developing and adult cerebral cortex 157
*Discussion* 170

**S. S. Scherer, L. J. Bone, S. M. Deschênes, A. Abel, R. J. Balice-Gordon** and **K. H. Fischbeck** The role of the gap junction protein connexin32 in the pathogenesis of X-linked Charcot-Marie-Tooth disease 175
*Discussion* 185

**N. J. Severs** Cardiovascular disease 188
*Discussion* 206

**C. Dasgupta, B. Escobar-Poni, M. Shah, J. Duncan** and **W. H. Fletcher** Misregulation of connexin43 gap junction channels and congenital heart defects 212
*Discussion* 222

**D. A. Goodenough, A. M. Simon** and **D. L. Paul** Gap junctional intercellular communication in the mouse ovarian follicle 226
*Discussion* 235

**H. Yamasaki, Y. Omori, V. Krutovskikh, W. Zhu, N. Mironov, K. Yamakage** and **M. Mesnil** Connexins in tumour suppression and cancer therapy 241
*Discussion* 254

Summary 261

Index of contributors 274

Subject index 276

# Participants

**D. Becker** Department of Anatomy and Developmental Biology, University College London, Gower Street, London WC1E 6BT, UK

**E. C. Beyer** Department of Pediatrics, Section of Pediatric Hematology/Oncology and Stem Cell Transplantation, University of Chicago Children's Hospital, 5841 South Maryland Avenue, M/C 4060, Chicago, IL 60637, USA

**R. Dermietzel** Abt Neuroanatomie und Molekulare Hirnforschung, Ruhr-Universität Bochum, Universitätstrasse 150, 44801 Bochum, Germany

**W. H. Evans** Department of Medical Biochemistry, University of Wales College of Medicine, Heath Park, Cardiff CF4 4XN, UK

**W. H. Fletcher** Department of Anatomy, Loma Linda University School of Medicine, 11201 Benton Street, Loma Linda, CA 92357, USA

**A. Forge** Institute of Laryngology and Otology, University College London, 330–332 Gray's Inn Road, London WC1X 8EE, UK

**C. Giaume** Collège de France, INSERM U 114, Chaire Neuropharmacologie, 11 Place Marcelin-Berthelot, 75231 Paris Cedex 05, France

**N. B. Gilula** (*Chair*) Department of Cell Biology, The Scripps Research Institute, 10550 North Torrey Pines Road, La Jolla, CA 92037, USA

**D. A. Goodenough** Department of Cell Biology, Harvard Medical School, 240 Longwood Avenue, Boston, MA 02115, USA

**C. R. Green** Department of Anatomy with Radiology, School of Medicine, University of Auckland, Auckland, New Zealand

**S. Jamieson** (*Bursar*) University Department of Surgery, Western Infirmary, 44 Church Street, Glasgow G11 6NT, UK

**J. Kistler** School of Biological Sciences, University of Auckland, Auckland, New Zealand

**N. M. Kumar** Department of Cell Biology, The Scripps Research Institute, 10550 North Torrey Pines Road, La Jolla, CA 92037, USA

**A. F. Lau** Cancer Research Centre Hawaii, Department of Molecular Carcinogenesis, 1236 Lauhala Street, Room 308, Honolulu, HI 96813, USA

**N. Lench** Molecular Medicine Unit, University of Leeds, St. James' University Hospital, Leeds LS9 7TF, UK

**C.W. Lo** University of Pennsylvania, Department of Biology, Philadelphia, PA 19104, USA

**L. Musil** Vollum Institute for Advanced Biomedical Research, Oregon Health Services University, Vollum L474, 3181 SW Sam Jackson Park Road, Portland, OR 97201, USA

**B. J. Nicholson** SUNY at Buffalo, Department of Biological Sciences, Buffalo, NY 14260, USA

**J. G. Parnavelas** Department of Anatomy and Developmental Biology, University College London, Gower Street, London WC1E 6BT, UK

**M. J. Sanderson** Department of Physiology, University of Massachusetts Medical School, 55 Lake Avenue North, Worcester, MA 01655, USA

**S. S. Scherer** Department of Neurology, University of Pennsylvania School of Medicine, Philadelphia, PA 19104, USA

**N. J. Severs** National Heart and Lung Institute, Imperial College of Science, Technology and Medicine, Royal Brompton Hospital, Sydney Street, London SW3 6NP, UK

**D. I. Vaney** Vision, Touch and Hearing Research Centre, Department of Physiology and Pharmacology, University of Queensland, Brisbane 4072, Queensland, Australia

**A. Warner** Department of Anatomy and Developmental Biology, University College London, Gower Street, London WC1E 6BT, UK

**R. Werner** University of Miami, School of Medicine, Department of Biochemistry and Molecular Biology, PO Box 016129, Miami, FL 33101, USA

**K. Willecke** Institut für Genetik, Abt. Molekulargenetik, Universität Bonn, Römerstrasse 164, D-53117 Bonn, Germany

**H. Yamasaki** Multistage Carcinogenesis Unit, International Agency for Research on Cancer, 150 Cours Albert Thomas, F-69372, Lyon Cedex 08, France

**M. Yeager** Department of Cell Biology, The Scripps Research Institute, 10550 North Torrey Pines Road, La Jolla, CA 92037, USA

# Introduction

Norton B. Gilula

*Department of Cell Biology, The Scripps Research Institute, 10550 North Torrey Pines Road, La Jolla, CA 92037, USA*

A couple of years ago, when I first heard about the planning for this symposium, my initial response was that a symposium on this topic would be premature, because at that time, our knowledge about the involvement of gap junction-mediated communication and disease was limited. However, there has been tremendous progress in this area recently, and today marks the beginning of a meeting to define exactly what we do know and what we do not know about this topical area of health-related research.

The predecessor to this symposium was held here in 1986, just after the first reports of sequences for connexin genes were made by David Paul, Nalin Kumar and Klaus Willecke (Ciba Foundation Symposium 1987). It was an exciting time because no one knew the extent of connexin diversity. Howard Evans, Anne Warner, Klaus Willecke, myself and a number of other colleagues participated in that symposium on junctional complexes of epithelial cells. It brought together a wide spectrum of people interested in junctional complexes, and it was chaired by Sir Michael Stoker, who made important contributions to understanding the biological significance of contacts between cells. The discussions that followed those presentations provided a perspective on the status of what we understood at that time, and the same will be true of this symposium. Nigel Unwin, who was also at that symposium, presented some structural information, and this symposium marks a new era in the context of understanding gap junction channel structure, well below the limit of resolution that was available at that time.

Loewenstein was amongst the first to propose that there is a close relationship between growth control and cell communication. He emphasized the contribution of direct communication between cells through specialized channels, and as we will review and consider in this symposium, some of his ideas on growth control are relevant not only to tumorigenesis and cancer, but also to developmental processes.

Gap junctions have a characteristic structural appearance that has attracted a number of investigators to this area of research. The gap junctional plaque contains a polygonal arrangement of intramembrane particles, and now our perspective about what these particles contain, in terms of molecular content, has been extremely well defined, although we still can't say much more about the particles other than that they contain connexin protein and lipids. Some of the pressing issues today relate directly to the diversity of the connexin multigene family, and how these different members are utilized to create functional diversity.

Based on the classic picture of negatively stained isolated gap junctions, it was proposed that an oligomeric arrangement of a principle protein integrated into a lipid bilayer structure created a continuous channel structure between two cells. At this symposium, we will learn that this is still an excellent model for integrating the recent molecular information that has emerged.

The connexin proteins form a multigene family, of which there are many members. The nomenclature adopted by many people in the field is that every connexin is named by its molecular weight, although difficulties with this can arise when one is trying to understand the differences between certain connexins, e.g. Cx31 and Cx31.1. In general, people who work on these different connexins appear to be comfortable with this system. However, it is also possible to classify connexins on a phylogenetic basis, i.e. into an $\alpha$ class and a $\beta$ class. This classification system appears to have some relevance for understanding how the different members of the connexin multigene family are utilized selectively to create a practical diversity for producing channels that have different properties between cells. Why there are so many members and how they are utilized by individual cells will be an important issue to address at this symposium. It is clear that most cells use more than one connexin at the same time, and it is also clear from several recent studies that multiple connexin gene products can be associated into an oligomer to form an individual channel. What remains poorly understood are the rules that govern the association properties of the connexins. In many cases when we examine the expression of connexin genes by individual cells, it is difficult to identify all the connexin genes that might be used by that cell at a given time, and whether at any time thereafter the cell is capable of utilizing other connexin genes. These issues are prominent to those of us who are trying to understand how connexin gene diversification might influence the functional considerations that take place during the development of an organism from an oocyte. Also, in a stable differentiated cell type what are the changes in utilization of connexin genes that contribute to the loss of certain control functions? Hopefully, by the end of this symposium we will have a better understanding of some of the principles involved in determining how the different members of this multigene family are utilized.

Much of the information that led to the genesis of this symposium came from the recent finding of mutations in connexin genes and their involvement in certain health conditions. There are a number of observations implicating connexins in certain disease conditions that have been derived from the identification of changes in connexin coding sequences that result from mutations or other events. In this context, targeted gene disruptions, i.e. knockouts, in animal model systems have been an effective way to try to understand how the gap junction genes are contributing as a general unit to tissue and organ function.

There are also issues of regulation that are important to understand when considering the relationship of gap junction-mediated communication to health and disease. There are two principle kinds of regulatory events, one of which is being studied by Alan Lau which is focused on trying to understand post-translational modifications and what happens in a particular cell type when those events are altered or defective. The second regulatory event concerns the role of bioactive messengers, i.e. small molecules that can promote the regulation of multicellular organization and function in development and in adult tissues, and what happens when these regulatory events are abnormal.

I would like to mention a couple of recent studies from our own laboratory that illustrate the progress that is being make in this health-related area. The first has been published recently by Gong et al (1997). When the gene encoding a Cx46 ($\alpha_3$) connexin product used in the lens fibre region is knocked out in mice, the homozygote mice develop a lens phenotype that can be characterized as a nuclear age-dependent cataract. Recently, in collaboration with Rick Mathias, we examined the electrical coupling properties in these lenses, and we observed that the nuclear region of these lenses lacks detectable electrical coupling. Therefore, in this experimental model system there is direct evidence for an abnormal phenotype associated with a region in which there is a complete loss of gap junctional physiological properties. This is not necessarily an example of what one would expect to find when a connexin gene is mutated because this is a targeted gene deletion. Nevertheless, it is giving us some clues about the molecular basis for changes in the organization and properties of the soluble proteins in the lens, in this case the crystallins, that occur as a result of disrupting the communication pathway.

Another approach for studying the contribution of connexins to health-related processes is to make site-specific mutations in connexin sequences, and to then reintroduce them back into cells or organisms. In general, this approach has demonstrated that different phenotypes exist that are dependent on the mutation. The phenotypes range from defects in junctional connexin assembly, to trafficking, to function at the cell surface. These phenotypes may have relevance to many of the results obtained *in vivo* that will be discussed at this symposium, including the Charcot-Marie-Tooth syndrome.

Bioactive regulatory substances are associated with many different cellular processes. One of these bioactive substances was identified on the basis of its sleep-induction properties by Cravatt et al (1995). This substance is referred to as a fatty acid amide or oleamide. When it is injected into animals, it induces normal sleep, as opposed to drug-induced sleep. Recently, we reported that this molecule, along with another related molecule, anandamide, identified by Christian Giaume (Venance et al 1995), can promote the blockage of gap junction channels (Guan et al 1997). In our hands oleic acid does not block the channels, although Janis Burt (Hirschi et al 1993) has reported that it can cause the uncoupling of heart cells. However, this is a long-term effect that appears to be caused by a dietary change in the plasma membrane of the heart cells that leads to changes in the cell-to-cell conductance properties. Oleamide and anandamide have amide groups and they also contain the *cis*-oriented hydrocarbon tail that appears to be crucial for the uncoupling properties. What is so exiting about some of these bioactive lipids, including the anandamide that is thought to be the natural ligand for the cannabinoid receptor, is that they do not necessarily block channels between all cells. This would be beneficial for those of us who have high circulating levels of these molecules in our bodies; in particular, to avoid cardiac arrest. On the other hand, they could have some profound regulatory influences under normal conditions. For example, in the nervous system these molecules could determine whether or not the multicellular associations between neurons and neurons, neurons and glial cells, or vice versa can take place in a normal manner. It is also interesting to note that molecules such as these are closely related chemically to molecules that are used today on humans as anti-convulsants. Therefore, there is considerable interest in defining whether these bioactive lipids are the natural ligands for modulating and influencing connections between cells in the nervous system under normal conditions, and whether under abnormal conditions they influence the resultant pattern of behaviour.

Over the next few days I am going to ask those of you who have been involved in some of these issues to contribute as best you can, and to create a constructive dialogue. We will be trying to understand any contradictions or inconsistencies that may exist, and to consider how to relate gap junction-mediated communication to normal conditions and disease processes. In the interest of developing an optimal environment for this constructive dialogue, it will be imperative for us all to leave our egos outside while we participate in the process.

## References

Ciba Foundation 1987 Junctional complexes of epithelial cells. Wiley, Chichester (Ciba Found Symp 125)

Cravatt BF, Prospero-Garcia O, Siuzdak G et al 1995 Chemical characterization of a family of brain lipids that induce sleep. Science 268:1506–1509

Gong X, Li E, Klier G et al 1997 Disruption of $\alpha_3$ connexin gene leads to proteolysis and cataractogenesis in mice. Cell 91:833–843

Guan X, Cravatt BF, Ehring GR et al 1997 The sleep-inducing lipid oleamide deconvolutes gap junction communication and calcium wave transmission in glial cells. J Cell Biol 139: 1785–1792

Hirschi KK, Minnich BN, Moore LK, Burt JM 1993 Oleic acid differentially affects gap junction-mediated communication in heart and vascular smooth muscle cells. Am J Physiol Cell Physiol 265:C1517–C1526

Venance L, Piomelli D, Glowinski J, Giaume C 1995 Inhibition by anandamide of gap junctions and intercellular calcium signaling in striatal astrocytes. Nature 376:590–594

# Molecular biology of the interactions between connexins

Nalin M. Kumar

*Department of Cell Biology, The Scripps Research Institute, 10550 North Torrey Pines Road, La Jolla, CA 92037, USA*

*Abstract.* The protein structural component of gap junctions is the connexin. Studies on the association properties of the connexins to form heteromeric connexons and heterotypic gap junctions are necessary for a complete understanding of the role of different connexins in gap junction function. The connexins are coded by a multigene family consisting of at least 16 members. Most cells express multiple types of connexin that can potentially associate to form gap junction channels containing more than one type of connexin. The permeability and gating characteristics of gap junction channels are dependent on the isoform and post-translational modifications present on the connexins and their association properties. Together with an observed selectivity in the association properties of the different connexins and the development of more specific perturbation approaches, these studies have provided insights into the significance of connexin diversity and the temporal expression patterns for connexins that have been determined *in vivo* in both developmental and differentiating systems.

*1999 Gap junction-mediated intercellular signalling in health and disease. Wiley, Chichester (Novartis Foundation Symposium 219) p 6–21*

It has been more than 30 years since the presence of a low resistance electrical pathway between adjacent crayfish axons was first described (Furshpan & Potter 1959). This was followed by the finding that these same pathways also existed in non-excitable cells, and that they allowed the passage of larger hydrophilic molecules without leakage into the extracellular space. These observations suggested that the channel responsible for these pathways must exist between the two cell membranes and be continuous. The membrane differentiation that was implicated in this pathway was ultimately identified as the gap junction.

## Connexin multigene family

The gap junction was first clearly observed in its present form by electron microscopy as a septalaminar structure with a 2–4 nm space or 'gap' between two opposing cells. In *en face* views, a polygonal lattice of about 7.5 Å subunits could be visualized by using electron-opaque tracers. This arrangement was also evident in freeze-fracture replicas in which the particles on the inner membrane half (fracture face P) and a complimentary arrangement of pits on the outer membrane half (fracture face E) were observed. These structural features of the gap junction are conserved over a wide range of organisms, with the exception of arthropods where the intramembrane particles are located on fracture face E.

The early electrophysiological and morphological studies established some basic criteria that would need to be met in order for a polypeptide to be considered as forming the gap junctional channel. These include:

(a) the presence of hydrophobic regions that would enable it to exist in the hydrophobic environment of the lipid bilayer;
(b) the presence of at least one membrane-spanning domain containing polar residues that would allow for the passage of hydrophilic molecules;
(c) the formation of a channel between coupled cells by a minimum of two polypeptides, each of which is found in the membrane of two adjoining cells; and
(d) a high degree of homology among the gap junction polypeptides in each of the cells since the gross structural features were the same in each cell.

In addition, at least one of the gap junction polypeptides has to:

(e) have the capacity to form channels capable of transferring molecules larger than ions but smaller than 1500 Da; and
(f) be capable of forming gap junction plaques that can be identified by electron microscopic criteria, including negative staining and freeze fracture.

Strategies developed for the isolation of enriched subcellular fractions of gap junctions suggested that the gap junctions were formed by the associations of a single major polypeptide. Significant progress was subsequently made in characterizing this polypeptide at a molecular level. This effort resulted in the identification of the gene family, connexins, that are the protein structural components of gap junctions. The evidence that connexins are the major integral membrane protein present at the gap junction include the following: (1) the existence of putative transmembrane regions within the sequence (Milks et al 1988, Kumar & Gilula 1992); (2) solubilization by detergents and resistance to extraction under biochemical conditions that do not substantially disrupt the bilayer (Hertzberg 1984); (3) accessibility of predicted extracellular and cytoplasmic regions to site-specific antibodies and proteases (Zimmer et al 1987, Milks et al 1988, Yeager & Gilula 1992); and (4) high resolution structural studies of recombinant connexins (Unger et al 1997).

Several lines of evidence were consistent with the connexins mediating the functions of the gap junction. First, connexins are found in tissues known to contain functional gap junctions (e.g. the liver and heart; Paul 1986, Kumar & Gilula 1986, Beyer et al 1987). Furthermore, electron microscope localization studies using antibodies against peptides derived from the amino acid sequence confirmed that connexins were a component of gap junctions (Milks et al 1988, Hülser et al 1997). Finally, expression of connexin cDNAs in heterologous expression systems resulted in the formation of gap junctions that could be demonstrated by structural and/or functional criteria (Swenson et al 1989, Eghbali et al 1990, Naus et al 1993, Kumar et al 1995, Elfgang et al 1995). Thus, this family of proteins satisfied the criteria given above, and the structural, biochemical and electrophysiological data conclusively demonstrated that the connexins are the protein structural component of the gap junction channel.

Following the identification of the first gap junction gene product, an entire multigene family of connexins has been described (Bruzzone et al 1996, Kumar & Gilula 1996). In this multigene family two broad classes, termed $\alpha$ and $\beta$, have been proposed (Kumar & Gilula 1992). Of the 16 connexin genes so far identified in mammals, nine belong to the $\alpha$ class and six to the $\beta$ class. The members of this multigene family are described either with Greek ($\alpha$ and $\beta$) designations or by the deduced molecular size of the gene products (Table 1).

In addition, connexin cDNA has been identified from a primitive vertebrate, the skate, that encodes a protein of deduced molecular mass 35 kDa, and is hence termed connexin 35 (Cx35; O'Brien et al 1996). Recently, the homologue to Cx35 has been identified in mouse neurons (Cx36), and codes for a protein of molecular mass 36 kDa (Condorelli et al 1998). In both cases amino acid alignment indicated that this connexin falls near the divergence point between the $\alpha$ and $\beta$ groups. The gene structure for these two connexins was also unusual since, unlike the other 16, it contained an intron within the coding region, whereas other connexin genes that have been characterized contain a single intron in the 5′ non-translated region. The unusual gene structure of the Cx36 gene, as well as its divergence from the $\alpha$ and $\beta$ classes of connexins, suggests that this connexin may belong to a third class of connexins (O'Brien et al 1996).

Connexins have complex tissue distribution patterns that are temporally and spatially regulated during development and differentiation. At present, the factor(s) that govern the distribution of connexins in various organs and cell types is unknown. Most cells express more than one connexin, leading to the possibility that heteromeric connexons (composed of two or more connexins) as well as heterotypic associations may exist *in vivo*. This possibility is supported by the observation that multiple connexins can coexist in the same junctional plaque (Nicholson et al 1987, Paul et al 1991, Risek et al 1994).

**TABLE 1 Connexin multigene family**

| *Greek letter nomenclature* | *Molecular mass nomenclature* | *Predicted molecular mass (kDa)* | *Human chromosome location* | *Examples of organs with expression* |
|---|---|---|---|---|
| $\alpha_1$ | Cx43 | 43 | 6q21-q23.2 | Heart |
| $\alpha_2$ | Cx38 | 37.8 | | Embryo |
| $\alpha_3$ | Cx46 | 46 | 13q11-q12 | Lens |
| $\alpha_4$ | Cx37 | 37.2 | 1p35.1 | Endothelium |
| $\alpha_5$ | Cx40 | 40.6 | 1q21.1 | Lung |
| $\alpha_6$ | Cx45 | 45.5 | | Heart |
| $\alpha_7$ | Cx33 | 32.9 | | Testes |
| $\alpha_8$ | Cx50 | 48.2 | 1q21.1 | Lens |
| $\alpha_9$ | Cx60 | 60 | | Ovary |
| $\alpha_{10}$ | Cx46.6 | 46.6 | 1q41-q42 | Cochlea |
| $\beta_1$ | Cx32 | 32 | Xq13.1 | Liver |
| $\beta_2$ | Cx26 | 26.2 | 13q11-q12 | Liver |
| $\beta_3$ | Cx31 | 31 | | Skin |
| $\beta_4$ | Cx31.1 | 31.1 | | Skin |
| $\beta_5$ | Cx30.3 | 30.3 | | Skin |
| $\beta_6$ | Cx30 | 30.4 | | Skin |

## Connexin structure

The deduced amino acid sequences suggest the presence of four transmembrane domains that would enable the polypeptide to exist in the hydrophobic environment of the lipid bilayer. Topological analysis of connexins using site-specific antibodies, protease digestions and structural considerations (Milks et al 1988, Yeager & Gilula 1992) indicates that for all members of this family, both amino and carboxyl termini are located on the cytoplasmic surface with two external loops, E1 and E2, between the transmembrane domains directed towards the extracellular space or gap. Further, one of these domains (M3) has an amphipathic character consistent with its potential contribution to the lining of the channel. These features of the gap junction channel are supported by high resolution electron microscopic analysis of truncated, recombinant Cx43 ($\alpha_1$; Unger et al 1997, 1999, this volume).

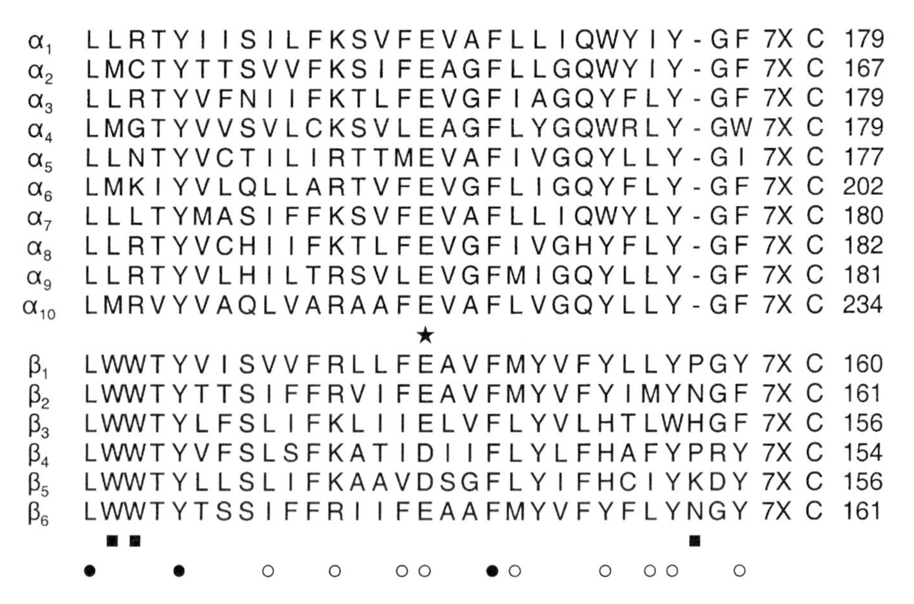


FIG. 1. Alignment of amino acid sequences of rodent connexins in the putative channel-forming domain, M3, extending into the second extracellular domain, E2. Closed squares represent residues conserved in the $\beta$ class of connexins; closed circles represent those residues conserved in all connexins; and open circles represent conservative substitutions. The star denotes the acidic residue present in M3.

Alignment of the amino acid sequence for the different connexins of the putative pore-forming domains (M3) extending into the second extracellular loop (E2) is shown in Fig. 1. A high degree of similarity can be observed in this region, as well as in the first extracellular region of the different connexins. Some notable differences are also seen, however, between the proposed classes. For example, the number of amino acids between the conserved acidic residue in M3 (indicated by a star) and the first cysteine residue in the second extracellular loop is one amino acid shorter in the $\alpha$ class when compared to the $\beta$ class of connexins. In addition, other conserved residues and motifs, such as the predominance of a GQ pair, in this region of the $\alpha$ class of connexins when compared to the $\beta$ class of connexins can also be discerned. The significance of these conserved residues in this region of the connexin and their role in the structure and function of gap junctions is not understood but may contribute to the selectivity in the association between the different connexins.

## Interaction of connexins

Perturbation and mutagenesis approaches have suggested that the two extracellular loops participate in the interactions that result in the gap junctional contact between two adjacent membranes (Dahl et al 1992, Meyer et al 1992, White et al 1994). Studies using heterologous expression systems have indicated that connexins show a specificity in their interactions to form a functional gap junction channel and that the specificity domain resides in both the second extracellular domain (E2) as well as in specific intracellular domains (Bruzzone et al 1994, White et al 1994, 1995, Haubrich et al 1996).

In one approach, for studying the oligomerization properties of connexin, rodent gap junction plaques have been analysed by scanning transmission electron microscopy. The results have been interpreted to indicate that homomeric connexons and heterotypic gap junction channels exist in mouse and rat liver (Sosinsky 1995). Some of these homomeric connexons may be derived from the differential distribution of these two connexins in rodent livers (Traub et al 1989).

Evidence for the oligomerization of different connexins into heteromeric connexons has also been reported (Stauffer 1995, Cascio et al 1995, Jiang & Goodenough 1996). In these studies, gap junctions that have been dissociated by detergents have been utilized as starting material. That heteromeric connexins exist requires a number of criteria to be satisfied, if connexons are used that have been prepared from solubilized (dissociated) gap junctions. These include the following: (a) the original interactions present within the connexon must be maintained, and subunit exchange must not occur under conditions of solubilization; (b) individual connexons, rather than paired connexons or aggregates, are present after solubilization; (c) heteromeric connexons can be unequivocally distinguished biochemically from homomeric connexons; and (d) there is no cross-reactivity of the antibodies under the conditions used.

The above criteria have been satisfied in an approach that we have used in our laboratory. In this approach beads to which connexin antibodies are attached have been used to separate mixtures of connexons derived either from *in vivo* sources or from expression of connexin cDNA in heterologous systems (N.M. Kumar, unpublished results 1998). For example, using rat liver as a source of solubilized gap junctions, both $\beta_1$ (Cx32) and $\beta_2$ (Cx26) connexin were recovered from material bound to the $\beta_1$ antibody beads. These observations are consistent with the formation of heteromeric connexons containing $\beta_1$ and $\beta_2$ connexin in the liver. Furthermore, a similar analysis with cells coexpressing $\alpha_1$ together with $\beta_1$ or $\beta_2$ connexin indicated that $\alpha_1$ does not form a hetero-oligomer with either $\beta_1$ or $\beta_2$ connexin. These results suggest that there may be specificity in the interactions of different connexins to form a connexon (Table 2).

**TABLE 2 Specificity in the formation of heteromeric connexons**

| *Connexin subtype* | $\alpha_1$ | $\alpha_3$ | $\beta_1$ | $\beta_2$ |
|---|---|---|---|---|
| $\alpha_1$ | Y | Y | N | N |
| $\alpha_3$ | Y | Y | N | N |
| $\beta_1$ | N | N | Y | Y |
| $\beta_2$ | N | N | Y | Y |

Y represents the interaction of connexins as determined by immunoaffinity approaches; and N represents the lack of interactions as determined by immunoaffinity approaches.

Further support for this specificity is provided by studies on the co-localization of different connexins within cells. For example, an immunofluorescence study of the thyroid gland indicated the differential distribution of $\beta_1$ and $\alpha_1$ connexin (Guerrier et al 1995). In this study, $\beta_1$ connexin was detected in lateral membranes whereas $\alpha_1$ connexin was detected in subapical regions of the cells, suggesting that there was no interaction between these two connexins. A similar conclusion can be drawn from studies examining the localization of $\alpha_1$ and $\beta_2$ connexins in a single cell (Spray et al 1991, Laird et al 1992).

## Dominant negative inhibition of connexins

The specificity in the interaction of connexins to form a connexin oligomer is also supported by studies using dominant negative connexin constructs. In one such study, certain mutants of $\beta_1$ connexin could act as dominant negative inhibitors of $\beta_2$, but not $\alpha_5$ (Cx40), function when analysed in a *Xenopus* oocyte system (Bruzzone et al 1994). In our laboratory, mutagenesis studies have also been performed in which site-directed mutations or truncation of $\alpha_1$ or $\beta_1$ connexin cDNA were generated and expressed in BHK cells. These studies defined domains that are important for the transport and function of these gap junction proteins. Non-conservative substitutions in the different domains of the connexins frequently led to defects in their intracellular transport. For example, truncation of $\alpha_1$ connexin at amino acid 263 still enabled the formation of functional gap junctions on the surface of the cell, whereas truncation at amino acid 231 led to accumulation in the cytoplasm of the cell. Thus, the region between amino acid 231 and amino acid 263 is critical for the transport of $\alpha_1$ connexin to the cell surface to produce functional channels.

Many of these transport-defective mutant connexins will act as dominant negative mutant proteins when coexpressed with normal, wild-type connexins. Analysis of doubly infected BHK cells containing wild-type $\alpha_1$ with $\beta_1$ or $\beta_2$ were found to express both sets of connexins on the cell surface (N. M. Kumar, unpublished observations 1998). Similarly, when the $\alpha_1$ truncated at amino acid 263 was coexpressed with wild-type $\alpha_1$ connexin, gap junctions could be detected on the cell surface. However, cell surface gap junctions were significantly reduced in frequency when the $\alpha_1$ connexin truncated at amino acid 231, was coexpressed with wild-type $\alpha_1$ connexin. These immunofluorescence results were obtained with two different antibodies and confirmed by freeze-fracture electron microscope analysis.

These results indicated that mutant connexins with transport defects can interact with wild-type connexins to prevent transport of both types of connexin to the cell surface. They appear to function as dominant negative mutations whose defects prevent normal connexins from forming functional channels. These dominant negative effects could occur by at least three mechanisms: (a) blockage of transport of wild-type connexin by oligomerization of the mutated connexin with the other connexins; (b) competition between wild-type and mutated connexons for pairing with wild-type connexons present in adjacent cells, thereby reducing the number of functional channels and; (c) formation of heteromeric connexons that contain the mutated connexin and that are transported normally to the cell surface where they may exist as functional or non-functional channels depending on the mutation.

## Summary

The molecular mechanisms responsible for the selective oligomerization of connexins to form a connexon have not yet been determined. One possibility is that selective association is an intrinsic property of the determined sequence of the connexin or related to its post-translational modifications. In ongoing studies, connexin chimeras are being generated and utilized to study these interactions (N. M. Kumar, unpublished results 1998). Preliminary findings indicate that the carboxyl half of $\beta_1$ contains a domain required for selective oligomerization. However, additional motifs in other regions of the connexin could act together with the carboxyl half to influence the heteromeric association of the connexins (Falk et al 1997).

The specific interactions between connexin subunits and the capacity of individual cells to express multiple connexins may lead to the formation of a variety of channels with differing properties. Indeed, the presence of heteromeric connexons in cells has been implicated by the determination of functional differences such as conductance and voltage sensitivity (Brink et al 1997) and permeability to biological signalling molecules (Bevans et al 1998). The potential incompatibility between certain pairs of connexons may have other consequences, such as the formation of communication barriers between groups of cells in contact. The selective association of connexins may provide an explanation for the diversity in gap junction genes and why certain connexins are utilized by specific cell types.

## References

Bevans CG, Kordel M, Rhee SK, Harris AL 1998 Isoform composition of connexin channels determines selectivity among second messengers and uncharged molecules. J Biol Chem 273:2808–2816

Beyer EC, Paul DL, Goodenough DA 1987 Connexin43: a protein from rat heart homologous to a gap junction protein from liver. J Cell Biol 105:2621–2629

Brink PR, Cronin K, Banach K et al 1997 Evidence for heteromeric gap junction channels formed from rat connexin43 and human connexin37. Am J Physiol 273:C1386–C1396

Bruzzone R, White TW, Paul DL 1994 Expression of chimeric connexins reveals new properties of the formation and gating behavior of gap junction channels. J Cell Sci 107:955–967

Bruzzone R, White TW, Paul DL 1996 Connections with connexins: the molecular basis of direct intercellular signaling. Eur J Biochem 238:1–27

Cascio M, Kumar NM, Safarik R, Gilula NB 1995 Physical characterization of gap junction membrane connexons (hemi-channels) isolated from rat liver. J Biol Chem 270:18643–18648

Condorelli DF, Parenti R, Spinella F et al 1998 Cloning of a new gap junction gene (Cx36) highly expressed in mammalian brain neurons. Eur J Neurosci 10:1202–1208

Dahl G, Werner R, Levine E, Rabadan-Diehl C 1992 Mutational analysis of gap junction formation. Biophys J 62:172–182

Eghbali B, Kessler JA, Spray DC 1990 Expression of gap junction channels in communication-incompetent cells after stable transfection with cDNA encoding connexin 32. Proc Natl Acad Sci USA 87:1328–1331

Elfgang C, Eckert R, Lichtenberg-Fraté H et al 1995 Specific permeability and selective formation of gap junction channels in connexin-transfected HeLa cells. J Cell Biol 129:805–817

Falk MM, Buehler LK, Kumar NM, Gilula NB 1997 Cell-free synthesis and assembly of connexins into functional gap junction membrane channels. EMBO J 16:2703–2716

Furshpan EJ, Potter DD 1959 Transmission at giant motor synapses of the crayfish. J Physiol (Lond) 145:289–325

Guerrier A, Fonlupt P, Morand I et al 1995 Gap junctions and cell polarity: connexin32 and connexin43 expressed in polarized thyroid epithelial cells assemble into separate gap junctions, which are located in distinct regions of the lateral plasma membrane domain. J Cell Sci 108:2609–2617

Haubrich S, Schwarz HJ, Bukauskas F et al 1996 Incompatibility of connexin 40 and 43 hemichannels in gap junctions between mammalian cells is determined by intracellular domains. Mol Biol Cell 7:1995–2006

Hertzberg EL 1984 A detergent-independent procedure for the isolation of gap junctions from rat liver. J Biol Chem 259:9936–9943

Hülser DF, Rehkopf B, Traub O 1997 Dispersed and aggregated gap junction channels identified by immunogold labeling of freeze-fractured membranes. Exp Cell Res 233:240–251

Jiang JX, Goodenough DA 1996 Heteromeric connexons in lens gap junction channels. Proc Natl Acad Sci USA 93:1287–1291

Kumar NM, Gilula NB 1986 Cloning and characterization of human and rat liver cDNAs coding for a gap junction protein. J Cell Biol 103:767–776
Kumar NM, Gilula NB 1992 Molecular biology and genetics of gap junction channels. Semin Cell Biol 3:3–16
Kumar NM, Gilula NB 1996 The gap junction communication channel. Cell 84:381–388
Kumar NM, Friend DS, Gilula NB 1995 Synthesis and assembly of human gap junctions in BHK cells by DNA transfection with the human $\beta_1$ cDNA. J Cell Sci 108:3725–3734
Laird DW, Yancey SB, Bugga L, Revel JP 1992 Connexin expression and gap junction communication compartments in the developing mouse limb. Dev Dyn 195:153–161
Meyer RA, Laird DW, Revel JP, Johnson RG 1992 Inhibition of gap junction and adherens junction assembly by connexin and A-CAM antibodies. J Cell Biol 119:179–189
Milks LC, Kumar NM, Houghten N, Unwin N, Gilula NB 1988 Topology of the 32-kD liver gap junction protein determined by site-directed antibody localizations. EMBO J 7:2967–2975
Naus CCG, Hearn S, Zhu D, Nicholson BJ, Shivers RR 1993 Ultrastructural analysis of gap junctions in C6 glioma cells transfected with connexin43 cDNA. Exp Cell Res 206:72–84
Nicholson B, Dermietzel R, Teplow D, Traub O, Willecke K, Revel JP 1987 Two homologous protein components of hepatic gap junctions. Nature 329:732–734
O'Brien J, Al-Ubaidi MR, Ripps H 1996 Connexin 35: a gap-junctional protein expressed preferentially in the skate retina. Mol Biol Cell 7:233–243
Paul D 1986 Molecular cloning of cDNA for rat liver gap junction protein. J Cell Biol 103:123–134
Paul DL, Ebihara L, Takemoto LJ, Swenson KI, Goodenough DA 1991 Connexin46, a novel lens gap junction protein, induces voltage-gated currents in nonjunctional plasma membrane of *Xenopus* oocytes. J Cell Biol 115:1077–1089
Risek B, Klier FG, Gilula NB 1994 Developmental regulation and structural organization of connexins in epidermal gap junctions. Dev Biol 164:183–196
Sosinsky G 1995 Mixing of connexins in gap junction membrane channels. Proc Natl Acad Sci USA 92:9210–9214
Spray DC, Moreno AP, Kessler JA, Dermietzel R 1991 Characterization of gap junctions between cultured leptomeningeal cells. Brain Res 568:1–14
Stauffer KA 1995 The gap junction proteins $\beta_1$-connexin (connexin-32) and $\beta_2$-connexin (connexin-26) can form heteromeric hemichannels. J Biol Chem 270:6768–6772
Swenson KI, Jordan JR, Beyer EC, Paul DL 1989 Formation of gap junctions by expression of connexins in Xenopus oocyte pairs. Cell 57:145–155
Traub O, Look J, Dermietzel R, Brümmer F, Hülser D, Willecke K 1989 Comparative characterization of the 21-kD and 26-kD gap junction proteins in murine liver and cultured hepatocytes. J Cell Biol 108:1039–1051
Unger VM, Kumar NM, Gilula NB, Yeager M 1997 Projection structure of a gap junction membrane channel at 7 Å resolution. Nat Struct Biol 4:39–43
Unger VM, Kumar NM, Gilula NB, Yeager M 1999 Electron cryo-crystallography of a recombinant cardiac gap junction channel. In: Gap junction-mediated intercellular signalling in health and disease. Wiley, Chichester (Novartis Found Symp 219) p 22–37
White TW, Bruzzone R, Wolfram S, Paul DL, Goodenough DA 1994 Selective interactions among the multiple connexin proteins expressed in the vertebrate lens: the second extracellular domain is a determinant of compatibility between connexins. J Cell Biol 125:879–892
White TW, Paul DL, Goodenough DA, Bruzzone R 1995 Functional analysis of selective interactions among rodent connexins. Mol Biol Cell 6:459–470
Yeager M, Gilula NB 1992 Membrane topology and quaternary structure of cardiac gap junction ion channels. J Mol Biol 223:929–948

Zimmer DB, Green CR, Evans WH, Gilula NB 1987 Topological analysis of the major protein in isolated intact rat liver gap junctions and gap junction-derived single-membrane structures. J Biol Chem 262:7751–7763

## DISCUSSION

*Nicholson:* You mentioned that in the liver a proportion of connexin26 (Cx26; $\beta_2$) was associated with Cx32 ($\beta_1$). Those results also suggest that there is a proportion that might not form heteromers. Is the mixing complete or partial? The latter would suggest that at least a fraction might exist in homomeric forms.

*Kumar:* This is a difficult question to answer because we don't yet know the affinities of each antibody.

*Scherer:* When Cx32 is coexpressed with Cx43 ($\alpha_1$), does Cx32 reach the plasma membrane?

*Kumar:* Yes, both Cx43 and Cx32 reach the membrane.

*Musil:* You concluded from your chimera studies that the carboxyl end might dictate whether or not different connexin species can co-oligomerize to form heteromeric connexons. Can you comment on Mathais Falk's *in vitro* studies (Falk et al 1997), in which he indicated that the amino terminus is responsible?

*Kumar:* There are several possible reasons for this apparent difference. First, they used an *in vitro* translation system, whereas our study utilized connexin expression inside the cell. Second, Falk et al (1997) showed that when there was a cleavage of the first extracellular loop there was a change in the association properties of the connexin, and they inferred from this that the amino terminus may be involved. In contrast, we made chimeras and saw no cleavage of the product, as far as we could tell. Finally, it is possible that the amino terminus can interact with the carboxyl terminus, such that both domains are responsible for heteromeric association.

*Gilula:* There is currently no conclusive understanding of the differences between what Mathias Falk has observed and what you observed. These are two completely separate sets of observations that have identified different sequence regions that might be critical for connexin association. Whether or not these different observations can be explained by the differences in the systems that are being used remains to be seen.

*Willecke:* I would like to mention that Valiunas et al (1998), have looked at wild-type hepatocytes and Cx32-deficient hepatocytes. They report that about 20% of the channels are heterotypic Cx32/Cx26 channels and <1% are homotypic Cx26 channels, but they did not find heteromeric channels. One would expect that they should have seen heteromeric channels, if these have distinct electrophysiological characteristics. If heteromeric Cx32/Cx26 channels behave like the parental channels, however, Valiunas et al (1998) would not have detected them. Therefore, it is possible that different results are obtained in the baculovirus system, where there is a forced over-expression of connexins, than in primary cells. Therefore, it is important to find out whether Cx32/Cx26 heteromeric channels exist in the murine liver and whether they are functionally different from heterotypic and parental connexin channels.

*Beyer:* Your results contrast with our Cx43 and Cx37 ($\alpha_4$) coexpression results in transfected N2A cells (Brink et al 1997). We picked these two connexins because their channel properties were so different to start with. Our data showed that there was a range of channel types, and we could rarely detect homomeric channels. It is most likely that the connexins like to mix in these cells. Indeed, for two connexins to be incompatible, one would have to propose a mechanism to keep them from mixing if they could. We would have to postulate a system that involves, for example, the biosynthesis of chaperones or protective molecules that are assembled or made in different places. My bet is that if two connexins are compatible and they can mix, then they will.

*Evans:* We have some evidence which indicates that homomeric connexons can form in the endoplasmic reticulum, but that additional factors present in the Golgi apparatus are required for the formation of heteromeric connexons. I will be presenting these data in my talk this afternoon (Evans et al 1999, this volume).

*Beyer:* Does the Cx43/Cx32 chimera make functional channels?

*Kumar:* We don't yet know this.

*Werner:* We have tried for several years to make various chimeras between Cx32, Cx43 and Cx38 ($\alpha_2$). We wanted to find out whether the first extracellular loop binds to the first or second extracellular loop of the other connexin. We knew that Cx38 does not form heterotypic channels with Cx32 and that it does with Cx43, so we envisaged a scheme in which the pairing of chimeras in one orientation would form functional channels, whereas the pairing in the opposite orientation would not. However, this experiment did not pan out because most of our chimeras did not form functional channels. The only ones that did were those in which we limited the exchanged part to the extracellular loop 1; we could not exchange the extracellular loop 2 without loss of function, even when we included the adjacent transmembrane segment. We also found that a hemichannel in which the first extracellular loop of Cx32 was replaced by that of Cx43 formed a leaky hemichannel, i.e. a hemichannel that was open (Pfahnl et al 1997).

*Kumar:* You say that you did not observe functional channels, but do you know if there were any transport defects?

*Werner:* We didn't test that.

*Beyer:* One often finds that constructs don't work. Most of them probably don't reach the surface. If you find a construct that acts as a bona fide dominant negative you would have found a gold mine, at least in terms of having a tool. Many non-functional connexin mutants are just misfolded or mistargeted, but they don't inhibit other connexins.

*Fletcher:* Nalin Kumar said that the wild-type Cx43 with the 231 truncation stays inside the cell, and he interprets this as a dominant negative effect. However, for a dominant negative effect to occur there has to be an interaction between the wild-type and the mutant protein that somehow traps the wild-type protein. I'm curious as to how this can occur when the mutant is so severely truncated.

*Kumar:* It is only about 10 amino acids shorter than Cx26, for example.

*Fletcher:* But from my perspective you have taken out virtually every regulatory domain.

*Kumar:* It still has four transmembrane domains.

*Kistler:* If you are trying to find out whether this truncated Cx43 blocks the trafficking pathway, you should check whether other proteins that are usually trafficked to the cell surface are still being trafficked normally so that you can rule out the possibility of toxic effects.

*Kumar:* What we have done is to co-transfect a truncated form of Cx32 with a truncated form of Cx43, and we have demonstrated that the truncated Cx32 does reach the surface, whereas truncated Cx43 is retained in the cytoplasm.

*Nicholson:* We have injected Cx43 with a 244 truncation into *Xenopus* oocytes and we find that it behaves normally. My guess is that a critical mass sticking out into the cytoplasm is required, and that folding issues are also important. There is a dramatic contrast in some of the results that Klaus Willecke and our lab have obtained with the same set of clones. Klaus has put the extracellular loops of Cx43 onto Cx40 ($\alpha_5$) to change its docking specificity (i.e. Cx40 does not pair with Cx43) and has found that, when these are transfected into HeLa cells, the Cx40 which has all its extracellular domains derived from Cx43 pairs with both Cx40 and Cx43. This indicates that the membrane domains have an influence on the extracellular docking site (Haubrich et al 1995). We observe the opposite, i.e. that this same construct in oocytes doesn't pair with Cx40 or Cx43. We know that the construct is fine because it does pair with *Xenopus* Cx38, probably because at that temperature in oocytes Cx38 is able to fold better. We are now doing these experiments at high and lower temperatures. If folding is critical, then it's possible that amphibian oocytes just don't work well for some mammalian proteins—especially chimeras that may have more folding problems.

*Gilula:* This is a good example of how many of you have personal experimental experiences related to various aspects of Nalin Kumar's presentation. You are using your experiences to find out what can be concluded from the results of using chimeras and mutational analysis in one cell type versus another. If we can succeed in generating some experimental information that can be used to understand functionally what's happening to assembly, trafficking or function at the cell surface, then we all benefit. In spite of these frustrations, it is still an important area of exploration because the potential gain is large.

*Warner:* One important issue is whether there are any general principles emerging in terms of particular point mutations and their functional consequence. For example, will a mutation in the extracellular domain always block intracellular trafficking? It is crucial to try to understand what effect the various mutations, and particularly those that occur naturally, have on function.

*Kumar:* In terms of general principles, our analyses suggest that mutations in the first 200 amino acids result in changes in transport to the cell surface. So far, we have selected residues that are conserved among different connexins. However, in Charcot-Marie-Tooth (CMT) disease, for example, there are many mutations throughout the molecule, and as far as I'm aware they are not clustered in any one region.

*Fletcher:* When the CMT story came out it looked like what was happening was that Cx32 was a 'don't mess with me' sort of protein, because any mutation, no matter how subtle, caused a problem in the affected families. I first thought that this might also be the case in Scott Cunningham's data on Cx43 in heterotaxia patients (Britz-Cunningham et al 1995). However, we now have three out of four biological parents of children who have identified mutations and also have mutations in serine codons in the same domain, but they have no clinical symptomology that we could find. So the question is, are only the $\beta$ connexins (Cx32, Cx26 etc.) 'don't mess with me' types? Or is it just because we don't have sufficient information?

*Gilula:* It's an interesting question. I don't think anybody has an answer. How much have we learned as a result of carrying out mutational analysis, both from our experimental efforts and from the human studies? What are the changes that people have identified and what are the consequent phenotypes? In the beginning a pattern of information emerged that was totally inconclusive because the phenotypes were either functional or structural, and it was difficult to work out whether or not the failure to make a channel was a result of a defect on trafficking, assembly or transport, or whether the protein at the cell surface had a conformational change that resulted in it not being able to associate with other connexins.

*Yamasaki:* We have taken a different approach to try to understand this. We found that Cx32 is often aberrantly localized in tumours, and we wondered whether this was due to a specific mutation in the Cx32 gene. We analysed the Cx32 gene sequences in 20–30 human liver tumours but we only found one mutation, which suggests that there must be other defects in the trafficking processes that result in the aberrant localization of Cx32.

*Green:* I may have misunderstood Nalin Kumar, but I thought that there are few *in vivo* instances where there are heteromeric or heterotypic channels.

*Kumar:* I stated was that I wasn't aware of a cell type that only had one connexin gene product.

*Green:* I'm not sure whether this is a slightly sweeping statement. There are a number of biological systems—including the brain, the heart and during development—where there are specific connexins expressed at specific times.

*Beyer:* Do you think those single cells only have one connexin gene product?

*Green:* In general, yes I do.

*Beyer:* I disagree. I believe that there are few cases where a cell contains only a single connexin. The harder we look the more likely we will find that multiple connexins are present.

*Gilula:* And during the life cycle of any given cell, which may become highly differentiated, it will become more challenging to find a case where there is unequivocal evidence that only one connexin gene has been expressed.

*Green:* I accept that during the life cycle of a cell there will be changes. My point is that I am not convinced that in a particular cell at a particular moment in time there will necessarily be a multitude of connexins being expressed. Where there is, some connexins may be expressed at such a low level that it may be almost insignificant.

*Beyer:* It is likely to be significant, even if only low amounts are present, because if there is free heteromeric mixing then the number of possible hexamers will be large. Even a relatively small amount of one connexin may alter the properties of mixed channels as compared to pure homomeric channels.

*Goodenough:* There are many arguments about the different types of connexins and how they may mix between cells. However, the functional relevance of this is going to require coming up with some endpoint assays for the functions of given gap junctions in different cells or in different tissues. Without such an endpoint assay we can't tell which connexins and which mixtures of connexins are important.

*Gilula:* Colin Green would argue that you could look at a cell that you know well and define a functional assay for that cell type. If you know which connexins are being used then you can understand the meaning of the functional assay in terms of the utilization of connexins in a particular cell type. The challenge is much greater for those of us who do modelling experiments involving, for example, the injection of different connexin sequences into oocytes.

*Becker:* I would like to change the subject slightly and ask people to comment on the recent reports of the ability of the invertebrate protein Shaking-B (Phelan et al 1998) to form functional channels between *Xenopus* oocytes. Do the Opus family of proteins, of which Shaking-B is a member, represent gap junction proteins?

*Gilula:* There is some reasonable concern as to whether or not any of the proteins that have been identified, including the Shaking-B protein, have a sequence that enables them to form a gap junction channel. It will be necessary to find this out, because the nematode and *Drosophila* sequences that have been proposed to be related to gap junction sequences have all been associated with some channel activity. However, the important question is whether they are facilitating the formation of a junction between cells, or whether they themselves are a product that makes the membrane component of the channel. Either way they are interesting proteins. People have been trying to identify arthropod gap junction gene products for a long time. Now that putative gene(s) have been identified in *Drosophila* and the nematode, the challenge is to understand which of these sequences encodes the protein product that is required for these interesting arthropod gap junction structures that Colin Green and a number of others have studied for a long time.

## References

Brink PR, Cronin K, Banach K et al 1997 Evidence for heteromeric gap junction channels formed from rat connexin43 and human connexin37. Am J Physiol 273:C1386–C1396

Britz-Cunningham SH, Shah MM, Zuppan CW, Fletcher WH 1995 Mutations of the connexin43 gap junction gene in patients with heart malformations and defects of laterality. N Engl J Med 332:1323–1329

Evans WH, Ahmad S, Diez J, George CH, Kendall JM, Martin PEM 1999 Trafficking pathways leading to the formation of gap junctions. In: Gap junction-mediated intercellular signalling in health and disease. Wiley, Chichester (Novartis Found Symp 219) p 44–59

Falk MM, Buehler LK, Kumar NM, Gilula N 1997 Cell-free synthesis and assembly of connexins into functional gap junction membrane channels. EMBO J 16:2703–2716

Haubrich S, Schwarz H-J, Bukauskas F et al 1995 The incompatibility of connexin 40 and 43 hemichannels in gap junctions between mammalian cells is determined by intracellular domains. Mol Biol Cell 7:1995–2006

Pfahnl A, Ahou X-W, Werner R, Dahl G 1997 A chimeric connexin forming gap junction hemichannels. Pflueg Arch Eur J Physiol 433:773–779

Phelan P, Stebbings LA, Baines RA, Bacon JP, Davies JA, Ford C 1998 *Drosophila* Shaking-B protein forms gap junctions in paired *Xenopus* oocytes. Nature 391:181–184

Valiunas V, Niessen H, Willecke K, Weingart R 1998 Electrophysiological properties of gap junction channels in hepatocytes isolated from connexin32 deficient and wild type mice, submitted

# Electron cryo-crystallography of a recombinant cardiac gap junction channel

Vinzenz M. Unger[1], Nalin M. Kumar*, Norton B. Gilula* and Mark Yeager*[2]†

**Department of Cell Biology, The Scripps Research Institute, 10550 North Torrey Pines Road, and †Division of Cardiovascular Diseases, Scripps Clinic, 10666 North Torrey Pines Road, La Jolla, CA 92037, USA*

*Abstract.* Gap junctions in the heart play an important functional role by electrically coupling cells, thereby organizing the pattern of current flow to allow co-ordinated muscle contraction. Cardiac gap junctions are therefore intimately involved in normal conduction as well as the genesis of potentially lethal arrhythmias. We recently utilized electron cryo-microscopy and image analysis to examine frozen–hydrated 2D crystals of a recombinant, C-terminal truncated form of connexin43 (Cx43; $\alpha_1$), the principal cardiac gap junction protein. The projection map at 7 Å resolution revealed that each 30 kDa connexin subunit has a transmembrane $\alpha$-helix that lines the aqueous pore and a second $\alpha$-helix in close contact with the membrane lipids. The distribution of densities allowed us to propose a model in which the two apposing connexons that form the channel are staggered by $\sim 30°$. We are now recording images of tilted, frozen–hydrated 2D crystals, and a preliminary 3D map has been computed at an in-plane resolution of $\sim 7.5$ Å and a vertical resolution of $\sim 25$ Å. As predicted by our model, the two apposing connexons that form the channel are staggered with respect to each other for certain connexin molecular boundaries within the hexamer. Within the membrane interior each connexin subunit displays four rods of density, which are consistent with an $\alpha$-helical conformation for the four transmembrane domains. Preliminary studies of BHK hamster cells that express the truncated Cx43 designated $\alpha_1$Cx263T demonstrate that oleamide, a sleep inducing lipid, blocks *in vivo* dye transfer, suggesting that oleamide causes closure of $\alpha_1$Cx263T channels. The comparison of the 3D structures in the presence and absence of oleamide may provide an opportunity to explore the conformational changes that are associated with oleamide-induced blockage of dye transfer. The structural details revealed by our analysis will be essential for delineating the molecular basis for intercellular current flow in the heart, as well as the general molecular design and functional properties of this important class of channel proteins.

*1999 Gap junction-mediated intercellular signalling in health and disease. Wiley, Chichester (Novartis Foundation Symposium 219) p 22–37*

[1]Current address: Max-Planck Institut für Biophysik, Abteilung Strukturbiologie, Heinrich-Hoffmann-Str. 7, D-60528 Frankfurt/Main, Germany.
[2]This chapter was presented at the symposium by Mark Yeager, to whom correspondence should be addressed.

The theme of this symposium is the role of gap junction intercellular signalling in health and disease. In terms of the heart, this is straightforward. Cardiac gap junctions play an important functional role by electrically coupling cells, thereby organizing the pattern of current flow to allow co-ordinated muscle contraction (Barr et al 1965, De Mello 1982). Cardiac gap junctions are therefore intimately involved in normal conduction as well as the genesis of potentially lethal arrhythmias (Spach 1983, 1997). Cardiac electrophysiology has provided tremendous insight into the mechanisms of cardiac conduction and the basis for intercellular current flow viã gap junctions (reviewed in Spooner et al 1997). Fundamental to a complete understanding of basic mechanisms of cardiac conduction and arrhythmias is detailed knowledge about the structure of cardiac gap junctions.

Gap junction channels are formed by the end-to-end docking of two hemichannels (connexons), each of which is formed by a hexameric cluster of protein subunits (connexins) that allow the intercellular exchange of molecules up to about 1000 Da (Loewenstein 1981). We recently reviewed the structure of cardiac gap junctions based on work in our laboratory as well as others (Yeager 1998). In summary, the channels have been localized to the intercalated disks between heart cells by immunofluorescence microscopy. Immunogold electron microscopy combined with protease cleavage were used to confirm a folding model based on hydropathy analysis in which each connexin is composed of four hydrophobic transmembrane domains with the N- and C-termini located on the cytoplasmic membrane face. Electron microscopy and image analysis of 2D crystals of rat heart gap junctions demonstrated that each connexon is formed by a hexameric cluster of subunits, analogous to liver gap junctions, and that the intercellular channel is $\sim 250$ Å thick, compared with $\sim 150$ Å for liver gap junctions. Circular dichroism spectroscopy showed that the $\alpha$-helical content is sufficient so that the four transmembrane domains may be folded as $\alpha$-helices. Most recently, recombinant connexins have been expressed in BHK hamster cells, in which they self-assemble into gap junctions that are indistinguishable from those formed in native tissues. Analysis of a C-terminal truncation mutant by electron cryo-crystallography has allowed visualization, for the first time, of transmembrane $\alpha$-helices. To our knowledge, this is the first example where a recombinant polytopic membrane protein has been expressed in a heterologous system and examined by structural methods. Here we will review the topological features and functional regions of connexin43 (Cx43; $\alpha_1$), the principal cardiac gap junction connexin, describe the projection structure determined by electron cryo-crystallography, present a model for connexon docking and conclude with work in progress to delineate the 3D structure of the intercellular channel. This research summary is based our previous research publications (Yeager & Gilula 1992, Unger et al 1997) and invited reviews (Yeager & Nicholson 1996, Yeager 1998).

FIG. 1. (a) Folding model for connexin 43 (Cx43; $\alpha_1$). The amino-acid sequence of Cx43 (Beyer et al 1987) was deduced from cDNA analysis, and the residues are coded as follows: hydrophobic in yellow, acidic in red, basic in blue and cysteine in green. Hydropathy analysis predicts four membrane-spanning domains, referred to as M1, M2, M3 and M4, proceeding from the N- to the C-terminus. The predicted locations of the extracellular and cytoplasmic regions were confirmed with site-directed antibodies (blue and yellow bars indicate cytoplasmic and extracellular epitopes, respectively). Note the three conserved cysteine residues (shown in green) located in each of the extracellular loops (designated E1 and E2) of the three connexins.

(b) Cx43 indicating the locations of various functionally important residues (His95) and domains 1 (yellow) and 2 (blue) as determined by mutagenesis and chimera studies. Indicated sites of covalent modification are based on consensus sequences, modification of synthetic peptides *in vitro* and mutagenesis studies. The significance of the functional domains (circled numbers) and specifically mutated residues are as follows. Domain 1 (yellow) is the predominant determinant for specificity of heterotypic interactions between Cx43, Cx50 ($\alpha_8$) and Cx46 ($\alpha_3$). This result was based on chimeras of Cx50 and Cx46. Domains 2 (blue) impart high sensitivity to pH gating of Cx43. Gating by pH is also influenced by the charge on His95. The SH3 domains of v-Src bind to proline-rich motifs and a phosphorylated tyrosine at position 265 of Cx43. The numbered references are as follows: 1, Beyer et al (1989); 2, Yancey et al (1989); 3, El Aoumari et al (1990); 4, Laird & Revel (1990); 5, Swenson et al (1990); 6, Kennelly & Krebs (1991); 7, Yeager & Gilula (1992); 8, Goldberg & Lau (1993); 9, Kanamitsu & Lau (1993); 10, Liu et al (1993); 11, Sáez et al (1993); 12, Ek et al (1994); 13, White et al (1994); 14, Loo et al (1995); 15, White et al (1995); 16, Kwak et al (1995); 17, Morley et al (1996); 18, Warn-Cramer et al (1996); 19, Kanemitsu et al (1997). Modified from Yeager & Nicholson (1996) by permission of Current Biology Ltd.

FIG. 2. Projection density map at 7 Å resolution of a recombinant cardiac gap junction channel determined by electron cryo-crystallography. This end-on view shows that each connexon is formed by a hexameric cluster of subunits with three major features: (i) a ring of circular densities centred at a radius of 17 Å, interpreted as $\alpha$-helices that line the channel; (ii) a ring of densities centred at a radius of 33 Å, interpreted as $\alpha$-helices that are most exposed to the lipid; and (iii) a continuous band of density at a radius of 25 Å, separating the two groups of $\alpha$-helices, which may arise from the superposition of projections of additional transmembrane $\alpha$-helices and polypeptide density arising from the extracellular and intracellular loops within each connexin subunit. Note that the inner and outer rings of $\alpha$-helices are staggered by 30°. The rank order of mass density is white > yellow > red > violet > blue. From Unger et al (1997) and reproduced here by permission of Nature Structural Biology.

FIG. 3. Schematic model for the packing of $\alpha$-helices, connexins and connexons in the gap junction intercellular channel. Each connexon is formed by six connexin subunits, and the complete intercellular channel is formed by the end-to-end docking of two connexons as shown at the right. Each connexin subunit is depicted as an arbitrary four-helix bundle, and the transmembrane $\alpha$-helices are depicted as straight cylinders. The superposition of the apposed connexons (left) is in accordance with the observed projection density map shown in Fig. 2 and predicts that the connexons within the channel will be rotationally staggered by 30°. The superposition of the $\alpha$-helices in different connexons and this rotation dictate that the $\alpha$-helices within a connexin subunit of one connexon (shown in blue) will be superimposed with $\alpha$-helices of two connexin subunits in the apposed connexon (shown in orange and yellow). However, it should be noted that this rotational stagger does not apply for all theoretically possible outlines of the molecular boundary of the connexin monomer. Nevertheless, common to all models is the superposition of $\alpha$-helices of one connexon with the $\alpha$-helices of at least two connexin subunits in the apposed connexon.

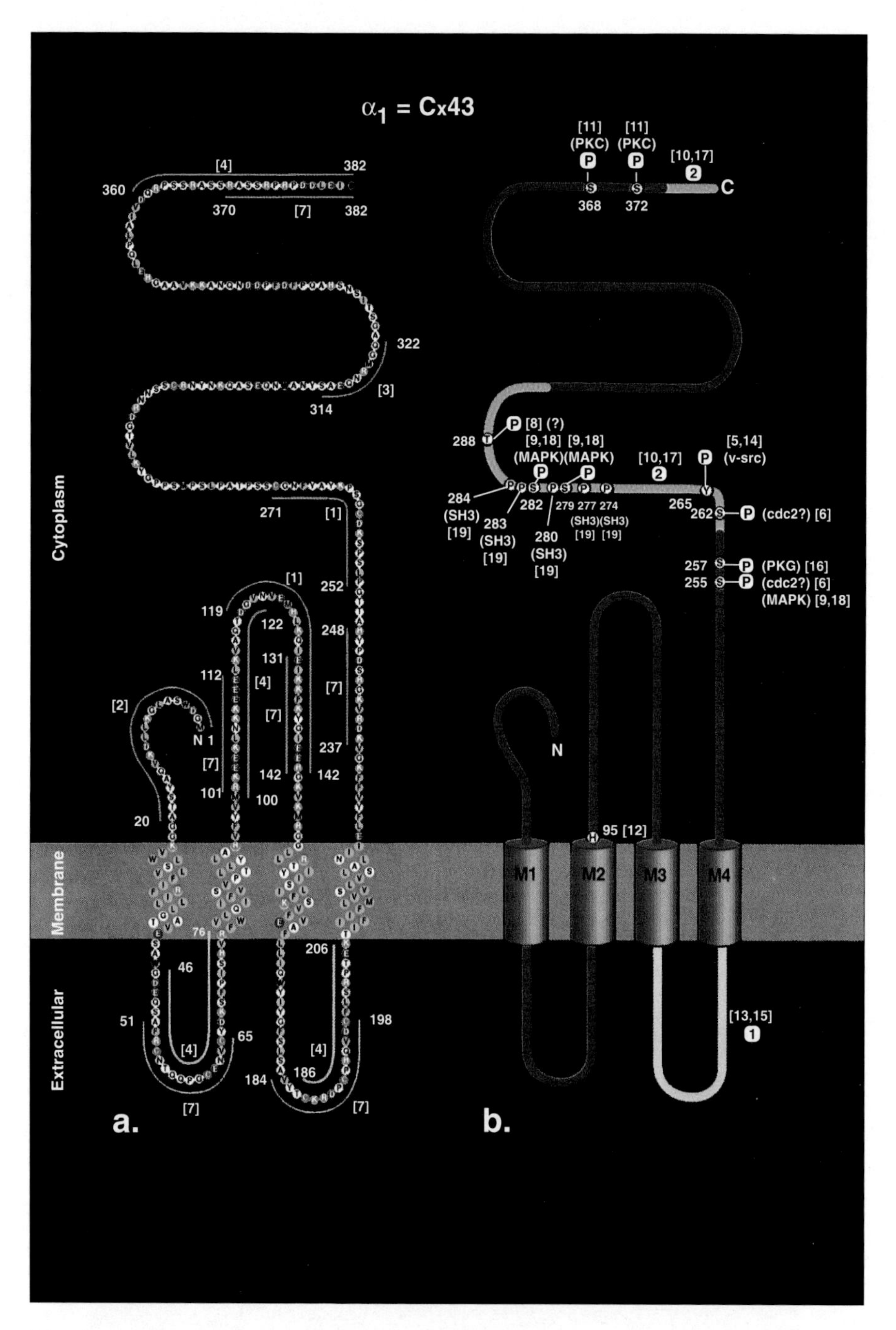

α1 = Cx43
Cytoplasm
Membrane
Extracellular
[4]
360
382
370
[7]
382
322
[3]
314
271
[1]
252
[1]
119
122
248
131
112
[4]
[7]
[2]
[7]
N 1
237
[7]
142
142
101
100
20
76
206
46
51
65
198
[4]
[4]
186
184
[7]
[7]
a.
[11]
(PKC)
[11]
(PKC)
[10,17]
C
368
372
288
[8] (?)
[9,18] [9,18]
(MAPK)(MAPK)
[10,17]
[5,14]
(v-src)
284
(SH3)
[19]
282
283
(SH3)
[19]
279 277 274
(SH3)(SH3)
[19] [19]
280
(SH3)
[19]
265
262
(cdc2?) [6]
257
(PKG) [16]
255
(cdc2?) [6]
(MAPK) [9,18]
N
95 [12]
M1
M2
M3
M4
[13,15]
b.


Fig. 1

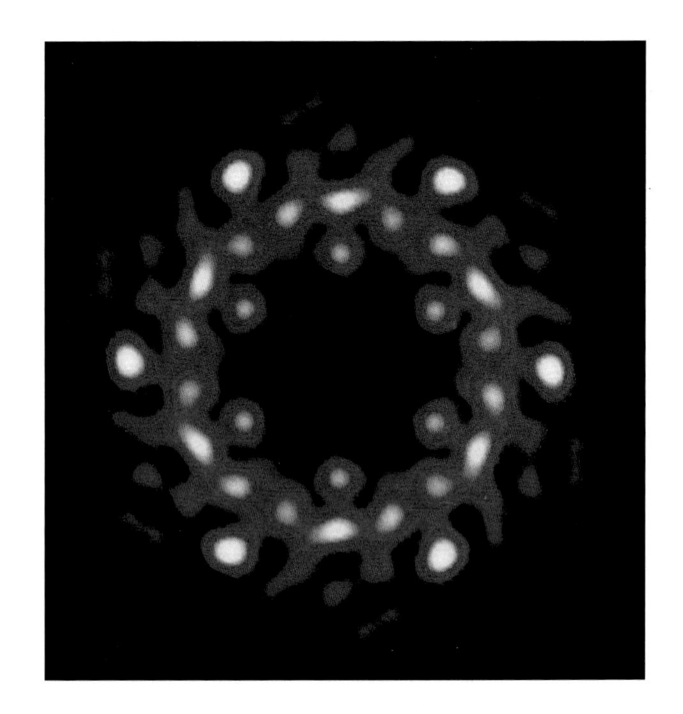

**Fig. 2**

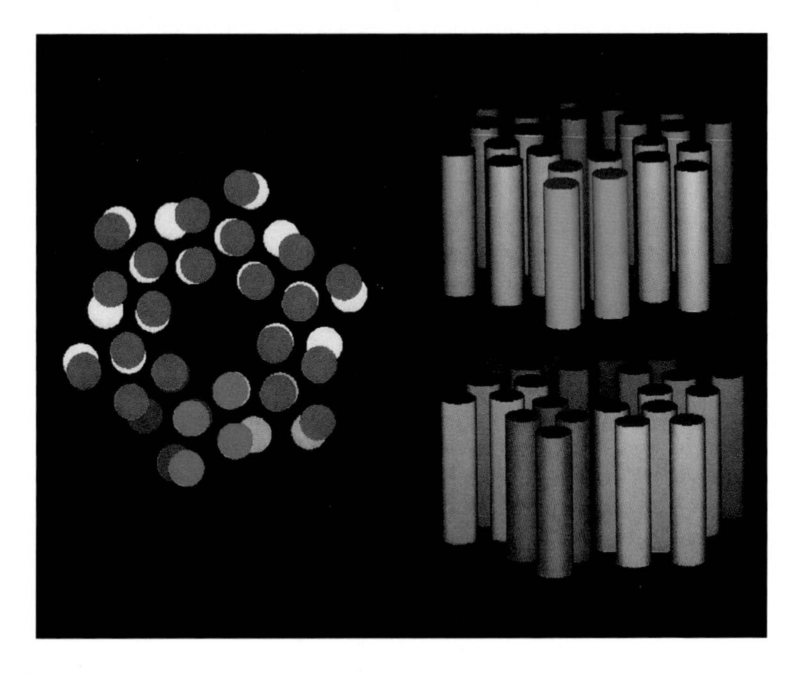

**Fig. 3**

## Membrane topology of connexin43

The membrane topology of Cx43 has been explored using peptide antibodies directed to several sites in the protein sequence and protease cleavage (Beyer et al 1989, Yancey et al 1989, Laird & Revel 1990, Yeager & Gilula 1992). In our laboratory, four affinity-purified peptide antibodies were directed to sites in cytoplasmic domains, and two antibodies were directed to the E1 and E2 extracellular loops (Fig. 1). Isolated gap junctions could not be labelled by the extracellular loop antibodies, consistent with the known narrowness of the extracellular gap region that would exclude macromolecular antibody probes. Immunoelectron microscopy demonstrates that gap junctions containing native Cx43 are heavily decorated on their cytoplasmic surfaces by the peptide antibodies directed to cytoplasmic domains. The N-termini of Cx43 and Cx32 ($\beta_1$) are homologous, and antibodies to the N-terminal sequence of Cx32 can bind to Cx43 (Zervos et al 1985). Yancey et al (1989) have also shown that the N-terminus of Cx43 is accessible on the cytoplasmic surface for labelling by peptide antibodies. Thus, the N-terminal domain, C-terminal domain and the M2–M3 loop define three polypeptide regions of Cx43 that are accessible on the cytoplasmic side of the membrane.

## Transmembrane α-helices visualized by electron cryo-crystallography of recombinant gap junctions

To explore the secondary structure and folding of cardiac gap junction channels, we sought to generate 2D crystals that are ordered to better than 10 Å resolution. Examination of biological specimens in the frozen–hydrated state offers the possibility for higher resolution structure analysis. Nevertheless, electron cryo-microscopy of native gap junctions isolated from tissues such as the liver have been limited to 15–20 Å resolution (Unwin & Ennis 1984, Gogol & Unwin 1988). This limiting resolution is presumably based on molecular heterogeneity that could arise from: (1) expression of multiple connexins within cells of the same tissue that could lead to formation of gap junctions that are assembled from different homomeric connexons, formation of heterotypic channels in which homomeric connexons are composed of different connexins and formation of heteromeric connexons in which the oligomers contain multiple connexin types (White et al 1995, Barrio et al 1991, Elfgang et al 1995, Sosinsky 1995, Stauffer 1995, Jiang & Goodenough 1996); (2) multiple oligomeric states within a connexon population (such as a small fraction of pentamers amongst the hexamers); (3) inherent flexibility in the native protein; (4) partial denaturation during specimen preparation; (5) the presence of non-connexin proteins that may disorder the lattice; (6) partial proteolysis; (7) different degrees of post-translational modification; (8) rotational flexibility of the connexons in the lattice; and (9) lipid heterogeneity that may prevent precise chemical interactions at the protein–lipid interface which may be necessary for 2D crystallization of the connexons.

Several of these problems can be overcome by expression of a single recombinant connexin. Progress has certainly been made in the over-expression of Cx32 in insect cells using a baculovirus vector (Stauffer et al 1991). The ability of over-expressed connexins to self-assemble into recognizable gap junction structures in infected cells is convincing evidence that the recombinant protein exhibits the critical structural elements of the native protein. Although initial 3D crystallization trials were encouraging (Stauffer et al 1991), crystals suitable for X-ray crystallography have not been forthcoming. Alternatively, detergent-solubilized and purified connexons may be amenable to *in vitro* reconstitution with lipids in order to grow 2D crystals (Kühlbrandt 1992, Jap et al 1992, Yeager et al 1999). However, this approach has not been successful in yielding high resolution 2D crystals thus far.

Recombinant connexins have now been expressed in a stably transfected BHK hamster cell line under the control of the inducible metallothionin promoter (Kumar et al 1995). Ultrastructural studies demonstrated that a truncated form of rat Cx43 that lacks most of the large C-terminal domain (truncation at Lys263; $\alpha_1$Cx263T) assembles into gap junctions having the characteristic septalaminar morphology. Freeze-fracture images revealed that the gap junctions form small 2D crystals, and their crystallinity and purity could be improved by extraction with non-denaturing detergents such as Tween20 and 1,2-diheptanoyl-sn-phosphocholine (DHPC; Unger et al 1997), an approach similar to that taken for crystallization of native rat heart gap junctions (Yeager 1994).

A projection density map based on the analysis of frozen–hydrated 2D crystals (Unger et al 1997) shows that the recombinant connexon has a diameter of ~65 Å and is formed by a hexameric cluster of connexin subunits (Fig. 2). At 7 Å resolution the map of the recombinant channel shows substantially more detail than the map of rat heart gap junctions at 16 Å resolution (Yeager & Gilula 1992). In particular, at 17 Å radius the channel is lined by circular densities with the characteristic appearance of transmembrane $\alpha$-helices that are oriented roughly perpendicular to the membrane plane (Unwin & Henderson 1975, Kühlbrandt & Downing 1989, Havelka et al 1993, Schertler et al 1993, Jap & Li 1995, Karrasch et al 1995, Mitra et al 1995, Walz et al 1995, Hebert et al 1997). A similar appearance for densities at 33 Å radius suggests the presence of $\alpha$-helices at the interface with the membrane lipids. The two rings of $\alpha$-helices are separated by a continuous band of density at a radius of 25 Å, which may arise from the superposition of projections of additional transmembrane $\alpha$-helices and polypeptide density arising from the extracellular and intracellular loops within each connexin subunit.

## Docking of connexons involves rotational stagger

A notable feature of the projection density map is the 30° displacement between the rings of $\alpha$-helices at 17 and 33 Å radius, which places constraints on possible structural models for the intercellular channel (Fig. 3). Assuming a roughly circular shape for the connexin subunit, the 30° displacement predicts that the two connexons forming the channel are rotationally staggered with respect to each other. The amount of rotational stagger will dictate whether six or 12 peaks are resolved in the outer ring of $\alpha$-helices at 33 Å radius. That is, models with less than 30° of rotational stagger between the connexons are not consistent with the map in Fig. 2 since there would be 12 rather than six peaks in the outer ring of $\alpha$-helices. In addition, the 30° displacement between the rings of helices is not consistent with models in which the $\alpha$-helices are co-linear through the centre of the channel. The model shown in Fig. 3 is in best agreement with the projection density map that shows superposition of the resolved $\alpha$-helices between the apposed connexins. Note that this model predicts that each subunit in one connexon will interact with two connexin subunits in the apposing connexon. Such an arrangement may confer stability in the docking of the connexons.

## Work in progress

### *3D structure analysis*

In order to assess the validity of the model shown in Fig. 3, we are recording images of tilted, frozen–hydrated 2D crystals. A preliminary 3D map has been computed that has an in-plane resolution of ~7 Å and a vertical resolution of ~25 Å. This preliminary map suggests that the two apposing connexons that form the channel are staggered by 30° with respect to each other for some of the possible subunit boundaries within the hexamer. Within the membrane interior there appear to be 24 rods of density per connexon, consistent with an $\alpha$-helical conformation for the four transmembrane domains within each connexin subunit. The close packing of $\alpha$-helices within and between connexin subunits stabilizes the hexameric connexon oligomer. The $\alpha$-helical rods merge with a tight ring and arcs of density in the extracellular gap, which would form a tight electrical and chemical seal to facilitate current flow between cells in the heart.

### *Conformational states of gap junction channels*

Preliminary functional studies of BHK cells that express $\alpha_1$Cx263T demonstrate that oleamide, a sleep-inducing compound, blocks *in vivo* dye transfer. As previously demonstrated for other connexins (Guan et al 1997), this behaviour is an indication that oleamide causes closure of $\alpha_1$Cx263T. These results have encouraged us to examine whether oleamide affects the structure of the channel. 2D crystals of $\alpha_1$Cx263T grown in the presence and absence of oleamide display reflections to ~11 Å resolution by optical diffraction and in many cases to ~7 Å resolution after correction for lattice distortions. The comparison of the 3D structures in the presence and absence of oleamide should allow us to explore the conformational changes that are associated with oleamide-induced blockage of dye transfer. The structural details revealed by our analysis will be essential for delineating the functional properties of this important class of channel proteins.

*Acknowledgements*

This work has been supported by grants from the National Institutes of Health (M.Y. and N.B.G.), the Lucille P. Markey Charitable Trust (N.B.G.), the Gustavus and Louise Pfeiffer Research Foundation (M.Y.), and the Baxter Research Foundation (M.Y.). V.M.U. was a recipient of a postdoctoral fellowship from the American Heart Association. During the course of this work M.Y. was an Established Investigator of the American Heart Association and Bristol-Myers Squibb, and is now the recipient of a Clinical Scientist Award in Translational Research from the Burroughs Wellcome Fund.

## References

Barr L, Dewey MM, Berger W 1965 Propagation of action potentials and the structure of the nexus in cardiac muscle. J Gen Physiol 48:797–823

Barrio LC, Suchyna T, Bargiello T et al 1991 Gap junctions formed by connexins 26 and 32 alone and in combination are differently affected by applied voltage. Proc Natl Acad Sci USA 88:8410–8414 (erratum 1992 Proc Natl Acad Sci USA 89:4220)

Beyer EC, Paul DL, Goodenough DA 1987 Connexin 43: a protein from rat heart homologous to a gap junction protein from liver. J Cell Biol 105:2621–2629

Beyer EC, Kistler J, Paul DL, Goodenough DA 1989 Antisera directed against connexin43 peptides react with a 43-kD protein localized to gap junctions in myocardium and other tissues. J Cell Biol 108:595–605

De Mello WC 1982 Intercellular communication in cardiac muscle. Circ Res 51:1–9

Ek JF, Delmar M, Perzova R, Taffet SM 1994 Role of histidine 95 on pH gating of cardiac gap junction protein connexin43. Circ Res 74:1058–1064

El Aoumari AE, Fromaget C, Dupont E et al 1990 Conservation of a cytoplasmic carboxy-terminal domain of connexin 43, a gap junctional protein, in mammal heart and brain. J Membr Biol 115:229–240

Elfgang C, Eckert R, Lichtenberg-Fraté H et al 1995 Specific permeability and selective formation of gap junction channels in connexin-transfected HeLa cells. J Cell Biol 129: 805–817

Gogol E, Unwin N 1988 Organization of connexons in isolated rat liver gap junctions. Biophys J 54:105–112

Goldberg GS, Lau AF 1993 Dynamics of connexin43 phosphorylation in pp60v-src-transformed cells. Biochem J 295:735–742

Guan X, Cravatt BF, Ehring GR et al 1997 The sleep-inducing lipid oleamide deconvolutes gap junction communication and calcium wave transmission in glial cells. J Cell Biol 139: 1785–1792

Havelka WA, Henderson R, Heymann JA, Oesterhelt D 1993 Projection structure of halorhodopsin from *Halobacterium halobium* at 6 Å resolution obtained by electron cryo-microscopy. J Mol Biol 234:837–846

Hebert H, Schmidt-Krey I, Morgenstern R et al 1997 The 3.0 Å projection structure of microsomal glutathione transferase as determined by electron crystallography of $p2^12^12$ two-dimensional crystals. J Mol Biol 271:751–758.

Jap BK, Li H 1995 Structure of the osmo-regulated $H_2O$-channel, AQP-CHIP, in projection at 3.5 Å resolution. J Mol Biol 251:413–420

Jap BK, Zulauf M, Scheybani T et al 1992 2D crystallization: from art to science. Ultramicroscopy 46:45–84

Jiang JX, Goodenough DA 1996 Heteromeric connexons in lens gap junction channels. Proc Natl Acad Sci USA 93:1287–1291

Kanemitsu MY, Lau AF 1993 Epidermal growth factor stimulates the disruption of gap junctional communication and connexin43 phosphorylation independent of 12-0-tetradecanoylphorbol 13-acetate-sensitive protein kinase C: the possible involvement of mitogen-activated protein kinase. Mol Biol Cell 4:837–848

Kanemitsu MY, Loo LWM, Simon S, Lau AF, Eckhart W 1997 Tyrosine phosphorylation of connexin 43 by v-Src is mediated by SH2 and SH3 domain interactions. J Biol Chem 272:22824–22831

Karrasch S, Bullough PA, Ghosh R 1995 The 8.5 Å projection map of the light-harvesting complex I from *Rhodospirillum rubrum* reveals a ring composed of 16 subunits. EMBO J 14:631–638

Kennelly PJ, Krebs EG 1991 Consensus sequences as substrate specificity determinants for protein kinases and protein phosphatases. J Biol Chem 266:15555–15558

Kühlbrandt W 1992 Two-dimensional crystallization of membrane proteins. Q Rev Biophysics 25:1–49

Kühlbrandt W, Downing KH 1989 Two-dimensional structure of plant light-harvesting complex at 3.7 Å (corrected) resolution by electron crystallography. J Mol Biol 207:823–828 (erratum 1989 J Mol Biol 210:243)

Kumar NM, Friend DS, Gilula NB 1995 Synthesis and assembly of human $\beta_1$ gap junctions in BHK cells by DNA transfection with the human $\beta_1$ cDNA. J Cell Sci 108:3725–3734

Kwak BR, Sáez JC, Wilders R et al 1995 Effects of cGMP-dependent phosphorylation on rat and human connexin 43: gap junction channels. Pflügers Arch 430:770–778

Laird DW, Revel J-P 1990 Biochemical and immunochemical analysis of the arrangement of connexin43 in rat heart gap junction membranes. J Cell Sci 97:109–117

Liu SG, Taffet S, Stoner L, Delmar M, Vallano ML, Jalife J 1993 A structural basis for the unequal sensitivity of the major cardiac and liver gap junctions to intracellular acidification: the carboxyl tail length. Biophys J 64:1422–1433

Loewenstein WR 1981 Junctional intercellular communication: the cell-to-cell membrane channel. Physiol Rev 61:829–913

Loo LWM, Berestecky JM, Kanemitsu MY, Lau AF 1995 $pp60^{src}$-mediated phosphorylation of connexin 43, a gap junction protein. J Biol Chem 270:12751–12761

Mitra AK, van Hoek AN, Wiener MC, Verkman AS, Yeager M 1995 The CHIP28 water channel visualized in ice by electron crystallography. Nat Struct Biol 2:726–729

Morley GE, Taffet SM, Delmar M 1996 Intramolecular interactions mediate pH regulation of connexin43 channels. Biophys J 70:1294–1302

Sáez JC, Nairn AC, Czernik AJ, Spray DC, Hertzberg EL 1993 Rat connexin 43: regulation by phosphorylation in heart. In: Hall JE, Zampighi GA, Davis RM (eds) Progress in cell research, vol 3: Gap junctions. Elsevier Science Publishers, Amsterdam, p 275–282

Schertler GF, Villa C, Henderson R 1993 Projection structure of rhodopsin. Nature 362:770–772

Sosinsky G 1995 Mixing of connexins in gap junction membrane channels. Proc Natl Acad Sci USA 92:9210–9214
Spach MS 1983 The role of cell-to-cell coupling in cardiac conduction disturbances. Adv Exp Med Biol 161:61–77
Spach MS 1997 Discontinuous cardiac conduction: its origin in cellular connectivity with long-term adaptive changes that cause arrhythmias. In: Spooner PM, Joyner RW, Jalife J (eds) Discontinuous conduction in the heart. Futura, Armonk, NY, p 5–51
Spooner PM, Joyner RW, Jalife J (eds) 1997 Discontinuous conduction in the heart. Futura, Armonk, NY
Stauffer KA 1995 The gap junction protein $\beta$1 connexin (Cx32) and $\beta$2 connexin (Cx26) can form heteromeric hemichannels. J Biol Chem 270:6768–6772
Stauffer KA, Kumar NM, Gilula NB, Unwin N 1991 Isolation and purification of gap junction channels. J Cell Biol 115:141–150
Swenson KI, Piwnica-Worms H, McNamee H, Paul DL 1990 Tyrosine phosphorylation of the gap junction protein connexin43 is required for the pp60v-src-induced inhibition of communication. Cell Regul 1:989–1002
Unger VM, Kumar NM, Gilula NB, Yeager M 1997 Projection structure of a gap junction membrane channel at 7 Å resolution. Nat Struct Biol 4:39–43
Unwin PNT, Ennis PD 1984 Two configurations of a channel-forming membrane protein. Nature 307:609–613
Unwin PNT, Henderson R 1975 Molecular structure determination by electron microscopy of unstained crystalline specimens. J Mol Biol 94:425–440
Walz T, Typke D, Smith BL, Agre P, Engel A 1995 Projection map of aquaporin-1 determined by electron crystallography. Nat Struct Biol 2:730–732
Warn-Cramer BJ, Lampe PD, Kurata WE et al 1996 Characterization of the mitogen-activated protein kinase phosphorylation sites on the connexin-43 gap junction protein. J Biol Chem 271:3779–3786
White TW, Bruzzone R, Wolfram S, Paul DL, Goodenough DA 1994 Selective interactions among the multiple connexin proteins expressed in the vertebrate lens: the second extracellular domain is a determinant of compatibility between connexins. J Cell Biol 125:879–892
White TW, Paul DL, Goodenough DA, Bruzzone R 1995 Functional analysis of selective interactions among rodent connexins. Mol Biol Cell 6:459–470
Yancey SB, John SA, Lal R, Austin BJ, Revel J-P 1989 The 43-kD polypeptide of heart gap junctions: immunolocalization, topology, and functional domains. J Cell Biol 108:2241–2254
Yeager M 1994 In situ two-dimensional crystallization of a polytopic membrane protein: the cardiac gap junction channel. Acta Cryst D50:632–638
Yeager M 1998 Structure of cardiac gap junction intercellular channels. J Struct Biol 121:231–245
Yeager M, Gilula NB 1992 Membrane topology and quaternary structure of cardiac gap junction ion channels. J Mol Biol 223:929–948
Yeager M, Nicholson BJ 1996 Structure of gap junction intercellular channels. Curr Opin Struct Biol 6:183–192
Yeager M, Unger VM, Mitra AK 1999 Three-dimensional structure of membrane proteins determined by two-dimensional crystallization, electron cryo-microscopy and image analysis. Methods Enzymol 294:135–180
Zervos AS, Hope J, Evans WH 1985 Preparation of a gap junction fraction from uteri of pregnant rats: the 28-kD polypeptides of uterus, liver, and heart gap junctions are homologous. J Cell Biol 101:1363–1370

## DISCUSSION

*Werner:* Do you have any idea which of the four transmembrane segments represents the lining of the pore?

*Yeager:* As you know, M3 in all the connexins has a hydrophilic stripe (Milks et al 1988), making it the best candidate for lining the aqueous channel. The data from your lab using cysteine scanning mutagenesis suggest that M1 is exposed to the aqueous channel (Zhou et al 1997). From our 3D map one can see that each of the six connexin subunits contributes two $\alpha$-helices to form the lining of the aqueous channel. However, the 7.5 Å resolution of the 3D map is not high enough to allow us to resolve amino acid side chains or the connecting loops between transmembrane $\alpha$-helices. Hence, the pore-lining $\alpha$-helices cannot be unambiguously assigned in the map.

*Werner:* We tried cysteine scanning mutagenesis on M3, but we had little luck with this, which is why we suggested M1 as a likely candidate because we found two amino acids in this region that were accessible from the pore lumen (Zhou et al 1997).

*Gilula:* A short-term challenge would be to do some chemical labelling.

*Nicholson:* One of the transmembrane segments seems to be more angular than the other. Not the one that lines the pore, but one that lies alongside. Are there any indications that helices other than the one you showed lining the pore have notable bends?

*Yeager:* All of the $\alpha$-helices are tilted with angles ranging from about 10 to 30 degrees. The $\alpha$-helix positioned at a radius of $\sim$33 Å from the centre of the channel is the straightest.

*Kistler:* The inner ring of rods seems to have kinks in the closed channel structure, but the pore size is still 17 Å. Is that sufficient to stop the flow of ions?

*Yeager:* The 3D map shows that the minimum pore size is about 15 Å. As noted, our map does not resolve amino acid side chains, which would add about 5 Å to the resolved boundaries of the $\alpha$-helices. The addition of side chains would therefore reduce the pore size to about 5 Å, which is small enough to block dyes such as Lucifer yellow but would not be small enough to block anhydrous ions. We are keen to find out if under these conditions the cells are electrically coupled.

*Gilula:* Jim Hall's electrophysiological analysis of connexin43 (Cx43; $\alpha_1$) expressed in glial cells clearly shows that they are electrically coupled (Guan et al 1997), so I would be surprised if these weren't also electrically coupled. There is going to be a difference between the size we theoretically propose will cause a block versus a functional block in reality.

*Giaume:* Do you have any data related to conformational changes induced by other uncoupling agents, such as alcohols or $\alpha$GA, which differ from oleamide in their structure and likely their mode of action?

*Yeager:* Not yet. We are currently comparing gap junction channels in the presence and absence of oleamide. Future experiments include the examination of possible conformational changes induced by pH, calcium and phosphorylation.

*Kistler:* How is the blockage of the channel caused by oleamide related to the voltage gating?

*Yeager:* That's a good question. The way I formulate this is that the connexons are built with exquisite versatility and have multiple mechanisms affecting channel closure, unlike voltage-regulated channels and ligand-gated channels, which presumably have one conformational pathway for gating. Some of the factors affecting gap junction channel closure include pH, phosphorylation, voltage, calcium and bioactive lipids such as oleamide. I envisage that these factors lead to a variety of structural rearrangements that trigger a final common pathway of conformational changes that cause channel closure.

*Vaney:* What is the channel size in the absence of oleamide?

*Yeager:* Our preliminary data in the absence of oleamide suggest that the α-helical packing is conserved. Certain sections of the maps in the presence and absence of oleamide show that the α-helical densities are virtually superimposable. Other sections of the map show intriguing shifts in the density that may be related to closure. The full 3D data set needs to be completed before we can be confident that such changes are significant.

*Gilula:* At these limits one has to be extremely conservative about interpreting such data.

*Kistler:* What is the estimated conformational change in terms of the distance that the α-helices move?

*Yeager:* I really can't answer your question with any confidence since the data in the absence of oleamide are not complete. Our goal is to record images from a sufficient number of tilted crystals so that we can divide each individual map in half and do statistical comparisons with itself and then cross-comparisons between the two maps in the presence and absence of oleamide.

*Dermietzel:* What is the difference between the old Nigel Unwin model (Unwin & Ennis 1984) and your model?

*Yeager:* It is important to note that the experiments are different. Nigel Unwin examined the structure of rat liver gap junctions in the presence and absence of calcium, whereas we are examining recombinant cardiac gap junction channels in the presence and absence of oleamide. An important difference is that Nigel's map at 20 Å resolution only resolved the molecular boundary of each subunit, whereas our map at 7.5 Å resolution allows us to delineate the four transmembrane α-helices within each subunit. We eventually hope to compute maps at lower resolution in the presence and absence of oleamide so that we can address whether the subunits slide and tilt during closure, as was suggested by Nigel.

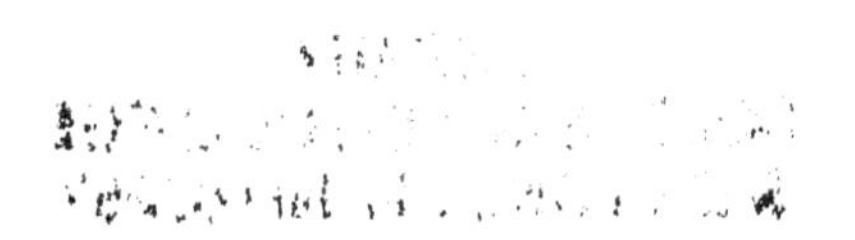

*Dermietzel:* Is your model more physiological than Nigel's? Because his treatment involved a deficit of calcium that was extremely non-physiological.

*Yeager:* I imagine that most gap junction physiologists would say that both Nigel's work and our work is non-physiological. The possibility exists that the conformational state of the channel may not be retained during isolation and crystallization of the gap junctions. I am not aware of any approach that will allow us to assay the functional state of the channel within the frozen–hydrated 2D crystals.

*Nicholson:* Our observations in oocytes suggest that when an oocyte is perfused intracellularly, neither pH, nor calcium alone are incapable of gating at least Cx43 channels. Cytosolic factors are required. So what Nigel was looking at may, or may not, be the gating of the channel.

*Warner:* The way in which calcium and oleamide interact with the connexins is going to be different because one is lipid soluble and will penetrate the membrane, whereas the other is not. This may account for the different crystallographic pictures, and both models may therefore turn out to be correct. However, there is a problem with all these models in that there is apparently still a gap in the closed conformation, even if it is only 7 Å or so. On a crystallographic map can you detect interactions across the channel wall between the different rods that line the channel? There could be a functional block that you might not see crystallographically if there were ionic interactions between the amino acids of the channel lining in localized regions of the connexin.

*Yeager:* An excellent example of this is Nigel Unwin's analysis of the acetylcholine receptor at a resolution of 9 Å (Unwin 1993, 1995). Leucine residues from the M2 transmembrane segments of the five receptor subunits are thought to form the gate within the receptor molecule that is ~150 Å in length. When you project the density along the full length of the channel into a plane parallel with the membrane, the pore appears to be in an open conformation even when the channel is in fact closed. The pore appears open because the five leucine residues only account for a thickness of a few ångströms, and the projected density is dominated by the aqueous channel which spans the full length of the receptor molecule.

*Warner:* If interactions between single amino acids, or small numbers of amino acids, cannot be distinguished crystallographically, then this should be borne in mind when discussing pore size, selectivity and channel permeability.

*Sanderson:* You state that oleamide affects the structure of the channel. Therefore, if you express Cx43 gap junctions in cell types in which they are not normally found won't the differences in background levels of membrane lipids affect the structure of the channel? You are putting Cx43 from cardiac cells into BHK hamster cells, and then you're using oleamide, based on findings that are related to glial neuron interactions, so will Cx43 look the same in glial cells, cultured cells and cardiac cells?

*Yeager:* That's a good point. We have to be clear about the system we're examining. I see the lipid bilayer as a solvent for the membrane protein, and depending on the solvent the conformation can certainly change.

*Kistler:* The problem is not too serious because you can do functional experiments on the cells from which you isolate the gap junctions.

*Sanderson:* It may not be so simple. For example, in the liver there are many gap junctions and, thus, a high level of conductance, so why do cardiac cells have Cx43 as opposed to Cx32 ($\beta_1$) to conduct current? There may be subtle differences in the ways in which connexins conduct current in different environments.

*Yeager:* It is notable that there are numerous examples in the literature on voltage-regulated and ligand-gated channels demonstrating that their functional properties can be influenced by fatty acids (reviewed by Ordway et al 1991).

*Goodenough:* Would it be possible to do similar experiments to those performed by Makowski et al (1984) using an electron-dense medium in your frozen specimens to demonstrate whether or not you had a completely open channel in either of the isolated states?

*Yeager:* That is a good suggestion. One possibility is to use aurothioglucose. The idea is that aurothioglucose would allow modulation of the solvent density. If the reagent can penetrate the channel, then you would expect to see an increased density in the aqueous regions all the way through the channel, whereas if the channels are in a closed state, then the reagent would not have access to the extracellular vestibule. Hence, one would see decreased solvent density in the extracellular vestibule compared with the cytoplasmic aqueous space.

*Nicholson:* I would like to show some of our recent data that fit well with the images Mark Yeager showed in his presentation. We are interested in the docking of connexins, and because we are better set up to do mutagenesis rather than to determine the structure directly, we focused on conserved cysteine residues in the extracellular loops in an effort to map which ones connect to which. It has previously been demonstrated by John & Revel (1991) and Rahman & Evans (1991), that disulfides form, but in an exclusively intramolecular manner. Subsequently, Dahl et al (1992) showed that if any of the cysteines are knocked out, the channel no longer works. Our strategy was that if any of the cysteines are moved the resulting channel would probably be non-functional. However, if the cysteine to which it is paired is also moved by a similar number of residues, then the disulfide would be reformed and function may be reconstituted. We focused on the first and the third cysteine in each extracellular loop. We moved the first cysteine 'back' towards the N-terminus and the third cysteine 'forward' towards the C-terminus. Within the loops, both of these movements should bring the cysteines closer to the membrane. We consistently found that if we moved the first cysteine in one loop back two residues and the third cysteine in the other loop forward two residues, then the function of the channel was

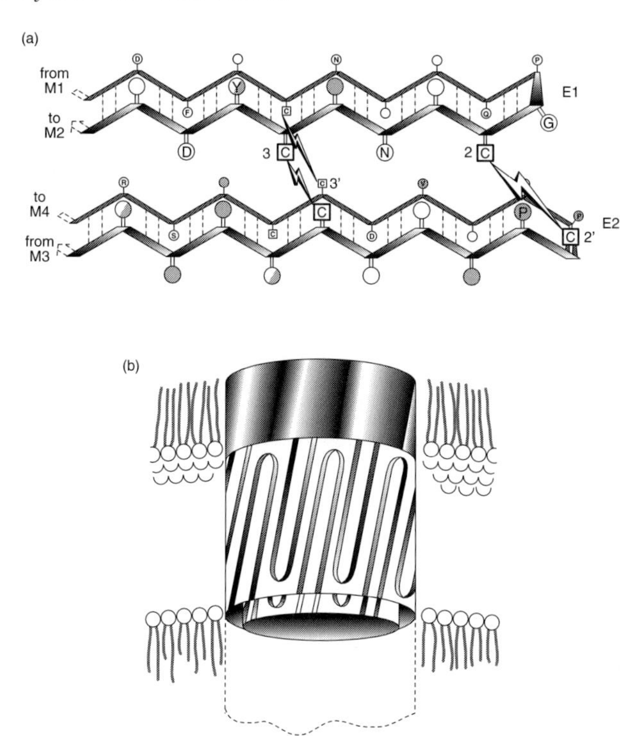


FIG. 1. A model of the docking interface between gap junctional hemichannels based on mutagenic mapping of the disulfides between the extracellular loops of connexin32 ($\beta_1$). (a) The two extracellular loops of a single connexin that are deduced to form stacked $\beta$-sheets held in place with three interloop disulfides. (b) The proposed means by which these loops could interdigitate during gap junction formation to form two concentric $\beta$-barrels (termed a $\beta$-zip) that would provide the electrically tight seal between hemichannels in the extracellular space. Reproduced from Foote et al (1998) with permission.

reconstituted. This was true for the first cysteine in E1 (first extracellular loop) and the third cysteine in E2 (second extracellular loop), or the third cysteine in E1 and first cysteine in E2. We found that periodicity was important, since movements of two worked, but movements of one and three did not. We also found that if we moved both first and third cysteines within one loop by two residues, then we could reconstitute channel function. Our explanation was that the loops have the ability to slide with respect to one another, and there was enough tolerance in the system that allowed for regeneration of channel function by sliding between the loops by one $\beta$-sheet repeat. If we moved the first and third cysteines within E2 four residues away from their original positions, function was not reconstituted. Perhaps this was because sliding between the loops by two $\beta$-sheet repeats was beyond the tolerance limits of channel docking. If this were true, and our model of inter-loop β-sheets disulfides is correct, then if we combine this mutant with movements of the first and third cysteines in E1 by two residues, the cysteines between loops would be closer together again and function would be reconstituted. This experiment worked, i.e. the movement of the first and third cysteines in E2 by four residues killed channel activity but if the first and third cysteines in E1 were also moved by two residues, channel activity was reconstituted—so a quadruple mutant worked better than a double mutant. We haven't touched the middle cysteines in each loop, but it seems likely that they may also form a disulfide. The model comprises two β-sheets that are stabilized by inter-loop disulfides, and we imagine that these will interdigitate with the β-sheets from a connexin in the apposed cell in a staggered pattern. We have termed this structure a β-zip (Fig. 1; Foote et al 1998). The result would be two concentric β-barrels, with one of the loops contributing to the inner barrel and the other loop contributing to the outer barrel. One is E1 and the other is E2, but we don't know which is which. An electrical seal is created because of the stable anti-parallel β-barrel. It would be difficult to disrupt these hydrogen bonds and hence neither water nor ions could readily traverse the channel wall, even in the absence of a hydrophobic membrane barrier. One problem with the model is that each of the two β-barrels have 24 strands, and therefore one cannot be continuous because it has a larger radius than the other. This fits well with Mark Yeager's images of the extracellular space, which show an inner complete ring and an outer ring with breaks in it. Links are also present between the outer and inner ring, which may represent disulfide bonds, or at least large hydrophobic residues. I am anxious to see the results of higher resolution experiments to see if this model fits.

*Yeager:* Your model appears to be consistent with our data, but to resolve β-sheets we will need to derive a map at higher resolution.

*Musil:* Dahl et al (1992) have speculated in the past that the disulfide bonding of the extracellular loops may change upon docking of a channel. Does your mutagenesis analysis give any insights into possible mutations that would affect channel docking but not connexon assembly?

*Nicholson:* It provides some indirect evidence. One could imagine for example that docking causes a switch from inter-loop to intra-loop disulfide bonds. When we move the two cysteines in the same loop by two residues we can still form functional channels. Our guess was that if we move the cysteines in both loops by two residues, the protein would be more likely to accommodate possible disulfide rearrangements, and this would enhance channel function. However, we found that this quadruple mutant didn't work as well as either of the double mutants. This result does not support the idea that there are changes in disulfide bonds. Other patterns, such as the formation of inter-molecular disulfide bonds, are possible, but our mutagenesis analysis would not address this.

*Gilula:* Have you looked at any phenotype other than channel activity? For example, have you looked at whether the products of these mutations assemble and make it to the cell surface?

*Nicholson:* Many of the mutants did not work, and we would have loved to have done that analysis but the oocyte is not particularly amenable to those studies, so we are unable to give any definitive answers. We have looked at them in an *in vitro* translation system, and with one exception all of the mutants are inserted into the membrane with the appropriate topology, and we can even detect disulfide formation. It is surprising that when we just move one cysteine, for example, we can no longer detect disulfide formation between the two loops, whereas in the double cysteine movements that are compatible with functional channels we can detect such disulfides. I suspect that the structure is balanced on a knife edge, such that if one of the pairs of disulfides are removed, inappropriate disulfide bonds are generated. Whether or not that results an accumulation of the protein inside the cell is not known.

## References

Dahl G, Werner R, Levine E, Rabadan-Diehl C 1992 Mutational analysis of gap junction formation. Biophys J 62:171–182

Foote CI, Zhou L, Zhu X, Nicholson BJ 1998 The pattern of disulfide linkages in the extracellular loop regions of connexin 32 suggests a model for the docking interface of gap junctions. J Cell Biol 140:1187–1197

Guan X, Cravatt BF, Ehring GR et al 1997 The sleep-inducing lipid oleamide deconvolutes gap junction communication and calcium wave transmission in glial cells. J Cell Biol 139:1785–1792

John SA, Revel J-P 1991 Connexin integrity is maintained by non-covalent bonds: intramolecular disulfide bonds link the extracellular domains in connexin 43. Biochem Biophys Res Commun 178:1312–1318

Makowski L, Caspar DL, Phillips WC, Goodenough DA 1984 Gap junction structures. V. Structural chemistry inferred from X-ray diffraction measurements on sucrose accessibility and trypsin susceptibility. J Mol Biol 174:449–481

Milks LC, Kumar NM, Houghten R, Unwin N, Gilula NB 1988 Topology of the 32-kd liver gap junction protein determined by site-directed antibody localizations. EMBO J 7:2967–2975

Ordway RW, Singer JJ, Walsh JV Jr 1991 Direct regulation of ion channels by fatty acids. Trends Neurosci 14:96–100

Rahman S, Evans WH 1991 Topography of connexin 32 in rat liver gap junctions. Evidence for an intramolecular disulfide linkage connecting the two extracellular peptide loops. J Cell Sci 100:567–578

Unwin N 1993 Nicotinic acetylcholine receptor at 9 Å resolution. J Mol Biol 229:1101–1124

Unwin N 1995 Acetylcholine receptor channel imaged in the open state. Nature 373:37–43

Unwin PNT, Ennis PD 1984 Two configurations of a channel-forming membrane protein. Nature 307:609–613

Zhou XW, Pfahnl A, Werner R et al 1997 Identification of pore lining segments in gap junction hemichannels. Biophys J 72:1946–1953

# General discussion I

## Channel rectification

*Werner:* Three different connexins are expressed in the heart, Cx43 ($\alpha_1$), Cx40 ($\alpha_5$) and Cx45 ($\alpha_6$). Ventricular fibrillation is usually caused by ectopic foci of cells that depolarize causing an action potential to move in the opposite direction. Is it possible that there is a need in heart muscle for a rectifying channel, in the sense that action potentials can only go from the node towards the apex of the heart? If the channels were all rectifying, then the action potential would not be able to return. Has anyone looked at the distribution of different connexins within myocytes to see whether they are asymmetrically distributed within a single cell?

*Severs:* We have quite a lot of information on connexin distribution in the heart. There are clear differences in the connexin profiles in different of tissues of the heart, which are associated with different conduction properties. Cx40 and Cx45 are present in the atrioventricular conduction system, and Cx43 is also expressed at the distal ends joining to working myocardium. However, we don't know the *in vivo* configurations of the individual channels, so we can't predict whether rectifying channels are present.

*Goodenough:* Is this in the mouse?

*Severs:* There are some exceptions, but many of the features of connexin distribution are common to a number of different species. Atrial myocardium typically has a lot of Cx40 plus Cx43, and working ventricular myocardium is almost entirely composed of Cx43, although there may be a slight trace of Cx45. A rectifying system is unlikely to operate in these tissues, but there is one place it might conceivably have a role, i.e. the sinoatrial node. A rectifying system, if it existed here, could protect the sinoatrial node from the hyperpolarizing influence of the surrounding atrial tissue.

*Warner:* Much functional work has been done on the patterns of flow of the action potential through the heart. There is no evidence for rectifying gap junctions, even at the sinoatrial node. Denis Noble and colleagues are building models that reproduce the electrical activity of the intact heart, and Noble & Winslet (1997) recently reconstructed transmission from the sinoatrial node to the atrium, including the contribution of gap junctions. They show that one-way transmission of the action potential from node to atrium is facilitated by the anatomical relationship between sinoatrial cells and atrial cells. Atrial cells have finger-like protrusions which interdigitate with node cells so that each atrial cell is surrounded by, and in gap junctional contact with, many sinoatrial cells (see Noble & Winslet 1997). This overcomes the low density of gap junctions in the sinoatrial node and, together with the refractory period of the atrial action potential, which prevents rapid re-excitation behind an invading action potential, ensures one-way transmission without local re-entry.

*Severs:* With regard to current flow in relation to connexin distribution, well-known examples are the differences in the electrical properties between, for example, the fast-conducting Purkinje fibres and the slowing down of the impulse through the atrioventricular node, and these properties do correlate with spatial differences in connexin distribution.

*Beyer:* In some of these cases, the presence of a particular connexin may not be the only crucial factor. In the generation of re-entrant arrhythmias, the cellular distribution of gap junctions, at least at the level of end-to-end versus side-to-side distribution, may result in large differences in the relative anisotropy of conduction between cells. These differences may be enough to make a major contribution to the generation of local re-entrant circuits, especially in cases such as infarct border zones where there may be a decreased abundance of gap junctions.

*Gilula:* Do we have any fundamental understanding of the structural basis for rectification?

*Nicholson:* Rectification of junctional conductance has been observed in Cx26/Cx32 ($\beta_2/\beta_1$) heterotypic pairs (Barrio et al 1991). There is a threefold difference in conductance recorded at transjunctional potentials of $-100$ mV and $+100$ mV. This can now be explained by connecting, in series, two channels with different ionic activities. If two channels with different preferences for anions and cations are connected, there are problems in trying to maintain equal currents in the two channels. With one polarity of pulses, cations and anions enter the respective ends of those channels faster than they can leave. The result is that ions accumulate inside the channel to the point when the diffusion barrier equalizes the currents in the two hemichannels. If the polarity is reversed, exit occurs at a faster rate than entry, so the channel is depleted of ions until the concentration gradients equalize the currents. Hence, a $+100$ mV field (Cx26 side positive) is applied across an effective ion concentration that is 50% higher than bulk cytoplasm, whereas a 100 mV field of the opposite polarity is applied across an ion concentration that is 50% lower than the cytoplasm, thus generating rectification. This is a moderate rectification, but it's not steep.

*Gilula:* That is in a model system with a heterotypic interaction. But what do we know about sites of rectification *in vivo* in terms of utilization and diversity of connexin genes?

*Fletcher:* Nedergaard (1994) has shown that in astrocyte neuron coupling, in which there is a 15–20 mV potential difference, there is rectification. If the signal affects the astrocyte, and the signal transduction mechanism affects a triply charged molecule, such as inositol-1,4,5-trisphosphate ($InsP_3$), then 85% of the $InsP_3$ will go to the neuron. On the other hand, if the signal affects the neuron, and $InsP_3$ is the second messenger, only 15% of the signal will go to the astrocyte. This also seems to work for smooth muscle endothelial cell interactions in culture.

*Giaume:* I would like to mention that the group of A. Campos de Carvalho and V. Moura Neto from Brazil have recently reported the occurrence of dye and electrical coupling between neurones and astrocytes in culture (M. Froes-Ferrao A. H. P. Correia, J. Garcia-Abreu, D. C. Spray, V. Moura Neto & A. Campos de Carvalho, unpublished paper, International Gap Junctions meeting, Key Largo, USA, 12–17 July 1997). In addition, a member of this group, M. Froes Ferrao, is now working in my laboratory and we observed that Lucifer yellow also diffuses in the reverse direction, i.e. from astrocytes to certain neurons.

*Gilula:* This is interesting, but again let me ask whether any observations have been made that would lead us to understand how the multigene connexin family is utilized to generate a gap junction with rectification properties. It seems that no-one here can answer this question quickly, so I propose that this area deserves some serious consideration. Investigators who are using models should give some thought as to how they could construct a model that would reflect what may not yet be known about the *in vivo* utilization of the multigene connexin family.

*Warner:* Rectification is an interesting functional property of gap junctions that has received relatively little attention. This may be because of technical difficulties, particularly with respect to ionic current flowing through the junction. In any multicellular system (not just the heart) the experiment has to be set up with the electrodes some distance apart to allow current rectification at the gap junction level to be detected. Even strong rectification may not be apparent. My suspicion is that few people have set up their experiments in a way that will answer the question.

*Sanderson:* We should not restrict this discussion to just current rectification. Large signalling molecules that go through gap junctions are probably equally important and can display rectification. It is relatively easy to detect signal rectification if you're looking at molecules such as $InsP_3$ or cAMP with imaging techniques. Just concentrating on current rectification, which may be important in the heart, may not reflect the kind of rectification that occurs in non-excitable cells.

*Giaume:* Caution should be taken when we talk about rectification at electrical synapses. Indeed, most of the time rectification for junctional currents is due to a difference in voltage-dependence properties of the two hemichannels forming junctional channels (Giaume et al 1987). However, in some cases rectification at electrical synapses is due to a difference in properties of the non-junctional membranes (Smith & Baumann 1969). For intercellular calcium signalling, rectification or asymmetry of the propagation of calcium waves between two cell types could also be due to differences in the environment rather than to junctional channels themselves. For example, the respective location of $InsP_3$ receptors, the level of filling of internal calcium pools or calcium buffering property of the two cytoplasms could be also critical elements that may account for a difference in calcium wave propagation between two different cell types.

## The structural diversity of connexins

*Gilula:* What do we know about this multigene family with respect to generating structural and functional diversity? We talk in detail about the molecular organization of the different members of the multigene family to generate an oligomer, but the only major piece of information we have is that the arthropod gap junction structure is dramatically different. However, to my knowledge this has not been exploited by people who are trying to understand the relationship of the gap junction genes to gap junction structure.

*Goodenough:* We don't have any data on this point at the crystallographic level. Results reported in the freeze-fracture literature suggest that there is much diversity with respect to the packing of connexons in junctional plaques, although there is a long-standing argument about what these packing differences mean, particularly changes in packing that occur as a result of uncoupling.

*Gilula:* Has there been any progress in understanding why different cell types use different connexins to make plaques with different organizations?

*Kistler:* I can only add that packing is probably influenced by the length of the carboxyl tail. In the lens there is an unordered and loose packing arrangement of connexons in the plaques, and when you do reconstitution experiments using the whole-length 'native' channel, then the unordered arrangement is reconstituted. In contrast, when you use channels cleaved either *in vitro* with calpain or *in vivo* by maturation-associated processing, a close hexagonal packing arrangement is observed. I believe this may be one of the reasons why Mark Yeager uses truncated Cx43.

*Yeager:* The truncated form is better ordered, but we do have some preliminary results with full-length Cx43.

*Gilula:* We have looked at other connexins in heterologous expression systems and found that some have a more ordered plaque arrangement when they are expressed as full-length molecules. Therefore, I would say we don't know enough yet to draw any conclusions.

*Musil:* I seem to remember that Nalin Kumar has shown that full-length Cx46 ($\alpha_3$) expressed in BHK hamster cells forms well-ordered crystalline arrays.

*Kumar:* Yes. When we expressed full-length Cx46 in BHK hamster cells it formed crystalline arrangements in the membrane that were as extensive as when the truncated form of Cx43 is expressed in these type of cells.

*Musil:* In which case it would seem that whether full-length Cx46 packs into unordered or into hexagonal arrays depends on the system in which it is expressed. One possibility is that the unusual lipid composition of lens fibre cell plasma membranes somehow interferes with ordered channel packing; if so, then the exogenous lipids used in reconstitution experiments must also have the same ability.

*Kumar:* Another question is, how do post-translational modifications affect packing? This question is relevant to Cx46 because post-translational modifications that occur in the lens do not occur when Cx46 is expressed in BHK cells.

*Goodenough:* Last summer at the International Gap Junctions meeting in Key Largo, USA (12–17 July 1997), Alan Lau presented some evidence for an interaction between ZO1 and Cx43, and I would like to ask if there are any recent updates on this?

*Lau:* We have evidence that strongly suggests that Cx43 interacts with the tight and adherans junction-associated protein, ZO1, and this interaction seems to be between the PDZ2 domain of ZO1 and the hydrophobic C-terminus of Cx43 containing isoleucine. Most connexins have hydrophobic C-termini, containing either valine, isoleucine or leucine, that seem to be important for their interaction with other proteins. We are in the process of determining whether connexins that lack these hydrophobic C-termini are unable to interact with the PDZ2 domain of ZO1. We are also interested in demonstrating the co-localization of Cx43 with ZO1 in cells that we have used for our biochemical studies of this interaction. However, I'm sure you're aware of the work from Yvonne Munari-Silem's laboratory reporting the co-localization of Cx43 with ZO1 in thyroid epithelial cells by laser confocal microscopy (Guerrier et al 1995). These cells, which coexpress Cx43 and Cx32, showed an intriguing differential distribution of Cx32 to lateral plasma membrane regions, whereas Cx43 gap junctions were superimposed with ZO1-containing tight junctions in subapical membrane regions. These data suggest a possible function for this putative interaction between Cx43 and ZO1, but of course, this more difficult question must be resolved in our future work.

*Gilula:* David Vaney is going to talk to us about the retina at this symposium, so I would like to ask him whether the current studies of the retina that are beginning to identify which connexins are being expressed by specific cell types will give us a better appreciation of how the utilization of the various members of the connexin multigene family contribute to the specified structural organizations characteristic of gap junctions between different cell types.

*Vaney:* It's extraordinary that the retina has been so well characterized in terms of the patterns and physiology of cellular coupling but we know comparatively little about the identity of the connexins in the mature retina. It's only recently that preliminary data have emerged on the identity of the connexins in retinal neurons as opposed to the connexins in glial cells and connective tissue. There seems to be only limited evidence for connexin diversity, which is odd because this is a tissue in which there is strong evidence for functional diversity in terms of permeability and modulation.

*Lo:* You mentioned that there is little connexin diversity in the retina, in which there are differences in terms of permeability etc., but what evidence do we have that different members of the gene family are absolutely required in the various tissues in which they have been shown to be expressed? In other words, can they be exchanged?

*Gilula:* These issues will clearly come out when we discuss some of the knockout data. When a gene is knocked out, and apparently there are no impacts on function, it will be necessary to identify what is actually providing that function. Another concern is when sequences have been mutated but the effect isn't dramatic enough to know exactly what's going on.

## References

Barrio LC, Suchyna T, Bargiello T et al 1991 Gap junctions formed by connexin 26 and 32 along and in combination are differently affected by applied voltage. Proc Natl Acad Sci USA 88:8410–8414 (erratum: 1992 Proc Natl Acad Sci USA 89:4220)

Giaume C, Kado RT, Korn H 1987 Voltage-clamp analysis of a crayfish rectifying synapse. J Physiol (Lond) 386:91–112

Guerrier A, Fonlupt P, Morand I et al 1995 Gap junctions and cell polarity: connexin32 and connexin43 expressed in polarized thyroid epithelial cells assemble into separate gap junctions, which are located in distinct regions of the lateral plasma membrane domain. J Cell Sci 108:2609–2617

Nedergaard M 1994 Direct signaling from astrocytes to neurons in cultures of mammalian brain cells. Science 243:1768–1771

Noble D, Winslet R 1997 Reconstructing the heart: network models of SA node–atrial interactions. In: Panfilov AV, Holden AV (eds) Computational biology of the heart. Wiley, Chichester, p 49–64

Smith TG, Baumann F 1969 The functional organization within the ommatidium of the lateral eye of Limulus. Prog Brain Res 31:313–349

# Trafficking pathways leading to the formation of gap junctions

W. H. Evans, S. Ahmad, J. Diez, C. H. George, J. M. Kendall and P. E. M. Martin

*Department of Medical Biochemistry, University of Wales College of Medicine, Heath Park, Cardiff, CF4 4XN, UK*

*Abstract.* This chapter reports the mechanisms resulting in the assembly of gap junction intercellular communication channels. The connexin channel protein subunits are required to oligomerize into hexameric hemichannels (connexons) that may be homo- or heteromeric in composition. Pairing of connexons in contacting cells leads to the formation of a gap junction unit. Subcellular fractionation studies using guinea-pig liver showed that oligomerization of connexins was complete on entry into Golgi, and that connexons showed heteromeric properties. The low ratio of connexin26 (Cx26; $\beta_2$) relative to Cx32 ($\beta_1$) in endomembranes compared to the approximately equal ratios found in plasma membranes and gap junctions suggest that Cx26 takes a non-classical route to the plasma membrane. Cultured cells, expressing connexin–aequorin chimeras, also provided evidence that Cx26 takes a more rapid non-classical route to the plasma membrane, because brefeldin A, a drug that disrupts the Golgi, had minimal effects on trafficking of Cx26 to the plasma membrane in contrast to its disruption of Cx32 trafficking. Finally, a cell-free approach for studying synthesis of connexons provided further evidence that Cx26 showed membrane insertion properties compatible with a more direct intracellular route to gap junctions. The presence of dual gap junction assembly pathways can explain many of the differential properties exhibited by connexins in cells.

*1999 Gap junction-mediated intercellular signalling in health and disease. Wiley, Chichester (Novartis Foundation Symposium 219) p 44–59*

Gap junctions are constructed from large numbers of aligned hexameric channels embedded in the plasma membrane at areas of cell–cell contact. They provide pathways allowing fast and unhindered intercellular communication and signalling, increasingly recognized to facilitate integrative behaviour in tissues and organs. The number of intercellular channels may be small, as observed in cultured cells, but in tissues the channels accrete into large plaques displaying a characteristic morphology much studied by microscopic methods. Gap junctions are constructed from a family of connexin proteins that oligomerize in the membrane to generate hexameric hemichannels, the connexons (Goodenough et al 1996). Key questions arising in considering the architecture of these membrane communication channels concern the type of connexin subunits used (since cells are likely to express two or more channel subunit proteins thereby allowing for the generation of homo- or heteromeric channels), the stoichiometry and arrangement of the subunits surrounding the channel (since this can potentially regulate channel permeability), and thus the transmitted messengers and the routes followed inside the cell by unassembled and assembled subunits as they are targeted to gap junctions. These fundamental general considerations are illustrated in Fig. 1.

This chapter describes some recent developments in the field of gap junction assembly. It describes three complementary techniques that were applied to trace the pathways followed by two connexins, connexin26 (Cx26; $\beta_2$) and Cx32 ($\beta_1$) as they generate gap junctions in liver tissue or cultured mammalian cells. A cell-free reconstitution system also provided inputs into how and where connexins are inserted into membranes. Connexins, of which 13 isoforms have been identified in rodents, display a conserved topography in membranes, with cytoplasmic carboxyl and amino termini, four membrane-traversing regions, two extracellular loops and one intracellular loop. Superficially, the major difference perceived between Cx32 and Cx26 is that Cx32 has a longer cytoplasmic carboxyl tail. The present account provides evidence that connexins appear to follow different trafficking pathways to gap junctions, although connexons and gap junctions constructed of both connexin subunits are generated in liver tissue.

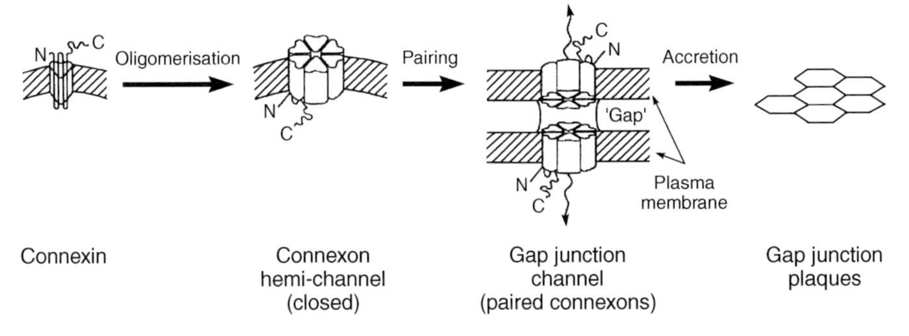


FIG. 1. Scheme showing the major stages in the assembly of gap junctions. Connexin protein subunits oligomerize to form hexameric connexon hemichannels; these are constructed of the same connexin (homomeric) or from two or more connexin isoforms (heteromeric). Docking of hemichannels at points of cell–cell alignment generates the gap junction channel, and large accretions of these form gap junction plaques, observed by morphological techniques as hexagonal arrays. Note that fusion of aequorin, a $Ca^{2+}$-dependent photoprotein, to the C-terminus of a connexin positions the reporter in the $Ca^{2+}$ environment at the cytoplasmic aspect of the endoplasmic reticulum and Golgi and immediately beneath the plasma membrane.

## Dual gap junction assembly pathways delineated in liver tissue

Subcellular fractionation of tissues or cells provides a method for dissecting out the major components of the trafficking pathways interconnecting organelles and endomembranes. The secretory pathway of eukaryotic cells delivers newly synthesized membrane proteins from the endoplasmic reticulum to the Golgi apparatus, which is credited with major roles in dictating membrane protein secretory trafficking, especially to the plasma membrane. Characterized subcellular fractions originating from specific stations on the secretory route can therefore be analysed to determine connexin composition at that point and the extent of their oligomerization into connexon hemichannels (Evans 1994). A range of anti-peptide antibodies generated to specific sequences on various connexins can be applied to probe their relative distribution and oligomeric status in different regions of the cell (Evans et al 1992).

Although extensive immunocytochemical studies of tissues such as liver (Kojima et al 1994) or heart (Severs et al 1993) identify prominent gap junction plaques at cell–cell contact areas, subcellular fractionation approaches indicate that substantial intracellular levels of connexins exist in rat liver (Rahman et al 1993). In addition to the high levels of Cx32 present in plasma membranes, appreciable amounts of the protein were detected using site-specific connexin antibodies in microsomal, Golgi and lysosomal fractions, and this was interpreted to reflect aspects of protein trafficking underlying the rapid synthesis and breakdown of gap junctions. Using guinea-pig liver, where approximately equal amounts of Cx32 and Cx26 are expressed, the relative distribution of both connexins was investigated in subcellular fractions enriched in endoplasmic reticulum-derived vesicles, Golgi and plasma membranes (Diez et al 1998). The results showed that although both connexins were present in equimolar amounts in isolated gap junctions and plasma membranes, Cx26, in contrast to Cx32, was barely detectable in microsomal and Golgi fractions (Fig. 2A). It was necessary to employ a higher titre antibody, generated to a different Cx26 epitope, to detect the relatively small amounts of this connexin in these endomembrane fractions (Fig. 2A). When the oligomeric state of Cx32 and Cx26 in the various fractions was examined, the results showed that both connexins were oligomeric, i.e. connexons were present in plasma membranes and Golgi membranes (Fig. 2B). In contrast, monomers and oligomeric connexin intermediates were detected in endoplasmic reticulum-derived vesicles, as well as a fraction containing vesicles derived from a location between the endoplasmic reticulum and Golgi (Schweitzer et al 1990). These results show that connexins oligomerize on entry into the Golgi apparatus. In view of the low Cx26 levels relative to Cx32 in the endoplasmic reticulum and Golgi sectors of the secretory pathway, we draw two conclusions. First, a high proportion of Cx26 arrives at the plasma membrane and

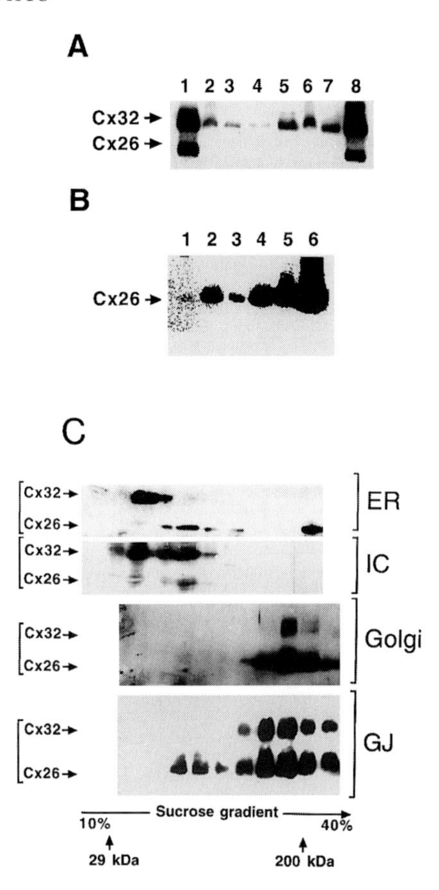


FIG. 2. Connexin26 (Cx26; $\beta_2$) and Cx32 ($\beta_1$) distribution and oligomeric status in gap junctions and subcellular fractions isolated from guinea-pig liver by SDS-PAGE and western blotting. (A) Lane 1, gap junctions; lane 2, endoplasmic reticulum; lane 3, endoplasmic reticulum–Golgi intermediate; lane 4, Golgi light fraction; lane 5, Golgi intermediate fraction; lane 6, Golgi heavy fraction; lane 7, sinusoidal plasma membranes; lane 8, lateral plasma membranes. The positions of Cx32 and Cx26 recognized by antibodies to each connexin are shown. (B) Lane 1, endoplasmic reticulum; lane 2, Golgi intermediate fraction; lane 3, Golgi heavy fraction; lane 4, sinusoidal plasma membranes; lane 5, lateral plasma membranes; lane 6, gap junctions. The proteins were identified by an anti-peptide Cx26 antibody of higher titre than used in A. (C) Subcellular fractions corresponding to endoplasmic reticulum (ER), endoplasmic reticulum–Golgi intermediate fraction (IC), Golgi and gap junctions (GJ) were solubilized in dodecyl maltoside detergent and the extracts fractionated on linear sucrose gradients. Each fraction was analysed by SDS-PAGE to determine whether the fractions consisted of connexins (migrating at the top end of the gradients), connexons (migrating at the bottom end of the gradients) or as intermediate products. Arrows show the position of Cx26 and Cx32. The results show that oligomerization is complete in Golgi and plasma membrane fractions, with connexin monomers and oligomeric intermediates present in the endoplasmic reticulum and the endoplasmic reticulum–Golgi intermediate fraction.

gap junction in guinea-pig liver by a route that bypasses the Golgi apparatus. Second, small amounts of Cx26 traffic through the Golgi apparatus, probably as heteromeric connexons that account for the presence of heteromeric gap junctions demonstrated by immunoprecipitation analysis of gap junctions (Diez et al 1998).

## Trafficking studies in live mammalian cells provide evidence for two gap junction assembly pathways

Membrane proteins with attached reporter groups, synthesized by recombinant technology, are proving useful in delineating intracellular trafficking pathways, especially to and from the Golgi apparatus (Lippincott-Schwartz 1998). Aequorin (Aeq), a $Ca^{2+}$-dependent photoprotein, was attached to the carboxyl termini of several connexin isoforms to allow measurement of the cytosolic calcium environments traversed by connexins during their assembly into gap junctions in transfected African green monkey COS kidney cells (Martin et al 1998). Connexin–Aequorin chimeras appear to traffic constantly from intracellular stores, where the $Ca^{2+}$ levels are 10–20-fold lower than at the plasma membrane (George et al 1998). The intracellular stores where connexins are located were shown, using antibodies to aequorin, to be present in 'distal' endoplasmic reticulum–Golgi apparatus regions of the cell by immunocytochemistry and by subcellular fractionation of COS cells transfected with cDNAs encoding the relevant connexin–Aequorin chimeras (C. H. George, unpublished work 1998). In cells expressing Cx26–Aeq, the aequorin reporter group was inactive, but chemiluminescent activity was retained when the tail of Cx26 was replaced with that of Cx43 ($\alpha_1$). This chimera is designated Cx26/43T–Aeq (George et al 1998). The aequorin moiety of the Cx–Aeq chimeras is reconstituted into active photoprotein *in situ* by binding the labile cofactor coelenterazine. Incubation of COS cells (expressing Cx–Aeq) in zero extracellular $Ca^{2+}$ retains the aequorin activity of plasma membrane-associated chimeras and, upon addition of 1 mM $Ca^{2+}$ to the medium, the influx of $Ca^{2+}$ triggers chemiluminescent light emission. These studies (C. H. George, unpublished work 1998) show that Cx32–Aeq takes about 10–15 min to transfer from intracellular stores to the plasma membrane. In contrast, Cx26/43T–Aeq arrived at the plasma membrane from less well-defined intracellular stores within 5 min. By carrying out these studies in the presence of inhibitors, details of the routes followed by the connexin chimeras to the plasma membrane emerge. Figure 3 shows that brefeldin A, a drug that disassembles the Golgi apparatus and redistributes its membrane vesicles to a perinuclear position, thereby interrupting the secretory pathway, severely restricted the replacement of Cx32–Aeq at the plasma membrane by precursors arriving from intracellular stores, but it had little effect on Cx26/43T–Aeq at the plasma membrane. Nocodazole, a microtubule-

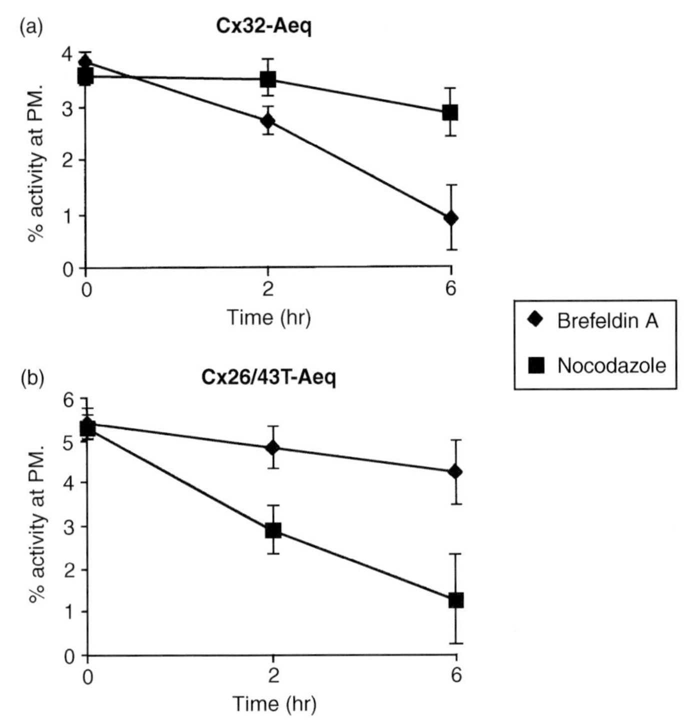


FIG. 3. Effects of brefeldin A (◆) or nocodazole (■) on trafficking of (a) connexin32 ($\beta_1$)-aequorin chimeras (Cx32–Aeq) and (b) a connexin26 ($\beta_2$)–aequorin chimera in which the tail of Cx26 was replaced with that of Cx43 ($\alpha_1$; Cx26/43T–Aeq) from intracellular pools to the plasma membrane in living COS African green monkey kidney cells. Cells expressing chimeras were incubated with either drug in the absence of extracellular $Ca^{2+}$. Plasma membrane-associated chimeras were selectively triggered upon the addition of 1.3 mM $Ca^{2+}$ to the perfusion medium (Marsault et al 1997). The amount of chimera at the plasma membrane was quantified using chemiluminescent activity of aequorin during influx of $Ca^{2+}$.

disrupting reagent, produced the opposite effect, severely inhibiting Cx26/43T–Aeq movement, but it had minimal effects on Cx32–Aeq levels at the plasma membrane.

In summary, studies with Cx–Aeq chimeras demonstrate that Cx32 traffics to the plasma membrane of COS cells via the secretory pathway, probably mediated by vesicular mechanisms that retain a high degree of functionality even after dissociation of microtubules. Cx26 trafficking to the plasma membrane, in contrast, appears to follow a route that, on the basis of its minimal disruption by brefeldin A, bypasses the Golgi. However, this non-classical route is critically dependent on the presence of intact microtubules in the cell.

## Evidence from cell-free studies that connexins show different membrane insertion and oligomerization properties

Most membrane proteins are inserted co-translationally into the endoplasmic reticulum by a series of orchestrated events that are at their most complex with multispanning proteins that include the connexins (High & Laird 1997). Connexins are proteins that lack a conventional signal sequence, and Cx32 and Cx26 are inserted co-translationally into endoplasmic reticulum-derived canine pancreatic microsomal membranes (Falk et al 1994, 1997). Cx26 is also inserted into membranes post-translationally (Zhang et al 1996, Ahmad et al 1998). The reasons underlying the post-translational insertion properties of Cx26 are unclear, although studies on the cystic fibrosis channel suggest that this property provides a second opportunity for insertion into the membrane of polytopic proteins lacking a signal sequence, and it involves topogenic determinants in the second transmembrane segment (Lu et al 1998).

Current work on the cell-free synthesis and oligomerization of various connexins shows that Cx26, unlike Cx32, is also inserted post-translationally into liver plasma membranes (S. Ahmad, unpublished work 1998). These results thus provide a mechanism for the direct and rapid trafficking of Cx26 to the plasma membrane via an intracellular pathway that need not implicate the Golgi apparatus. The cell-free analysis approach has also demonstrated that, in the presence of endoplasmic reticulum-derived microsomes, homomeric oligomers were the main forms, as shown by immunoprecipitation studies using antibodies to each connexin (Fig. 4). In contrast, supplementation of the *in vitro* reaction mixture with liver Golgi membranes resulted in the formation of heteromeric connexons, i.e. they were constructed of Cx26 and Cx32. Thus, unknown factors present in the Golgi apparatus catalyse the assembly of heteromeric connexons. However, the formation of homomeric connexons is initiated in the endoplasmic reticulum. Thus, the cell-free systems used appear to reinforce the general conclusions emerging from the other approaches.

## Synthesis

To study the mechanisms facilitating the assembly of connexin protein subunits into connexon hemichannels (Fig. 1), we pursued three independent and complementary experimental approaches. They all show that although the Golgi apparatus occupies a central position in the secretory trafficking route that is followed by connexins (and where post-translation modifications such as phosphorylation may occur; Laird et al 1995), Cx26 is also able to traffic to the plasma membrane and to gap junctions with minimal involvement of the Golgi. Combining the subcellular fractionation approach that provides a 'snapshot' view

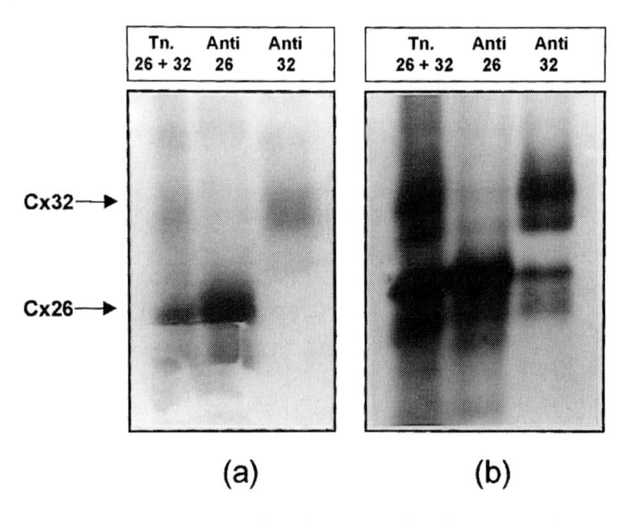


FIG. 4. Immunoprecipitation studies demonstrating homo- or heteromeric assembly of connexin26 (Cx26; $\beta_2$) and Cx32 ($\beta_1$) translated in a cell-free system. SDS-polyacrylamide gels showing in first lane of (a) and (b) Cx32 and Cx26 translated *in vitro* and supplemented with (a) canine pancreatic microsomal membranes and (b) in addition, liver Golgi membranes. The second and third lanes show the radiolabelled immunoprecipitated products using antibodies to Cx26 (second lane) or Cx32 (third lane). The figure shows that in the presence of Golgi membranes both antibodies immunoprecipitate two connexins, indicating heteromeric protein–protein interactions as would exist in connexons. In the absence of Golgi membranes, each antibody only immunoprecipitates its respective connexin, indicating that homomeric connexons were generated.

of trafficking with studies in which the movement of chimeric connexins to the cell surface in living cells is followed provided concordant conclusions regarding the presence of dual pathways leading to gap junction assembly. However, these pathways are not exclusive, for a small proportion of the Cx26 also traffics through the Golgi, thus accounting for the heteromeric connexons constructed of Cx26 and Cx32 that were detected by the immunoprecipitation approach, a powerful method of detecting protein–protein interactions. The two intracellular trafficking routes followed by the two connexins and the position in a generalized mammalian cell where the connexon hemichannels first form are shown in Fig. 5.

Mutational analysis approaches are also beginning to shed light on mechanisms of intracellular targeting of connexins and their assembly into functional gap junctions. In X-linked Charcot-Marie-Tooth disease, a form of hereditary motor and sensory neuropathy, studies on a number of mutations in Cx32 are providing information on specific amino acids and intracellular domains crucial for

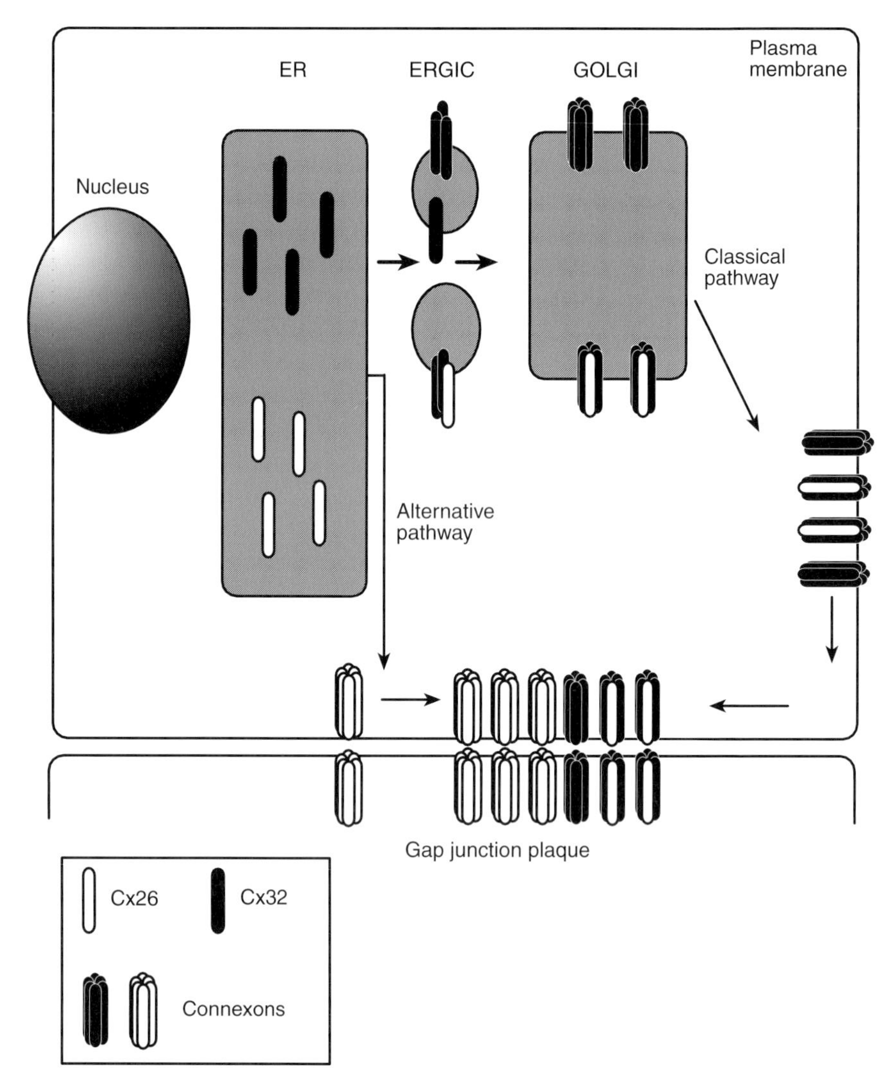


FIG. 5. Intracellular trafficking pathways of connexin26 ($\beta_2$) and Cx32 ($\beta_1$) in a generalized mammalian cell. Cx32 follows a classical pathway, with oligomerization into hexamers complete in the Golgi apparatus. A minor proportion of Cx26 also follows this route, and these form heteromeric connexons in which Cx26 exists at a low stoichiometric ratio to Cx32. The major route followed by Cx26 is via a pathway to the plasma membrane that bypasses the Golgi and leads to predominantly homomeric Cx26 connexons. ER, endoplasmic reticulum; ERGIC, endoplasmic reticulum–Golgi intermediate fraction.

trafficking to the plasma membrane (Deschênes et al 1997). This structure/function approach has identified amino acids at the amino terminus that are necessary for exit from the endoplasmic reticulum, and others that facilitate efficient oligomerization (P. E. M. Martin, unpublished work 1998).

This chapter illustrates facets of gap junction assembly on the assumption that connexins are the sole protein building blocks. Further work will, in due course, reveal the nature of any associated factors, e.g. $Ca^{2+}$-binding proteins such as calmodulin (Torok et al 1997), or other adhesion proteins, e.g. cadherins (Jongen et al 1991, Frenzel & Johnson 1996), that ensure the fidelity and rapidity of assembly and turnover. Gap junctions proteins, like other channel proteins (e.g. cystic fibrosis transmembrane regulator; Lukacs et al 1994), have half-lives that are approximately 10-fold shorter than the average membrane protein (Laird et al 1995). This rapid turnover of the channel subunits enables the extent and nature of intercellular cross-talk across gap junctions to be constantly modified.

*Acknowledgements*

This work was supported by a Medical Research Council Programme Grant and the Welsh Office for Research and Development.

## References

Ahmad S, Diez J, George CH, Evans WH 1998 *In vitro* synthesis and assembly of connexins into functional gap junction hemichannels, submitted

Deschênes S, Walcott JL, Wexler TL, Scherer SS, Fishbeck KH 1997 Altered trafficking of mutant connexin32. J Neurosci 17:9077–9084

Diez JA, Ahmad S, Evans WH 1998 Biogenesis of liver gap junctions. In: Werner R (ed) Gap junctions. IOS Press, Amsterdam, p 130–134

Evans WH 1994 Assembly of gap junction intercellular communication channels. Biochem Soc Trans 22:788–792

Evans WH, Carlile G, Rahman S, Torok K 1992 Gap junction communication channel: peptide and anti-peptide antibodies as structural probes. Biochem Soc Trans 20:856–861

Falk MM, Kumar NM, Gilula NB 1994 Membrane insertion of gap junction connexins: polytopic channel forming membrane proteins. J Cell Biol 127:345–355

Falk MM, Buehler LK, Kumar NM, Gilula NB 1997 Cell-free synthesis and assembly of connexins into functional gap junction membrane channels. EMBO J 16:2703–2716

Frenzel EM, Johnson RG 1996 Gap junction formation between cultured embryonic lens cells is inhibited by antibody to N-cadherin. Dev Biol 179:1–16

George CH, Kendall JM, Campbell AK, Martin PEM, Evans WH 1998 Reporting the calcium environment in trafficking pathways leading to gap junction biogenesis using connexin–aequorin chimerae. In: Werner R (ed) Gap junctions. IOS Press, Amsterdam, p 140–144

Goodenough DA, Goliger JA, Paul DL 1996 Connexins, connexons and intercellular communication. Ann Rev Biochem 65:475–502

High S, Laird V 1997 Membrane protein biosynthesis — all sewn up? Trends Cell Biol 7:206–210

Jongen WMF, Fitzgerald DJ, Asamoto M et al 1991 Regulation of connexin 43-mediated gap junctional intercellular communication by $Ca^{2+}$ in mouse epidermal cells is controlled by E-cadherin. J Cell Biol 114:545–555
Kojima T, Sawada N, Oyamada M, Chiba H, Isomura H, Mori M 1994 Rapid appearance of connexin 26-positive gap junctions in centrilobular hepatocytes without induction of mRNA and protein synthesis in isolated perfused liver of female rat. J Cell Sci 107:3579–3590
Laird DW, Castillo M, Kasprzak L 1995 Gap junction turnover, intracellular trafficking and phosphorylation of connexin 43 in brefeldin A-treated rat mammary tumor cells. J Cell Biol 131:1193–1203
Lippincott-Schwartz J 1998 Unravelling Golgi membrane traffic with GFP chimeras. Trends Cell Biol 8:16–20
Lu Y, Xiong X, Helm A, Kimani K, Bragin A, Skach WR 1998 Co- and post-translational translocation mechanisms direct cystic fibrosis transmembrane conductance regulator N terminus transmembrane assembly. J Biol Chem 273:568–576
Lukacs GL, Mohamed A, Kartner N, Chang X-B, Riordan JR, Grinstein S 1994 Conformational maturation of CFTR but not its mutant counterpart (ΔF508) occurs in the endoplasmic reticulum and requires ATP. EMBO J 13:6076–6086
Marsault R, Murgia M, Pozzan T, Rizzuto R 1997 Domains of high $Ca^{2+}$ beneath the plasma membrane of living A7r5 cells. EMBO J 16:1575–1581
Martin PEM, George CH, Castro C et al 1998 Assembly of chimeric connexin–aequorin proteins into functional gap junction channels. Reporting intracellular and plasma membrane calcium environments. J Biol Chem 273:1719–1726
Rahman S, Carlile G, Evans WH 1993 Assembly of hepatic gap junctions. Topography and distribution of connexin32 in intracellular and plasma membranes determined using site-specific antibodies. J Biol Chem 268:1260–1265
Schweitzer A, Fransen JAM, Mattew K, Kreis T, Ginsel L, Hauri H-P 1990 Identification of an intermediate compartment involved in protein transport from the endoplasmic reticulum to the Golgi apparatus. Eur J Cell Biol 131:185–196
Severs NJ, Gourdie RG, Harfst E, Peters NS, Green CR 1993 Intercellular junctions and the application of microscopical techniques: the cardiac gap junction as a case model. J Microsc 169:299–328
Torok K, Stauffer K, Evans WH 1997 Connexin32 of gap junctions contains two cytoplasmic calmodulin-binding domains. Biochem J 326:479–483
Zhang J-T, Chen M, Foote CI, Nicholson BJ 1996 Membrane integration of *in vitro* translated gap junctional proteins: co- and post-translational mechanisms. J Cell Biol 7:471–482

## DISCUSSION

*Musil:* I have a question about connexin26 (Cx26; $\beta_2$) in the subcellular fractions. Can you rule out the possibility that the reason you don't see Cx26 in the intracellular fractions is because it is unstable in those fractions?

*Evans:* We have looked extensively for Cx26, and possible precursors and degradation products of Cx26 in those fractions by using various antibodies to Cx26 and the various peptides used to generate the antibodies. The ratio of Cx26 to Cx32 ($\beta_1$) in the intracellular membrane fractions is always extremely low, so we conclude that the bulk of Cx26 is taking an alternative non-classical pathway to the cell surface, especially when we consider the other lines of evidence that I presented. When one looks in the literature there is a lot of evidence for proteins that take non-classical secretory pathways, such as fibroblast growth factors and viral proteins that interact with the membrane (Cleves 1997). In yeast there is evidence for four pathways from the endoplasmic reticulum to the plasma membrane for membrane proteins and secretory proteins (Titorenko et al 1997), suggesting that the secretory pathway is not the exclusive route from the endoplasmic reticulum to the plasma membrane for membrane proteins. The evidence I presented indicates that although Cx32 and Cx43 ($\alpha_1$) do take this pathway, a large proportion of Cx26 takes an alternative pathway to the plasma membrane.

*Musil:* Can you find Cx26 in any other fractions? In other words, if it's not in the endoplasmic reticulum or Golgi fractions, where is it?

*Evans:* It is present at low levels in the endoplasmic reticulum. A further possibility is that it is associated with polysomes, and that Cx26 is inserted directly into the plasma membrane, but further work is required to explore this. The observation that in a cell-free analysis system it is possible to replace canine pancreatic microsomes by rat liver plasma membranes suggests that this post-translational insertion, which is an unusual feature of membrane proteins, does fit into this array of unusual properties exhibited by Cx26. There are many other properties of Cx26 that distinguish it from all the other connexins, e.g. it doesn't have an extended carboxyl tail and it is not phosphorylated.

*Gilula:* Several people have observed post-translational insertion of Cx26, but there is incomplete agreement.

*Nicholson:* We have seen post-translational insertion consistently (Zhang et al 1996). The general conclusion is that co-translational insertion occurs preferentially to post-translational insertion, but if the co-translation mechanism is eliminated, then relatively efficient post-translation is observed, at least in the microsome system. If this happens *in vivo*, one would conclude that there might be Cx26 present in the soluble fraction.

*Gilula:* Do you have evidence that the integration you are seeing is relevant topologically? For example, is it possible that membranes which are not authentic plasma membranes but are present in the plasma membrane fraction as impurities or contaminants can accept the product of the Cx26 transcript?

*Evans:* The oligomerization assays suggest that a proportion of Cx26 migrates at the 9S position, indicating that there is some evidence for oligomerization. This will have to be tested using the liposome assay system, which can address whether channels are introduced into liposomes. These are in some ways preliminary results, but the plasma membranes are of high purity. There are a number of control experiments that still need to be done, one of which addresses the possibility that there are specific factors associated with the plasma membrane fraction. We have extracted the plasma membrane fractions with sodium chloride or alkali-stripped the membranes with sodium carbonate to test the specificity of the association. However, we still obtain post-translational insertion into alkali-stripped membranes.

*Green:* In Charcot-Marie-Tooth disease there is a gene defect in Cx32 throughout the nervous system but the phenotypic defect appears specific to demyelination. Is it feasible that the Schwann cells lack the non-classical pathway, so that at that point the Cx32 defect becomes crucial? Prior to that Cx26 may be playing a role via the non-classical pathway.

*Evans:*This may be more applicable to genetic defects in tissues that express Cx26, e.g. the inner ear. The bulk of Cx26 expressed in transfected African green monkey COS kidney cells and in the endogenous liver tissue system appears to take a brefeldin A-insensitive pathway. There could be some degree of compensation if, for example, a connexin incorporated into a heteromeric connexon, Cx26 could then traffic through the secretory pathway.

*Goodenough:* Could you clarify how you did the localization analysis with different concentrations of calcium. Is the idea that the concentrations of intracellular calcium are high near the plasma membrane then there is a gradient such that the concentration decreases towards the centre of the cell?

*Evans:* Yes, there is a high calcium concentration beneath the plasma membrane. The Cx–Aeq chimeras report up to 5 $\mu$M (Martin et al 1998, George et al 1998)

*Goodenough:* Is it possible that brefeldin-A changes the distribution of calcium in these cells?

*Evans:* The brefeldin A experiments were performed for 3, 6 and 12 h. The drug does not change bulk cytoplasmic $Ca^{2+}$ levels. The cells looked in good shape and the effects of brefeldin A were reversible; they also respond to agonists, e.g. bradykinin. One problem with using cell culture systems is that we have to incubate the cells for periods in the presence of EGTA, and certainly the cells are not at their healthiest under those conditions.

*Gilula:* The different transit times of the Cx–Aeq chimeras are interesting because they suggest that the larger the protein the slower the transit time. Is it possible to calculate the transit times without the chimeras being associated with aequorin?

*Evans:* It is necessary to have aequorin (the reporter) attached to the molecule. The next step is to use different reporters, and there are now experiments in progress with green fluorescent protein attached to Cx43 and Cx26 to determine this. We favour the interpretation that Cx43 takes a different time to Cx32 to get from the store to the plasma membrane by postulating that they are residential in stores located in different parts of the Golgi apparatus. Unfortunately, immunofluorescence microscopy doesn't have a sufficiently high resolution to tell us whether they are in the *cis*, *medial* or *trans* parts of the Golgi apparatus. We need higher resolution studies to find out exactly where they are, but the subcellular fractionation evidence I presented confirmed that the connexin stores are in the Golgi region.

*Gilula:* Do you envisage that each connexin will have different processing and assembly routes and will therefore have different transit times?

*Evans:* No. Cx32 and Cx43 are taking the same pathway and probably moving at the same speed, but they're coming from a different starting point. This is supported by the observation that Cx43–Aeq reports a different cytoplasmic calcium level (Martin et al 1998).

*Nicholson:* There is a tool that could be useful in trying to distinguish this in an intact cell system. When we saw *in vitro* post-translation insertion my hopes were that this would be an *in vivo* reality. Mathias Falk has engineered glycosylation sites in the first extracellular loop to prevent signal-peptidase cleavage of M1 in the *in vitro* system (Falk et al 1997). The prediction from this model is that if you followed this Cx32 mutant through the system, then it would be glycosylated, but that if you put the equivalent Cx26 mutant through the system it would not be glycosylated. It would be interesting to try this approach in your system.

*Kistler:* If I understand correctly, a small proportion of Cx26 follows a classical pathway and the bulk of it follows an alternative pathway. Can we exclude the possibility that the alternative pathway is an aberration? Have you blocked the cells with brefeldin A such that transport via the classical pathway is excluded? Because if you observed that all the Cx26 followed the alternative route, this would suggest that it is unlikely that the alternative pathway is an aberration.

*Evans:* Those are experiments for the future. I should also emphasize that both brefeldin A and nocodazole have differential effects on trafficking, which reinforces the differences between the trafficking routes of Cx26 and Cx32 and Cx43 to the plasma membrane.

*Goodenough:* Your experiment is not a pulse–chase experiment. You're looking at the steady-state situation, so it would be interesting to find out whether there is a different rate of connexin turnover under those conditions. Also, can you rule out that Cx26 is not internalized from the plasma membrane and thus undergoes a recycling loop that doesn't involve synthesis?

*Evans:* We are currently performing experiments to address the issue of degradation. We can measure the removal of all chimeras from the plasma membrane. We've shown that lactacystin, which blocks the proteosome pathway, has a 70% inhibitory effect on the breakdown of the chimeras, whereas leupeptin, which blocks the lysosomal system, has a 30% inhibitory effect. In this sense our results are in agreement with those of Laing et al (1997), which indicate that the majority of connexin degradation, including the chimeric ones, occurs via a proteosomal pathway. This is against the grain in some ways because most people believe that proteosomes operate between the endoplasmic reticulum and the Golgi, and that they deal mainly with misfolded connexins. Therefore, when you attach reporter groups to connexins then you run the risk that the connexin will have difficulty in folding and oligomerizing efficiently. However, it does look as though connexin degradation occurs, perhaps underneath the plasma membrane, and that it is affected by proteosomal inhibitors.

*Goodenough:* My concern is not whether it is degraded, but rather whether the 30% that escapes degradation can be accounted for by what looks like a rapid return of Cx26 to the plasma membrane, i.e. Cx26 that has already been in the plasma membrane is internalized and then returned intact to the plasma membrane. Therefore, it appears as though new Cx26 is reaching the plasma membrane, but in fact it's Cx26 that has already gone through the classical pathway, has reached the surface and is now recycling at the surface.

*Evans:* The aequorin moiety of the connexin chimeras targeted to the plasma membrane is triggered by extracellular calcium and therefore, if recycled, it is inactive. The new peaks of aequorin chemiluminescence indicate arrival at the plasma membrane of chimeras with untriggered aequorin. Therefore, it is unlikely to be a recycling phenomenon and we believe this approach tells us how long it takes for the Cx–Aeq chimeras to move from intracellular depots to the plasma membrane.

*Beyer:* Our most recent data suggest that both the proteosome and the lysosome have roles in the degradation of surface gap junctions, or at least gap junctions that are forming plaques and possibly also channels at the surface. The proteosome is also likely to be involved in proof-reading or degradation at the level of the endoplasmic reticulum. There are still major topological problems that have to be solved, i.e. how can a hydrophobic protein oligomer be pulled apart, removed from the membrane and moved to a place where it can be degraded?

*Musil:* You are using intracellular transport inhibitors in cultured cells and you're also doing connexon assembly assays. If the intermediate compartment between the endoplasmic reticulum and the Golgi is the site of assembly, then one might expect to see connexon assembly in cells that are treated at 15 °C because this temperature blocks the transport of newly synthesized secretory proteins within the intermediate compartment in mammalian cells but still allows connexon assembly in *Xenopus* oocytes in which transport through the secretory pathway continues. Have you looked at whether there is any oligomerization in your tissue culture cells at 15 °C of pulse-labelled connexin that would not be able to exit the compartment that you're saying is the assembly compartment?

*Evans:* We haven't yet done those experiments. Heteromeric connexons are formed preferentially if liver Golgi membranes are added to the *in vitro* translation system, whereas in the absence of Golgi only homomeric connexons are formed, suggesting that a factor(s) is resident in the secretory pathway that catalyses the formation of heteromeric connexons. We plan to sub-fractionate the

Golgi membranes to find out whether it is a soluble factor, a loosely attached factor or whether it is a property of the membrane itself.

*Beyer:* I wonder whether your data give us clues regarding the stoichiometry of heteromeric mixing. Your immunoprecipitation experiments with anti-Cx32 antibodies followed by western blotting with anti-Cx26 antibody looked to me as though the intensities of the bands were comparable, which would argue that there is as much Cx26 mixed with Cx32 as there is total Cx26, i.e. that all the Cx26 is mixed with Cx32. However, if your alternative pathway hypothesis is correct, most of the Cx26 should reach the surface by itself.

*Gilula:* I would like to clarify why you think that some of Cx26 goes through the classical Golgi pathway.

*Evans:* First, because by subcellular fractionation of liver tissue we can detect small amounts of Cx26. Second, we have to explain why heteromeric connexons are detected in liver gap junctions.

*Gilula:* Does this give you an explanation for the use of Cx26 in heteromeric associations?

*Evans:* Yes, it accounts for why gap junctions are constructed of heteromeric connexons in a tissue that has mainly homomeric connexons.

## References


Cleves AE 1997 Protein transport: the nonclassical ins and outs. Curr Biol 7:318–320

Falk MM, Buehler LK, Kumar NM, Gilula NB 1997 Cell-free synthesis and assembly of connexins into functional gap junction membrane channels. EMBO J 16:2703–2716

George CH, Kendall JM, Campbell AK, Evans WH 1998 Connexin aequorin chimerae report calcium environments along trafficking pathways leading to gap junction biogenesis in living cells. J Biol Chem 273:29822–29829

Laing JG, Tadros PN, Westphale EM, Beyer EC 1997 Degradation of Cx43 gap junctions involves both the proteosome and the lysosome. Exp Cell Res 236:482–492

Martin PEM, George CH, Castro C et al 1998 Assembly of chimeric connexin–aequorin proteins into functional gap junction channels. Reporting intracellular and plasma membrane calcium environments. J Biol Chem 273:1719–1726

Titorenko VI, Ogrydziak DM, Rachubinski RA 1997 Four distinct secretory pathways serve protein secretion, cell surface growth and peroxisome biogenesis in the yeast *Yarrowia lipolytica*. Mol Cell Biol 17:5210–5226

Zhang JT, Chen MA, Foote CI, Nicholson BJ 1996 Membrane integration of *in vitro* translated gap junctional protein: co- and post-translation mechanisms. Mol Biol Cell 7:471–482

# Interactions between growth factors and gap junctional communication in developing systems

Anne Warner

*Department of Anatomy and Developmental Biology, University College London, Gower Street, London WC1E 6BT, UK*

*Abstract.* In the vertebrate limb bud fibroblast growth factor (FGF) 4 secreted by cells of the posterior apical ectodermal ridge controls digit pattern, which is directed by polarizing cells in the posterior mesenchyme at the tip of the bud. FGF4 also controls the expression of gap junctions in the limb. Both chick and mouse limb bud mesenchyme express connexin 32 (Cx32; $\beta_1$) and Cx43 ($\alpha_1$), although not in the same gap junction plaques. Quantitative analysis reveals two gradients of gap junctions: from posterior to anterior in the subapical mesenchyme and from distal to proximal along the bud. The highest gap junction density is associated with the polarizing region. Micromass cultures of chick and mouse posterior and anterior mesenchyme cells were used to assess the ability of FGF4 to modulate gap junctional communication. Posterior mesenchyme (polarizing region) cells express a population of gap junctions that are highly sensitive to FGF4, whereas gap junctions between anterior mesenchyme cells are completely insensitive to FGF4. FGF4 doubles gap junction density, intercellular communication and the polarizing capacity of posterior mesenchyme cells, restoring polarizing capacity to *in vivo* levels. We conclude that gap junctional communication and polarizing capacity are intimately linked. Interactions between signalling molecules and junctional communication may play an important role in controlling development.

*1999 Gap junction-mediated intercellular signalling in health and disease. Wiley, Chichester (Novartis Foundation Symposium 219) p 60–75*

The idea that direct cell–cell communication enables the transmission of regulatory signals between developing cells, through a pathway that we now know to be mediated by gap junctions, was proposed by Potter et al (1966), after discovering that all cells in the early squid embryo were able to communicate with each other. There is now a wealth of data confirming that gap junctional communication between all cells is a universal feature of early embryos. Although direct cell–cell communication is observed at times when important developmental interactions are taking place, proof that gap junctional communication plays a part in controlling development remains elusive.

## Gap junctions and patterning in the amphibian embryo

The most direct evidence for an integral developmental role for gap junctional communication remains the demonstration that antibodies to gap junction proteins that block cell–cell communication generate embryos with developmental defects when introduced into cells of the amphibian embryo (Warner et al 1984, Warner 1986), mouse embryo (Lee et al 1987, Becker et al 1995) and the chick limb bud (Allen et al 1990). Such experiments reveal a requirement for gap junctional communication during development, but not why and how it is important. However, work in the amphibian embryo has shown (see Warner 1985, 1986) that injecting blocking antibodies into vegetal pole cells at the eight-cell stage generates embryos that lack overall pattern. Despite severe defects of the embryonic axis, the embryos nevertheless contain differentiated mesoderm derivatives, such as muscle and notochord cells. The antibodies remain effective well into the gastrula stage (see Warner 1986) so that intercellular communication will have been blocked during the blastula stages when endoderm cells induce the mesoderm. These results suggest that a primary role of gap junctional communication is to ensure normal pattern formation rather than induction of specific differentiated cell types. This conclusion is reinforced by the parallel finding (Warner & Gurdon 1987) that endoderm cells prevented from taking part in communication via gap junctions are nevertheless able to induce actin gene expression, a marker for differentiated muscle, in ectoderm caps.

A link between embryonic patterning and direct cell–cell communication is reinforced by the realization that gap junctions between cells in different regions of the embryo are not necessarily identical. In the amphibian embryo there is now strong evidence that the future dorsoventral axis is characterized by a gradient of gap junction properties from dorsal to ventral sides of the early embryo (Guthrie et al 1988). As early as the 32-cell stage, the discrimination between molecules moving through junctions between cells in future dorsal regions is not the same as that between ventral cells: junctions between future dorsal cells have different properties from junctions between future ventral cells. Lucifer yellow provides the clearest indicator of this gradient, with dorsal cells exchanging Lucifer yellow with much greater efficiency than ventral cells. The dorsoventral gradient of gap junction properties at the 32-cell stage is an early and accurate predictor of the future dorsoventral axis (Nagajski et al 1989). In embryos ventralized by u.v. irradiation of the vegetal pole early in the first cleavage cycle all gap junctions assume the ventral characteristic of inefficient exchange of Lucifer yellow. By contrast, after lithium treatment at the eight- to 16-cell stage, which induces dorsalization of the embryo, gap junctions in ventral regions of the 32-cell embryo assume the dorsal characteristic of highly efficient exchange of Lucifer yellow. In both cases the gradient is abolished. This has now been found to be the case also after injection of reagents such as $\beta$-catenin, *Wnt1* and activin, which dorsalize the ventral side of the axis, and improve Lucifer yellow transfer between ventral cells at the 32-cell stage (Guger & Gumbiner 1995, Olson & Moon 1992). The recent demonstration (Cao et al 1998) of connexin specific permeability differences between Cx45 ($\alpha_6$), Cx32 and Cx26 ($\beta_2$) expressed in HeLa transfectants and *Xenopus* oocytes suggests that there is yet much to learn about the functional properties of gap junctions.

## Gap junctions and patterning in the developing limb

Although the mechanism underlying the correlated effects on gap junction properties and the dorsoventral axis has yet to be defined, the recognition that growth factors and second messenger molecules are part of the cascades that control developmental patterning raises the possibility that the modulation of gap junction properties, either directly or indirectly, is part of the way in which such agents direct development. The control of patterning of the digits in the developing limb bud has been much used for studies on embryonic patterning. The system has two major advantages: first, a variety of approaches, both *in vivo* and *in vitro*, can be used; second, the molecules and signalling mechanisms that operate in the limb bud are similar to those found in other developing systems, so that greater understanding of the limb bud should have wide application.

Anteroposterior pattern in the limb is controlled by a small group of posterior mesenchyme cells at the tip of the limb bud (Saunders & Gasseling 1968). When transplanted to the anterior margin of the bud, these cells reprogramme the anterior mesenchyme and generate duplicated digit patterns. Preventing communication between signalling (polarizing posterior mesenchyme) and responding (anterior mesenchyme) cells with blocking polyclonal antibodies reduces substantially the duplication of the digits observed when polarizing cells are grafted into the anterior mesenchyme (Allen et al 1990), directly implicating gap junctional communication in the signalling process. Recent work has revealed the central role of growth factor signalling, particularly by the fibroblast growth factors (FGFs), in maintaining the ability of polarizing cells to respecify the anterior mesenchyme (e.g. Niswander et al 1993). We therefore examined how FGF4 might influence gap junctional communication between both posterior (signalling) mesenchyme cells and anterior (responding) mesenchyme cells (Makarenkova et al 1997).

The diagram in Fig. 1 shows the organization of the vertebrate limb bud at the time of our experiments; the digit pattern in normal limb buds and limb buds in which the polarizing region of one limb bud has been grafted into the distal anterior mesenchyme of a second are shown below. The apical ectodermal ridge

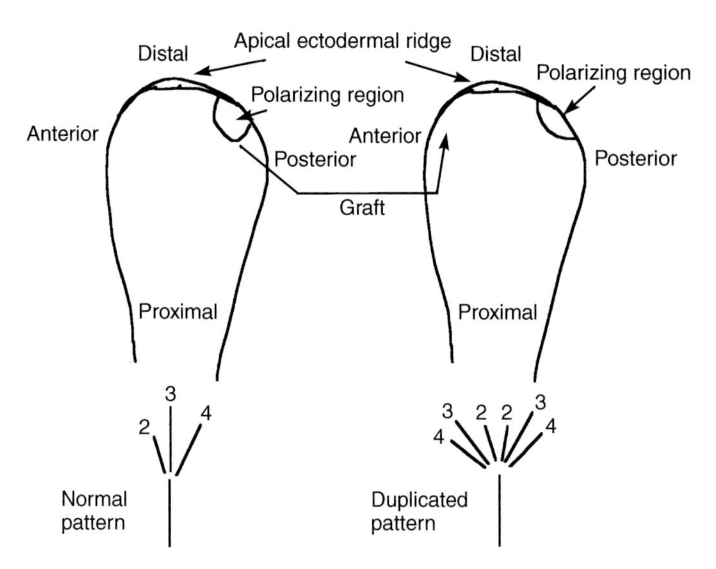


FIG. 1. Diagram to show the vertebrate limb bud at the time when the digit pattern is being set up. The digit pattern for the normal limb and the limb bud after grafting the polarizing region into the anterior mesenchyme of a second limb bud is shown below. The pattern observed for complete duplication of the digits is illustrated. This is equivalent to 100% respecification of the anterior mesenchyme (see text).

(AER) lies at the tip of the bud and is known to be crucial for normal outgrowth and pattern formation within the limb (Saunders 1948, Summerbell 1974).

### *Gap junctions are asymmetrically distributed in the limb bud*

Connexin-specific antipeptide antibodies detected Cx26, Cx32 and Cx43 (see Becker et al 1995). Cryosections of chick and mouse limb buds were taken when the digit pattern is being set up, stained with the appropriate anti-connexin antibody and examined in the confocal microscope (for details see Makarenkova et al 1997). In both species limb bud mesenchyme cells expressed Cx32 and Cx43 in gap junctions. Cx26 was not detected in chick limbs, but was found in the dorsal epidermis of the mouse. We therefore directed our subsequent analysis to Cx32 and Cx43. Most mesenchyme cells expressed both connexins, with little overlap, suggesting that the two connexins exist in different gap junction plaques, giving the opportunity for their expression to be differentially controlled. Quantitative estimates (see Makarenkova et al 1997) of the number of labelled junctions in different regions of the bud (posterior distal [polarizing region]; anterior distal; posterior proximal; and anterior proximal) revealed two gradients: from posterior to anterior and distal to proximal (Fig. 2), with both connexins

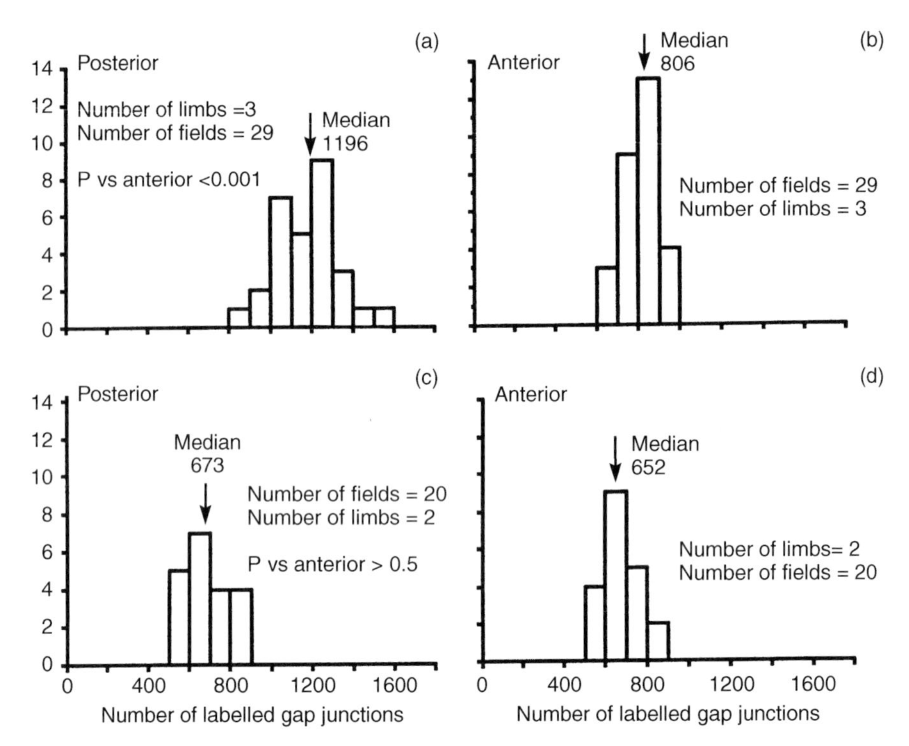


FIG. 2. The density of gap junctions in different regions of the chick limb bud. Cryosections were stained with antipeptide antibodies specific for connexin43 ($\alpha_1$). Labelled gap junctions were counted within 3600 $\mu m^2$ boxes placed over the sections. Ordinates: number of fields. Abscissae: number of labelled gap junctions/3600 $\mu m^2$. (a,b) Subapical mesenchyme at the tip of the bud beneath the apical ectodermal ridge. Note significantly more gap junctions were detected in the posterior subapical mesenchyme, which includes the polarizing region, than in the anterior subapical mesenchyme. (c,d) Proximal mesenchyme. Note that gap junction densities in posterior and anterior mesenchyme are identical.

following the same distribution. The density of gap junctions in the posterior distal mesenchyme, where the polarizing region is located, was significantly higher (50–70%) than in the distal anterior mesenchyme. The density of junctions in distal anterior mesenchyme was higher (about 20%) than in the proximal anterior mesenchyme, which expressed the same density of gap junctions as the proximal posterior mesenchyme. The greatest density of gap junctions was closely linked to the region of polarizing activity. The reduction in gap junction density and absence of a posterior to anterior gradient in proximal regions matches the decline in polarizing capacity from distal to proximal within the bud. Removing the AER, which plays an essential part in controlling pattern in the mesenchyme, removed the gradients, and gap junction density decreased to a level close to that normally found in proximal regions. Thus, the intact limb bud displays gradients of gap junctions that correlate well with the distribution of polarizing mesenchyme cells.

Gap junctions join transcripts for molecules such as *Sonic hedgehog* (*Shh*) in being asymmetrically distributed across the limb bud, with the highest levels in the polarizing region. The distribution of *Shh* transcripts is controlled by FGF4 secreted by ectoderm cells in posterior regions of the AER, and one possibility is that FGF4 also controls the asymmetric distribution of gap junctions. It proved to be technically impossible to test this hypothesis in the intact bud by grafting beads soaked in FGF4 into limb buds lacking an AER because the beads interfered with frozen sections. However, FGF4 maintains polarizing activity in cultured mouse posterior mesenchyme (Vogel & Tickle 1993), and we turned to short-term, micromass cultures of posterior and anterior mesenchyme from both chick and mouse limb buds for our experiments.

### *Gap junctions between polarizing cells are regulated by FGF4*

The outcome was clear. Gap junction density between undifferentiated posterior mesenchyme cells of both mouse and chick was highly sensitive to FGF4. By contrast, gap junction density between undifferentiated anterior mesenchyme cells was completely insensitive to FGF4. Mouse cultures retained expression of both Cx32 and Cx43, as in the intact limb bud, and both connexins were similarly responsive to FGF4. Cultures from chick limb buds retained Cx43, but lost the capacity to express Cx32. FGF4 did not restore Cx32 expression.

These results are summarized in Figs 3 & 4, compiled from single experiments in which all four possible combinations were examined. Figure 3 compares the density of gap junctions between distal posterior mesenchyme cells maintained in culture for 24 h in the absence (a,b,c) and presence (d,e,f) of FGF4 for chick Cx43 (a,d), mouse Cx43 (b,e) and mouse Cx32 (c,f). In all cases the density of gap junctions is significantly increased between cells maintained in culture in FGF4. This result was obtained on all occasions tested (for complete data see Makarenkova et al 1997). The complementary observations for distal anterior mesenchyme cells are shown in Fig. 4. In all cases FGF4 had no effect on gap junction density. In the experiment of Fig. 4c,d, which examines Cx32 in mouse anterior mesenchyme, the median value for FGF4-treated cultures was higher than in the control. However, the ranges are no different and other experiments did not confirm this indication: control and FGF4-treated Cx32 distributions were identical.

How completely was the normal posterior to anterior gradient abolished in cultures maintained in the absence of FGF4? For Cx43, both in the chick and mouse, the gradient was reversed: there was a greater density of gap junctions between anterior mesenchyme cells than between posterior mesenchyme cells

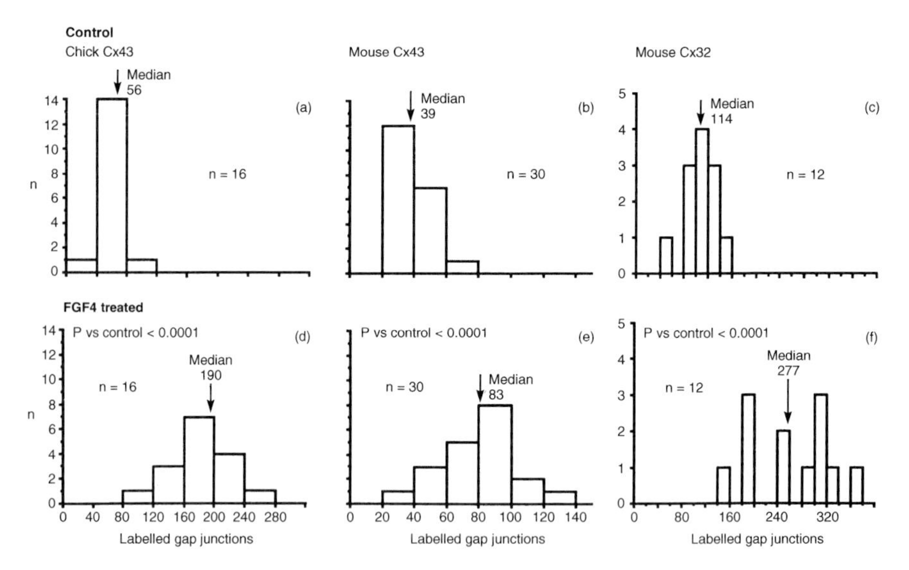


FIG. 3. Gap junctions between chick and mouse posterior (polarizing) mesenchyme cells are sensitive to fibroblast growth factor (FGF) 4. Micromass cultures of chick or mouse mesenchyme plated at $4 \times 10^4$ cells/10 $\mu$l. Ordinates: number of fields. Abscissae: number of labelled gap junctions/3600 $\mu m^2$. Data taken from single experiments in which cultures of posterior mesenchyme were plated in the presence and absence of FGF4 and gap junction density counted after 24 h in culture. Data for all experiments in Makarenkova et al (1997). Note in all cases FGF4 increases density of gap junctions.

(compare control medians of Figs 3 & 4). However, the Cx32 gap junctions retained in mouse mesenchyme cultures consistently retained a higher gap junction density in the posterior mesenchyme, even in the absence of FGF4 (compare control medians in Figs 3 & 4).

The density of gap junctions between cells within the aggregates of condensed cells scattered through the monolayers of undifferentiated cells in both posterior and anterior mesenchyme were not affected by FGF4. This may be linked to the loss of polarizing activity when mesenchyme cells leave the progress zone at the tip of the limb bud and begin to prepare for differentiation and the disappearance of the anteroposterior gradient in proximal regions of the intact bud.

### *Communication between mesenchyme cells is regulated by FGF4*

Quantitative estimates of gap junction density were matched by functional estimates of gap junctional communication made by microinjection of Lucifer yellow into single chick mesenchyme cells and determining dye transfer into the

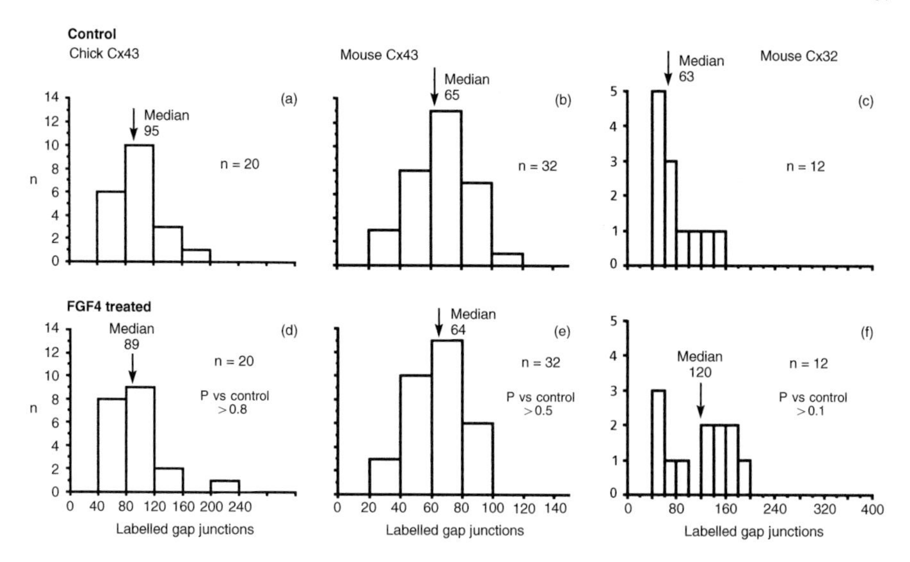


FIG. 4. Gap junctions between chick and mouse anterior mesenchyme cells are not sensitive to fibroblast growth factor (FGF) 4. Micromass cultures of chick or mouse mesenchyme plated at $4 \times 10^4$ cells/10 $\mu$l. Ordinates: number of fields. Abscissae: number of labelled gap junctions/3600 $\mu m^2$. Data taken from single experiments in which cultures of anterior mesenchyme were plated in the presence and absence of FGF4 and gap junction density counted after 24 h in culture. Data for all experiments in Makarenkova et al (1997).

culture. Dye transferred between anterior mesenchyme cells with an incidence of 57% ($n = 33$; $N = 2$) in the absence of FGF4 and 49% ($n = 39$; $N = 3$) in the presence of FGF4. In posterior mesenchyme cultures without FGF4 the failure rate was only 40% ($n = 68$; $N = 5$). This rose to 97% ($n = 59$; $N = 5$) in the presence of FGF4, indicating a substantial increase in the frequency of gap junctional communication between posterior mesenchyme cells. This was paralleled by the extent to which dye transferred. Anterior mesenchyme cells transferred dye to few cells in the absence of FGF4 (median of distribution 2 cells), with a similar extent in cultures maintained in FGF4 (median of distribution 3 cells; $P$ vs. control $> 0.8$). Posterior mesenchyme cells displayed equivalent, poor dye spread in the absence of FGF4 (median 2 cells), which rose significantly ($P < 0.0001$) to a median of 7 when FGF4 was available.

These results suggest strongly that gap junctions in the posterior mesenchyme are under the control of FGF4. Surprisingly, gap junctions between anterior mesenchyme cells are completely insensitive to FGF4, whether assessed by dye transfer or gap junction density. In the intact limb bud FGF4 is secreted by posterior ectoderm cells, which lie over the polarizing region, and even if gap junctions between all cells in the distal mesenchyme were sensitive to FGF4, an asymmetric distribution of gap junctions in the bud would be expected. The finding that gap junctions between anterior mesenchyme cells are not responsive to FGF4 suggests that the link between FGF4 and gap junctional communication is specific to polarizing cells.

### *FGF4 maintains the polarizing capacity of cultured posterior mesenchyme at in vivo levels*

In the long term we need to understand how the signalling mechanisms that operate within the limb bud interact and are regulated to generate limb bud pattern. A first step is to determine how gap junctions, FGF4 and the polarizing capacity of posterior mesenchyme cells are related. Such an analysis requires that the influence of FGF4 on polarizing capacity and gap junctional communication be determined under conditions that allow direct comparison. We also need to learn more about the way in which FGF4 influences polarizing capacity. Does polarizing capacity fall in the absence of FGF4 because of a decline in the number of polarizing cells, without any change in the ability of individual cells to signal, or because of an influence on all polarizing cells that modifies their ability to generate a polarizing signal? To gain insight into these issues, we determined how cell density and polarizing capacity are related in the absence and presence of FGF4 under the conditions used to analyse the behaviour of gap junctions. Micromass cultures of chick posterior mesenchyme cells were plated at the densities used for the analysis of gap junctions, either with or without FGF4, and fragments of treated or untreated cultures grafted into the distal anterior mesenchyme of host chick limb buds 24 h later (for details see Makarenkova et al 1997). Polarizing capacity was assessed by the ability to reprogramme the anterior mesenchyme and generate duplication of the digits. For each limb bud the digit pattern was scored according to the most posterior digit duplicated, with an additional digit 2 scored as 25% respecification, digit 3 as 50% respecification and digit 4 as 100% respecification (see Fig. 1). The average for all limb buds analysed in each group was used to calculate an average per cent respecification (see Allen et al 1990).

Figure 5 plots plating cell density against per cent respecification produced by cells grafted after culture for 24 h in the absence or presence of FGF4 and shows that cell density is logarithmically related to polarizing capacity. A single data point from the experiments of Vogel & Tickle (1993), which were done in the same way on mouse mesenchyme, falls onto the same relationship as the chick data, showing that chick and mouse cells behave identically. The best logarithmic fit to the two data sets is extremely good ($R > 0.99$ for both), but not the same. For control cultures the slope is 22, but for cultures maintained in FGF4 the slope is 44. The change in slope implies that the signalling mechanism alters in the absence of FGF4. If the number of cells available to polarize fell in the absence of FGF4, with no change in mechanism, the plot would shift along the axis with no change in slope. This leads to an important new conclusion: withdrawal of FGF4 does not reduce the number of cells available to polarize, but changes the mechanism by which polarizing cells signal. The ability to distinguish alterations in polarizing capacity derived from changes in numbers of polarizing cells from those arising from a change in signalling mechanism will greatly assist the mechanistic analysis of limb bud signalling.

It is important to know whether FGF4 maintains the polarizing capacity of cultured cells at *in vivo* levels. Estimates of normal polarizing capacity can be obtained from experiments in which untreated polarizing cells were grafted as pellets shortly after removal from donor limb buds (Tickle 1981, Allen et al 1990). When analysed as in Fig. 5 their results give slopes of 40 ($R = 0.93$; Tickle 1981) and 42 ($R = 0.96$; Allen et al 1990). The slope of 44 obtained from posterior mesenchyme cultures maintained in FGF4 shows that FGF4 is able to maintain polarizing capacity at *in vivo* levels for at least 24 h.

Comparison between the effects of FGF4 on gap junction density, gap junctional communication and polarizing capacity reveals that at the same cell density, FGF4 approximately doubles all three parameters. Given the difference in the methods used to obtain each set of measurements, such congruence seems unlikely to be fortuitous. It suggests that polarizing capacity and communication through gap junctions are intimately linked.

What about connexin knockouts? Mice with a single knockout of either Cx43 (Reaume et al 1995) or Cx32 (Nelles et al 1996) show no major defects in limb patterning. This is entirely consistent with our findings since chick and mouse limb buds normally express both Cx43 and Cx32, and our experiments suggest that they are equivalently sensitive to FGF4. Absence or defects in both connexins may be necessary to interfere with limb bud patterning.

## Conclusions

The most important conclusion to emerge from our experiments is that the signalling mechanisms which operate within the limb bud to control its development interact. Communication through gap junctions provides one of these signalling pathways. Secreted growth factors such as FGF4 provide another. Our results show that polarizing cells, but not other mesenchyme cells in the bud, express a population of FGF4-sensitive gap junctions. *Shh* expression also is controlled by FGF4 and limited to polarizing posterior mesenchyme cells. Are *Shh* expression and gap junctional communication independently or coordinately regulated in the bud? Is gap junctional communication part of the

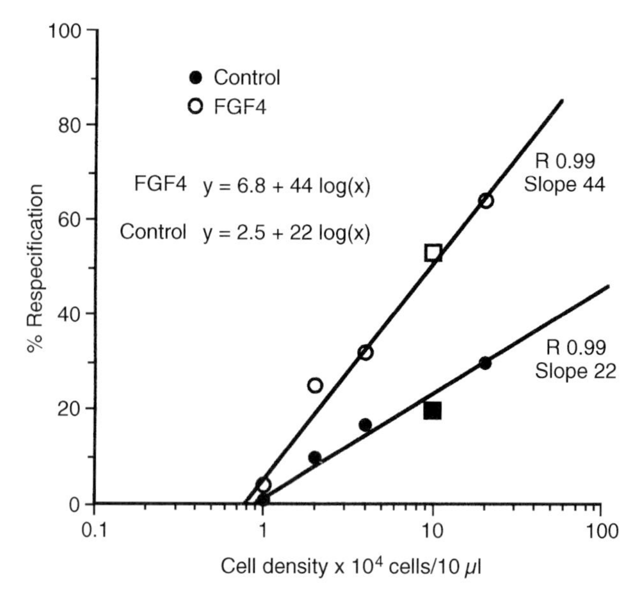


FIG. 5. Fibroblast growth factor (FGF) 4 maintains polarizing capacity in cultured chick posterior mesenchyme cells. Ordinate: per cent respecification achieved by fragments of micromass cultures after grafting into the anterior mesenchyme of host chick limb buds. Abscissa: plating cell density plotted on a logarithmic scale. Data based on at least 10 grafts for each condition tested. Note that polarizing capacity is logarithmically related to plating cell density. The slope of the relationship doubles when FGF4 is available to maintain polarizing capacity.

mechanism that regulates *Shh*? Such interactions open up completely new avenues for exploring the cellular mechanisms that underlie pattern formation in the limb bud. Our results could have wide applicability since FGF4/*Shh* interactions (Bueno et al 1996) are a feature of a number of developing systems in which cells are linked by gap junctions.

The cellular interactions that contribute to embryonic development are complex and it is perhaps not surprising that signalling mechanisms interact with each other. Evidence from both the amphibian embryo and the developing limb bud suggests that one way of understanding how gap junctional communication contributes to early development may be to explore interactions between signalling molecules such as growth factors and gap junctional communication.

*Acknowledgements*

The work described in this chapter was made possible by grants from the Medical Research Council and the Wellcome Trust. I thank the Royal Society for their support.

## References

Allen FLG, Tickle C, Warner A 1990 The role of gap junctions in patterning of the chick limb bud. Development 108:623–634

Becker DL, Evans WH, Green CR, Warner AE 1995 Functional analysis of amino acid sequences in connexin43 involved in intercellular communication through gap junctions. J Cell Science 108:1455–1467

Bueno D, Skinner J, Abud H, Heath JK 1996 Spatial and temporal relationships between *Shh*, *Fgf-4* and *Fgf-8* gene expression at diverse signalling centers during mouse development. Dev Dyn 207:291–299

Cao F, Eckert R, Elfgang C et al 1998 A quantitative analysis of connexin-specific permeability differences of gap junctions expressed in HeLa transfectants and *Xenopus* oocytes. J Cell Science 111:31–43

Guger KA, Gumbiner BM 1995 Beta-catenin has Wnt-like activity and mimics the Nieuwkoop signaling center in *Xenopus* dorsalventral patterning. Dev Biol 172:115–125

Guthrie H, Turin L, Warner AE 1988 Patterns of junctional communication during development of the early amphibian embryo. Development 103:769–783

Lee S, Gilula NB, Warner AE 1987 Gap junctional communication and compaction during preimplantation stages of mouse development. Cell 51:851–860

Makarenkova H, Becker DL, Tickle C, Warner AE 1997 Fibroblast growth factor 4 directs gap junction expression in the mesenchyme of the vertebrate limb bud. J Cell Biol 138:1125–1137

Nagajski DJ, Guthrie SC, Ford CC, Warner AE 1989 The correlation between patterns of dye transfer through gap junctions and future developmental fate in *Xenopus*: the consequences of uv irradiation and lithium treatment. Development 105:747–752

Nelles E, Bützler C, Jung D et al 1996 Defective propagation of signals generated by sympathetic nerve stimulation in the liver of connexin32-deficient mice. Proc Natl Acad Sci USA 93:9565–9570

Niswander L, Tickle C, Vogel A, Booth I, Martin GR 1993 FGF-4 replaces the apical ectodermal ridge and directs outgrowth and patterning of the limb. Cell 75:579–587

Olson DJ, Moon RT 1992 Distinct effects of ectopic expression of *Wnt*-1, activin B and bFGF on gap junctional permeability in 32-cell *Xenopus* embryos. Dev Biol 151:204–212

Potter DD, Furshpan EJ, Lennox ES 1966 Connections between cells of the developing squid as revealed by electrophysiological methods. Proc Natl Acad Sci USA 55:328–336

Reaume AG, de Sousa PA, Kulkarni S et al 1995 Cardiac malformation in neonatal mice lacking connexin43. Science 267:1831–1834

Saunders JW 1948 The proximo-distal sequence of origin of limb parts of the chick wing and the role of the ectoderm. J Exp Zool 108:363–404

Saunders JW, Gasseling MT 1968 Ectodermal–mesodermal interactions in the origin of limb symmetry. In: Fleischmajer R, Billingham RE (eds) Epithelial–mesenchymal interactions. Williams & Wilkins, Baltimore, p 78–97

Summerbell D 1974 A quantitative analysis of the effect of excision of the AER from the chick limb bud. J Embryol Exp Morphol 35:241–260

Tickle C 1981 The number of polarising region cells required to specify additional digits in the developing chick wing. Nature 289:295–298

Vogel A, Tickle C 1993 FGF-4 maintains polarising activity of posterior limb bud cells *in vivo* and *in vitro*. Development 119:199–206

Warner AE 1985 The role of gap junctions in amphibian development. In: Slack JMW (ed) Early amphibian development (British Society for Development Biology symposium). Cambridge University Press, Cambridge, p 365–380

Warner AE 1986 The pattern of communication through gap junctions during formation of the embryonic axis. In: Bellairs R, Ede DA, Lash JW (eds) Somites in developing

embryos (Proceedings of NATO Advanced Research Workshop). Plenum Press, New York, p 91–103
Warner AE, Gurdon JB 1987 Functional gap junctions are not required for muscle gene activation by induction in *Xenopus* embryos. J Cell Biol 104:557–564
Warner AE, Guthrie SG, Gilula NB 1984 Antibodies to gap-junctional protein selectively disrupt junctional communication in the early amphibian embryo. Nature 311:1–5

## DISCUSSION

*Musil:* You showed that the level of connexin 43 (Cx43; $\alpha_1$)-containing and Cx32 ($\beta_1$)-containing gap junction plaques was increased after a 24 h treatment with fibroblast growth factor (FGF) 4. This could be due to increased transcription, in which case it would be interesting that both Cx32 and Cx43 are affected, or it could instead reflect a post-translational event, such as increased plaque assembly. If you treated cells for shorter periods of time you might be able to distinguish between the two possibilities.

*Warner:* We don't yet know whether these changes stem from increased transcription or post-translational events, and we need to follow the time course in more detail. However, there are differences between Cx43 and Cx32, which suggest that the answer may not be straightforward. For example, in cultures of mouse mesenchyme, which retain expression of both Cx32 and Cx43, in the absence of FGF4 there are more Cx43-containing gap junctions between anterior mesenchyme cells than between posterior mesenchyme cells, the complete reverse of the *in vivo* situation. By contrast, Cx32-containing gap junctions remain at greater density between posterior mesenchyme cells than between anterior mesenchyme cells, as observed *in vivo*. The contrasting behaviour of Cx43- and Cx32-containing gap junctions may reflect nothing more than differences in turnover rate in response to withdrawal of FGF4 and have no significance for development. Both connexins behave in the same way in the limb bud.

*Lo:* Did you check the cell numbers in the FGF4-treated mesenchyme graft?

*Warner:* Yes, we counted nuclear density in posterior mesenchyme cultures maintained in the absence and presence of FGF4 and found that cell density did not change. We could therefore be confident that we were grafting the same number of cells in both cases. To explain the difference in polarizing capacity between untreated and FGF4-treated posterior mesenchyme on the basis of number of cells grafted, we would have to say that there are 100-fold more cells in the grafts of FGF4-treated micromasses, which is too great to be explained by errors in grafting technique. The slope of the cell density/polarizing capacity relationship for posterior mesenchyme cells *in vivo* can be estimated by cutting out posterior mesenchyme cells from the limb bud, determining cell density and immediately grafting the appropriate number of cells into the anterior mesenchyme of a host. This *in vivo* estimate is closely matched by the results obtained from cultured posterior mesenchyme cells maintained for 24 h in FGF4, suggesting that FGF4 maintains posterior mesenchyme close to the *in vivo* state. However, we are only at the beginning of the analysis and there are many aspects to explore; for example, the influence of FGF4 on *Sonic hedgehog* (*Shh*).

*Gilula:* Could you expand on your comments about gap junction quantity, because I'm concerned that the correlations may be fortuitous.

*Warner:* We were surprised that the numbers came out as similar as they did. The methods of the assays are different, so one doesn't know whether or not the correlation is fortuitous. I would be perfectly prepared to accept that it is fortuitous. However, when one does find a strong correlation, which doesn't happen often, one has to raise the possibility that it is meaningful. Even if the quantitative correlation does prove to be coincidental, in the absence of FGF4-polarizing capacity, dye transfer and gap junction density both fall back to basal levels, and for all three parameters, a property linked to gap junctions is modified by FGF4.

*Kistler:* How do you quantitate gap junction density? Do you just count spots or do you also take into account gap junction size?

*Warner:* We counted the number of spots using a confocal microscope. One has to be cautious about interpreting changes in spot size because they depend on how many sites are present that can bind antibody. Larger spots could reflect plaques being closer together, rather than individual junction plaques being larger. We did measure spot size, but we did not see any clear indications of change.

*Kistler:* Did you also integrate in three dimensions?

*Warner:* The data presented are based on single optical slices. One can do multiple slices and integrate over three dimensions, but this simply gives larger numbers and one runs into the problem of overlap between different sections.

*Becker:* I would like to mention some results that follow on from Anne Warner's presentation. Anne has shown the importance of Cx43 in limb development and the relationships between FGF4 expression, Cx43 expression and the polarizing ability of limb bud mesenchyme. Interestingly, the Cx43 knockout mouse shows no limb defects, suggesting that some other connexin compensates for its absence. Colin Green and myself have taken a different approach to this system by applying antisense to the limb in order to knockdown the levels of Cx43 in the limb bud transiently, just prior to patterning. We see two main effects from this: a truncation of the limb bud; or, more striking, a divided limb bud. We have gone on to examine these limb buds for the expression of other genes involved in patterning, such as *FGF4* and *Shh*. Their expression pattern indicates that the divided limb buds are not duplications but a splitting of the bud. *Shh* is only expressed in the posterior bud, and only the anterior bud possesses an apical ectodermal ridge (AER). There is normally a strong association between FGF4 expression and Cx43 expression in the AER of the normal limb bud and in the

Cx43 knockdown limb buds we see a reduction FGF4 expression in this region indicating an interaction in their expression.

*Goodenough:* How did you introduce the antisense into the cells?

*Becker:* We have developed a novel delivery system using Pluronic gel which we applied just to the developing limb bud area. The gel sets in place above 4°C and acts as a mild surfactant and reservoir for the antisense which slowly leaches out thereby remaining active in the tissue for a long period of time. In this way we can generate a knockdown of Cx43 protein within about 4–8 h (depending upon the cell type) which lasts for 24–48 h.

*Fletcher:* What is the incidence of the double limb bud?

*Becker:* We observe two effects on the limb bud. One is a truncation of the limb bud (Cx43 expression is required in the AER to maintain outgrowth) and the other is a splitting of the limb bud. Their combined incidence is about 35%.

*Willecke:* We have looked at the effects of Cx43 and Cx32 double knockouts in developing thyroid gland, teeth and limbs, where Cx43 and Cx32 are coexpressed in wild-type cells of all three tissues (Houghton et al 1999). We did not observe an immediate effect. The embryos develop to term and then die, just like the Cx43-deficient mice. We did not see any deviations in the number of digits of the fore- and hindlimbs in the double deficient mice.

*Goodenough:* What controls have you performed to check for non-specific RNA ablation?

*Becker:* We have examined other connexins expressed in this system by treatment with antisense oligonucleotides, sense oligonucleotides and random oligonucleotides. None of the treatments had any effect on the normal expression of Cx32 or Cx26 ($\beta_2$) and only the antisense oligonucleotides reduced the expression of Cx43.

*Goodenough:* We have injected antisense Cx38 ($\alpha_2$) into *Xenopus* oocytes and we have performed northern analysis to look at Cx38 mRNA degradation, using fibronectin as a loading control. To our surprise, the injection of antisense Cx38 also resulted in a decreased concentration of fibronectin mRNA. Therefore, an experiment that involves measuring the levels of other connexins won't necessarily rule out that you are, for example, inducing apoptosis.

*Becker:* We cannot rule out other effects but we are still limited in our knowledge of how gap junctional communication integrates with other signalling mechanisms. With respect to apoptosis, in the normal embryo we observe a down-regulation of Cx43 expression in the interdigit region of the limb bud just prior to the onset of apoptotic cell death in that region. A similar down-regulation is seen in the developing retina just prior to the onset of cell death. It is possible that apoptotic cell death may be a response to reduced Cx43 expression in some tissues. Alternatively, cells that are about to die may stop communicating first.

*Goodenough:* The gold standard would be to do a rescue experiment, i.e. add Cx43 back after the RNA ablation step to see if this will rescue the phenotype.

*Becker:* It would be good to do that, but it would be a difficult experiment to do.

*Gilula:* I would like to ask Alan Lau to comment on the growth factor effects of these experiments.

*Lau:* These effects are different from the ones we have been studying. Our epidermal growth factor (EGF) studies are short-term studies. We treat the cells for only a few hours, and we see a transient disruption of communication within 1–2 h after the addition of EGF. We don't generally don't see changes in the levels of Cx43, but we do see marked changes in its phosphorylation state.

*Lo:* Have you monitored coupling? And I would like to extend this question to Klaus Willecke in relation to the Cx43/Cx32 double knockout experiments. Because one always worries about compensatory changes, and this is a situation where an antisense experiment may give a different result to a knockout experiment.

*Becker:* We have not monitored coupling. We have, however, measured the time course of the down-regulation and the recovery of Cx43 protein expression. The expression of Cx32 and Cx26 were not affected.

*Willecke:* We have not monitored coupling either. We just looked at morphology.

*Lo:* Have you tried putting in connexin constructs into the micromass cultures or have you used retroviral vectors to manipulate connexin expression to address how coupling may be directly related to polarizing activity?

*Warner:* Those experiments are in hand. We don't know anything about the pathways that link FGF4, coupling through gap junctions and polarizing activity. We're looking at the outcome of a multistep process, in which we have to dissect out the individual steps. We may find that there are amplification steps, or we may find that the pathway is relatively linear.

## Reference

Houghton ED, Thönnissen E, Kidder GM et al 1999 Doubly mutant mice, deficient in connexin32 and 43, show normal prenatal development of organs where the two gap junction proteins are expressed in the same cells. Dev Gen, in press

# Biological functions of connexin genes revealed by human genetic defects, dominant negative approaches and targeted deletions in the mouse

Klaus Willecke, Susanne Kirchhoff, Achim Plum, Achim Temme, Eva Thönnissen and Thomas Ott

*Institut für Genetik, Abt. Molekulargenetik, Universität Bonn, Römerstrasse 164, D-53117 Bonn, Germany*

*Abstract.* Gap junction channels in mammalian organs can be built up of at least 13 different connexin proteins, most of which are expressed in only few cell types, although many cells express more than one connexin protein. Recently, the consequences of missing or defective connexin proteins were studied in human patients with defects in connexin32 (Cx32; $\beta_1$; X-linked Charcot-Marie-Tooth disease) or in Cx26 ($\beta_2$; non-syndromic sensorineural deafness), and in mice with targeted deletions in the Cx26, Cx32, Cx37 ($\alpha_4$), Cx43 ($\alpha_1$), Cx46 ($\alpha_3$) or Cx50 ($\alpha_8$) genes. Some effects of dominant negative mutations in connexin genes have been characterized in *Xenopus* oocytes and transfected mammalian cells in culture. Here we review results of these different experimental approaches and report new findings regarding the characterization of Cx40 ($\alpha_5$)- and Cx31 ($\beta_3$)-deficient mice. The phenotypic alterations, caused by different defective connexin genes in mice or humans, are divergent, although in most known cases the viability is not affected. When more than one connexin gene, coexpressed in the same cell, is inactivated, development or maturation can be more severely affected at an earlier stage. Some connexin proteins, if present in the same cell, can partially replace each other in certain functions. Thus, the diversity of connexin proteins in mammalian cells may provide functional overlap and complementation.

*1999 Gap junction-mediated intercellular signalling in health and disease. Wiley, Chichester (Novartis Foundation Symposium 219) p 76–96*

Intercellular gap junction channels are present in almost all animal tissues and cell types investigated. After the connexin protein subunits of these channels had been detected, it became obvious that they were coded for by a family of at least 13 genes in the mouse genome. So far, in all cases investigated, human homologues of the mouse connexin genes have been detected. Each connexin protein has a specific cell type expression pattern, some being expressed in several tissues and others in few.

Furthermore, most cell types express more than one connexin. These findings raise the possibility that different connexin proteins have a different biological function(s). Sequence comparisons between members of the connexin gene family suggest that functional differences may be determined by the intracellular domains of the proteins. The connexin polypeptide chains transverse the plasma membrane four times. The extracellular domains are highly conserved among different connexins, whereas the central cytoplasmic loop and the C-terminal region are largely different (for reviews see Kumar & Gilula 1996, Goodenough et al 1996). This insight into the structural diversity of connexins, however, did not answer questions concerning functional diversity. Why does mammalian physiology require more than a dozen different protein subunits to form intercellular gap junction channels? Is there functional redundancy and/or complementation among connexin proteins? Are connexin channels differently modulated by components of signal transduction pathways?

As with other complex biological systems, answers to these questions can be expected from the characterization of spontaneous or targeted mutants. During the last five years, considerable progress has been made in this area of molecular genetics: two known genetic diseases caused by defective connexin genes have been identified; transdominant negative connexin mutants have been expressed in *Xenopus* oocytes and cultured mammalian cells; and, to date, eight mouse mutants have been generated with targeted deletions in different connexin genes. Here, current results of these different approaches are reviewed in search for answers to the above questions.

## Inherited human genetic diseases due to defective connexin genes

The first human genetic disorder shown to be caused by a mutated connexin gene was the X-linked form of Charcot-Marie-Tooth disease (CMTX; Bergoffen et al 1993). CMTX disease, one of the most common inherited myelination disorders of the peripheral nervous system, is characterized by progressive wasting of distal muscles of the limbs, combined with decreased sensation in peripheral extremities. CMTX disease can be due to defects in different genes (cf. Patel & Lupski 1994), including mutations in the connexin32 (Cx32; $\beta_1$) gene, located on the mammalian X chromosome. Cx32 defects cause demyelination of Schwann cells, which serve to insulate peripheral nerves, leading to reduced nerve conduction velocity. Cx32-containing gap junctions in Schwann cells connect the different layers of the same cell wrapped around an axon (i.e. 'reflexive' gap junctions). It has been pointed out that the distance for diffusion of metabolites or ions through gap junction channels between the adaxonal area and the nucleus of the same Schwann cell is much shorter than for diffusion following the circumferential pathway through the cytoplasm (Deschênes et al 1996). Low molecular weight compounds whose exchange through Cx32-containing gap junction channels is apparently interrupted in Schwann cells of CMTX patients have not yet been identified. Surprisingly, besides symptoms of peripheral neuropathology, no gross systemic dysfunctions of other organs have been described in human CMTX patients, although the Cx32 protein is also expressed in human liver, and likely in the pancreas, the brain and other organs. In most of these cases, Cx32-containing gap junctions are formed between adjacent cells of the same cell type. This could mean that Cx32-containing gap junctions in human cells other than Schwann cells may have no individual biological role or, more likely, that their biological function can be compensated for by gap junctions composed of other connexins.

The second example of an inherited human disorder due to defective gap junctions was recently reported (Kelsell et al 1997). Hereditary non-syndromic sensorineural deafness can be caused by mutations in different genes, such as the Cx26 ($\beta_2$) gene expressed in the cochlea of the inner ear. It has been suggested that deafness in patients with mutated Cx26 genes may be caused by defective gap junctions between sensory hair cells and supporting cells near the basal membrane. These gap junctions may allow the recycling of $K^+$ ions from hair cells through supporting cells back to the endolymph (Kikuchi et al 1995), although diffusion of other essential metabolites through these gap junctions is also likely. Recently, several more Cx26 mutants in patients suffering from hereditary deafness have been published (Denoyelle et al 1997, Zelante et al 1997). It is not yet clear whether other phenotypic alterations are associated with this defect, since Cx26 protein is probably expressed in skin, liver, pancreas and other human tissues. Perhaps the Cx26 defect is either not essential in these organs or can be compensated for by the expression of other connexins.

Another human disease that may be caused by defective gap junctions is the visceroatrial heterotaxia syndrome, which can be treated by heart transplantation in young children. Britz-Cunningham et al (1995) found single nucleotide exchanges in PCR-amplified Cx43 ($\alpha_1$) genes of affected hearts. Transfection of the mutated Cx43 coding sequence into cultured mouse cells deficient in gap junctional communication yielded differences in dye transfer, compared to wild-type Cx43. No electrophysiological confirmation of these findings was reported. The authors suggested that defects in the Cx43 gene expressed at high levels in ventricular myocardiocytes, and at low levels in several other tissues, may affect permeability through Cx43-containing gap junctions in patients suffering from visceroatrial heterotaxia. It is not clear whether these mutations were inherited or occurred spontaneously. More recently, it has been reported that no mutations in the Cx43 gene are found in other cases of visceroatrial heterotaxia (Gebbia et al 1996, Penman-Splitt et al 1997).

## Dominant negative effects of mutated connexin genes

Bruzzone et al (1994) expressed mutated Cx32 RNA from human CMTX patients in paired *Xenopus* oocytes. They observed that the mutated Cx32 RNA inhibited the channel-forming ability of coexpressed Cx26, but not of Cx40 ($\alpha_5$). Omori et al (1996) showed that transfected human Cx32 mutants inhibited dye transfer in cultured HeLa cells expressing wild-type Cx32, consistent with the notion of a dominant negative effect. Recently, Duflot-Dancer et al (1997) described dominant negative effects of mutated Cx26 genes on growth control in cultured HeLa cells expressing wild-type Cx26.

So far, dominant negative effects of mutated connexin genes have not been demonstrated in humans or mice. If these effects were strong enough to functionally inactivate expression of the second allele of a given connexin gene, this could lead to dominant inheritance of connexin defects. Possibly, dominant negative effects could also lead to non-functional heteromeric or heterotypic connexin channels, where more than one type of connexin subunit contributes to gap junctional hemichannels or total channels, respectively. Currently, it is not possible to predict which type of connexin mutant would be expected to show dominant negative effects on the same or other types of connexin. Furthermore, these effects could be strongly dependent on the expression level of the mutated connexin protein. Finally, it has been reported that Cx33 ($\alpha_7$), which is expressed in the mammalian testis, shows inhibitory effects on Cx37 ($\alpha_4$) channels when expressed in *Xenopus* oocytes (Chang et al 1996), although it does not form functional homotypic channels under these conditions. Thus, Cx33 could be a naturally occurring, dominant-acting connexin, whose biological role could be to inhibit other connexins. These speculations need to be verified experimentally.

## Targeted connexin deletion mutants in the mouse

The inactivation of genes by targeted deletions in the mouse genome (‘knockout’ technology) has led to many insights into the biological functions of these genes. The conclusions are limited, however, if gene deletions result in embryonic lethality or if the lost gene product can be compensated by the product of another gene, leading to an apparently normal phenotype of the mutant mouse. Currently, more than half of the known connexin genes have been deleted in the mouse genome. In the following paragraphs, we have tried to summarize these results as far as they relate to biological functions of these genes in mice and humans. Since the experimental technique of gene targeting is similar in all published cases, technical details are not reviewed here.

## Connexin43

Inactivation of the Cx43 gene leads to death of pups shortly after birth due to a morphological defect of the heart. The blood in these animals cannot be properly oxygenized, since the right ventricular outflow tract is blocked (Reaume et al 1995). This result shows that Cx43-containing gap junctions must fulfil an essential, but currently unclear, role during heart development. It is likely that neural crest cells which contribute to heart development are already affected, but it is obscure whether migration, positioning or other processes related to myocardial morphogenesis are altered in Cx43-deficient embryos. Interestingly, it has also been reported that overexpression of the Cx43 gene in the neural crest lineage of transgenic mice led, with incomplete penetrance, to embryonic cranial neural tube defects and heart malformations (cf. Lo 1996). Since Cx43 appears to be one of the earliest connexins expressed during mammalian development, it was surprising that Cx43-deficient embryos did not show any further signs of gross morphological abnormality or early embryonic lethality. If functional gap junctions play an essential role during early development, then the role of Cx43 must be compensated for by expression of other connexins. No up-regulation of Cx40 or Cx45 ($\alpha_6$) transcripts, which are present in the wild-type heart, has been found in Cx43-deficient embryos (Reaume et al 1995), although the amount and cellular location of the corresponding proteins have not been analysed in detail. It is noteworthy that the hearts of Cx43-deficient embryos are beating. This suggests that expression of the Cx43 protein cannot be essential for heartbeat. Recently, a detailed electrophysiological analysis of adult mice heterozygous for the Cx43 deletion revealed reduced ventricular conduction velocity compared to wild-type mice (Guerrero et al 1997). This indicates a role of Cx43-containing gap junctions in heart physiology. It is necessary to generate mice with a conditionally inactivated Cx43 gene to test Cx43 functions in the adult heart.

## Connexin32

Surprisingly, Cx32-deficient mice are viable and fertile, although they weigh approximately 17% less than wild-type littermates. More detailed investigations revealed that glucose mobilization from hepatic glycogen stores is decreased by 78% upon electrical stimulation of sympathetic nerves in Cx32-deficient livers compared to wild-type livers (Nelles et al 1996). Electrical stimulation is supposed to release noradrenaline from the endings of sympathetic nerves that innervate mouse liver only scarcely. Noradrenaline triggers the intracellular release of a second messenger, presumably inositol-1,4,5-trisphosphate, which stimulates glucose mobilization from glycogen. It has been suggested that signal propagation via gap junctions to contacting hepatocytes, mediated by these second messenger molecules, may be diminished in Cx32-deficient liver. Cultured primary Cx32-deficient hepatocytes showed 85% decreased intercellular electrical conductance and exhibited only Cx26 homotypic channels (Valiunas et al 1998). Cx26 protein is coexpressed in murine wild-type hepatocytes with Cx32 protein. In Cx32-deficient livers the Cx26 protein is down-regulated, although the total amount of Cx26 mRNA is not altered compared to wild-type liver. Thus, for unknown reasons, the Cx32 protein appears to stabilize Cx26 protein in the liver. In this context, it is interesting to note that Valiunas et al (1998) did not find heteromeric Cx32/Cx26 channels in primary embryonic and adult hepatocytes from wild-type mice, rather only heterotypic Cx32/Cx26 channels. Thus, it is uncertain whether stabilization of the Cx26 protein by Cx32 could occur by formation of heteromeric functional channels.

Cx32-deficient mice up to four months of age do not show any signs of peripheral neuropathy (Nelles et al 1996), but they develop morphological abnormalities (i.e. onion bulb structures of peripheral myelin) afterwards (Anzini et al 1997, Scherer et al 1998). Similar alterations are seen in peripheral nerves of Cx32-deficient human CMTX patients (cf. Martini 1997). Thus, Cx32-deficient mice can be considered as an animal model for the human inherited Cx32 defect, although the symptoms of the human CMTX disease are only mildly apparent in the mouse. This could be due to the much shorter length of peripheral axons in the mouse, which may weaken the effects of demyelination on peripheral nerve conduction.

We noticed that Cx32-deficient mice develop spontaneous and chemically induced liver tumours at higher frequencies than wild-type controls. After one year, we found eight- and 25-fold more spontaneous liver tumours in female and male Cx32-deficient mice, respectively, than in wild-type mice (Temme et al 1997). Furthermore, intraperitoneal injection of the carcinogen diethylnitrosamine led, after one year, both to more liver tumours in Cx32-deficient mice than in controls and to accelerated tumour growth. Apparently, the Cx32 defect affects proliferation of tumorigenic hepatocytes to a greater extent than initiation by the carcinogen. We noticed that labelling of S phase nuclei by bromodeoxyuridine was four- and 14-fold higher in male and female Cx32-deficient livers, respectively, compared to wild-type liver (Temme et al 1997). Apparently, proliferating Cx32-deficient hepatocytes can express more spontaneous or induced mutations in growth control genes and, thus, overcome the restraints of growth control.

We have looked at whether liver regeneration is altered after partial hepatectomy in Cx32-deficient and wild-type mice. We did not find any significant differences in time- or size-dependent regeneration of liver mass, although the mortality after partial hepatectomy in Cx32-deficient mice was 2.5-fold higher than in wild-type mice. Furthermore, the number of bromodeoxyuridine-labelled hepatocytic nuclei was significantly lower in male Cx32-deficient than in wild-type mice after partial hepatectomy (Temme et al 1998). These findings suggest that the Cx32 defect in liver may not directly cause abnormal control of proliferation.

Increased frequencies of liver tumours in Cx32-deficient human CMTX patients have not been described to date. The CMTX defect and hepatocellular carcinomas are both rare in western human populations. Thus, it is possible that the combined effects have not been noticed to date. Alternatively, the physiological consequences of the Cx32 defect in mouse and human livers may be different. For example, the density of liver innervation is much higher in the human liver than in the mouse liver.

Since Cx32 is also coexpressed with Cx26 protein in pancreatic acinar cells, we tried to induce pancreatic tumours by intraperitoneal injection of azaserine, a known pancreatic carcinogen in the rat. After 12 months, we did not observe any pancreatic tumours in Cx32-deficient and in wild-type mice. Instead, we noticed an increased occurrence of liver tumours (T. Ott, A. Temme, F. Dombrowski & K. Willecke, unpublished results 1998), suggesting that azaserine acted as liver carcinogen in the mouse, but failed to induce acinar tumours in the pancreas. Chanson et al (1998) observed a 2.5-fold increased secretion of amylase from Cx32-deficient pancreatic acini compared to wild-type mice. This directly confirms the older hypothesis (Meda et al 1987) that exocrine secretion in the pancreas is inversely proportional to gap junctional intercellular communication among acinar cells.

## Connexin26

Cx26 is the second gap junction protein expressed in mammalian glands, such as liver, pancreas and other tissues. When we generated Cx26-deficient mice, we noticed that these mice died around 11 days post coitum (dpc) *in utero* (Gabriel et al 1998). The expression of Cx26 at 8.5 dpc in the developing labyrinth region of the mouse placenta serves to complement and then to replace the nutrition of the embryo via the yolk sac. Cx26 is expressed in gap junctions between syncytiotrophoblasts I and II, which separate maternal from fetal blood (Shin et al 1996). We showed that following injection of radioactive methylglucose into the tail vein of pregnant mice, the accumulation of this non-metabolizable glucose analogue decreases by 60% at 10 dpc in homozygous Cx26-deficient embryos compared to heterozygous and wild-type littermates. This result suggests that the uptake of nutrients such as glucose, or the exchange of metabolic waste products through the labyrinth region of the mouse placenta, is dependent on Cx26-containing gap junctions. No other connexin protein (Cx31 [$\beta_3$], Cx32, Cx37, Cx40, Cx43 or Cx45) was detected in the labyrinth region of the mouse placenta. The embryonic lethality of homozygous Cx26-deficient mice precluded studies of adult Cx26-deficient mice. Conditional inactivation of the Cx26 gene will be necessary in order to study whether adult mice suffer from deafness, similar to human Cx26-deficient patients. The human placenta contains only one gigantic syncytiotrophoblast cell layer which separates maternal from fetal blood. Thus, there is no functional need for gap junctions in this tissue.

## Connexin37

Expression of Cx37 proteins has been observed in many endothelial cells of blood vessels in several organs. Cx37-deficient mice are viable, but females are infertile (Simon et al 1997). Cx37-containing gap junctions are present between oocytes and granulosa cells in wild-type females. Cx37-deficient mice fail to ovulate, and they develop several inappropriate corpora lutea. Oocyte development is arrested before meiotic competence is achieved. Simon et al (1997) showed that transfer of Lucifer yellow from the microinjected oocyte to surrounding granulosa cells is blocked in $Cx37^{-/-}$ but not in $Cx37^{+/+}$ follicles. This is consistent with the notion that an unknown inhibitory signal may be transferred from the oocyte through Cx37-containing gap junctions to the granulosa cells, resulting in the prevention of corpus luteus formation. The type of connexin in granulosa cell membranes (of the cumulus oopherus) that interacts with Cx37 in the oocyte membrane is not yet known. No effects on endothelial cells of Cx37-deficient mice have been reported to date.

## Connexin40

Besides Cx37, Cx40 is the second gap junctional protein that is expressed in many endothelial cells. In the mouse Cx40 protein is also present in myocardial gap junctions, i.e. in the atrium and in the conductive system of ventricular His bundle and Purkinje fibres. Recently, the generation and characterization of targeted Cx40-deficient mice has been independently achieved by Simon et al (1998) and, in our laboratory, by Kirchhoff et al (1998). In both cases, these mice were viable and fertile. No alteration in the amount and cellular location of other myocardial connexins (i.e. Cx43, Cx45 and Cx37) were noticed in adult Cx40-deficient mice compared to control mice. The frequency of heartbeat was not altered, but the QRS complex of the electrocardiogram of Cx40-deficient mice was significantly prolonged, suggesting that ventricular conduction in Cx40-deficient heart tissue was reduced. In addition, Kirchhoff et al (1998), reported that the P wave and PQ interval of Cx40-deficient mice was significantly longer than in wild-type mice, suggesting that the conduction velocity in the atrium was also reduced, consistent with the known expression pattern of Cx40 in the mouse myocardium. Furthermore, we found that six out of 31 Cx40-deficient mice developed arrhythmias, whereas normal sinus rhythm was found in 61 heterozygous and wild-type mice (Kirchhoff et al 1998). Thus, arrhythmias can arise as a consequence of the Cx40 defect in mouse heart. Since expression of Cx40 is similar in mouse and human hearts, it is tempting to speculate that a Cx40 defect in humans may also cause alterations in the electrocardiogram and even arrhythmias.

**Connexin46**

Cx46 ($\alpha_3$) is one of the major gap junction proteins expressed between lens fibre cells of mammals. Gong et al (1997) have characterized Cx46-deficient mice, which are viable and fertile but develop nuclear lens cataracts at two to three weeks of age. Slight effects on the expression pattern of Cx50 ($\alpha_8$) are also apparent in the Cx46-deficient lens, but more obvious is the presence of a proteolytic cleavage product of $\gamma$-crystallin, which is exclusive to the Cx46-deficient lens. Gong et al (1997) concluded that Cx46-containing gap junction channels in the lens may allow intercellular diffusion of metabolites and ions that regulate activity of an unknown protease and maintain the suprastructural organization of lens proteins responsible for lens transparency. A hereditary form of human cataract has been mapped to the same region on human chromosome 13, where the Cx46 gene is located (Mackay et al 1997). Currently it is not known whether this form of human cataract is due to a defect in the Cx46 gene.

**Connexin50**

Cx50 is coexpressed with Cx46 in lens fibre cells. White et al (1997) reported that targeted, homozygous Cx50-deficient mice are viable and fertile but that their eyes are smaller, and the lenses weigh only 56% of those in wild-type littermates. In addition, Cx50-deficient mice developed nuclear opacity after one week of age. Thus, similarly as reported for Cx46 (Gong et al 1997), functional Cx50-containing gap junction channels appear to determine lens transparency. The underlying molecular mechanism and the influence of Cx50 gap junction channels on eye size are not understood at present.

**Connexin31**

Cx31 is coexpressed with Cx43 at the blastocyst stage of the mouse embryo. During implantation, Cx31 becomes restricted to tissues derived from trophectoderm. In adult mice Cx31 is expressed in the suprabasal layer of mouse epidermis and in the testis. Preliminary characterization of Cx31-deficient mice in our laboratory (Plum et al 1998) showed that these mice are viable and fertile. Expression of the *lacZ* reporter gene, replacing the deleted Cx31 reading frame in Cx31-deficient mice, confirmed that Cx31 is expressed in extraembryonic ectoderm, ectoplacental cone (at 7.5 dpc), developing hindbrain (8.5 dpc) and branchial arches (9.5 dpc), as well as in adult skin. Interbreeding of heterozygous Cx31-deficient mice results in 10% homozygously mutated animals, genotyped three weeks after birth, instead of the 25% expected for Mendelian inheritance. At the preimplantive blastocyst stage (3.5 dpc) no distortion of the expected Mendelian ratio is found. This suggests that one or more limiting steps must be overcome during preimplantation or postimplantive development of Cx31-deficient embryos. We are currently trying to identify this developmental step and searching for molecular mechanisms that can explain the role of Cx31 during development.

## Characterization of mice lacking two different connexin genes

Since most cell types express more than one connexin protein, it is possible that different connexins can in part replace or compensate one another for biological functions. Compensatory or complementary effects of different connexin proteins may become more apparent when these connexins are lacking in the same cell type. This can be achieved by interbreeding connexin-deficient mice to obtain heterozygously or homozygously mutated progeny deficient in both connexins.

As a first example, we recently generated mice that were homozygously deficient for both the Cx43 and Cx32 genes. These double mutant pups die, like Cx43-deficient mice, shortly after birth. A survey of the major organ systems of the doubly mutated fetuses, including the thyroid gland, developing teeth and limbs, where Cx43 and Cx32 are coexpressed in wild-type mice, failed to reveal any morphological abnormality not already seen in Cx43-deficient fetuses. Thus, at least during prenatal life, normal development of these organs lacking Cx43 cannot be explained by the additional presence of Cx32, and vice versa. If gap junctional communication is necessary for development, it may be provided by yet another connexin when both Cx32 and Cx43 are absent (Houghton et al 1999).

We also noticed (S. Kirchhoff, unpublished results 1998) that mice which lack Cx40 and are heterozygously mutated in the Cx43 gene die around birth, although homozygous Cx40-deficient mice and heterozygous Cx43-deficient mouse mutants are viable and fertile. This could be due to complementary functions of both connexin proteins that are coexpressed, for example, in atrial myocardiocytes. Undoubtedly, several other combinations of the above mentioned connexin defects will be studied in the near future in order to clarify whether coexpression of different connexins in the same or neighbouring cells results in functional complementation.

## Conclusions and prospects

Identification of human connexin mutants and characterization of connexin-deficient mouse mutants has led to considerable insights into the function of different connexin genes. Since genetics and physiology may differ between humans and mice, it is necessary to compare the details of each connexin defect. So far, all connexin defects in the mouse genome have been generated by targeted deletion, whereas many of the mutants identified in the human Cx32 and Cx26 genes are due to single base exchanges. It should be possible to express human connexin mutants in the mouse in order to study possible dominant negative interactions of mutated and non-mutated connexin proteins. Preliminary results suggest that different connexins, when coexpressed in the same cells, may complement each other for the function of this cell type in the whole organ. Thus, one can possibly expect to dissect the contribution of each connexin type to total organ physiology, when all connexins coexpressed in this organ are inactivated. The diversity of connexins may have evolved, at least in part, due to the need for functional complementation. Future 'knock-in' experiments, where the coding sequence of one connexin gene is replaced by another one in transgenic mice, may reveal the extent of compensation between different connexins. Each connexin probably shares some redundant function(s) with other coexpressed connexins and can also fulfil one or more complementary functions. It will be important to identify metabolites and ions that pass through different connexin channels and mediate compensatory or complementary functions.

*Acknowledgements*

Work in our laboratory on gap junctions is supported by grants of the Deutsche Forschungsgemeinschaft (SFB 284), the Dr. Mildred Scheel Stiftung, the Fonds der Chemischen Industrie and the Biomed Program of the European Community.

## References

Anzini P, Neuberg DH, Schachner M et al 1997 Structural abnormalities and deficient maintenance of peripheral nerve myelin in mice lacking the gap junctional protein connexin32. J Neurosci 17:4545–4561

Bergoffen J, Scherer SS, Wang S et al 1993 Connexin mutations in X-linked Charcot-Marie-Tooth disease. Science 262:2039–2042

Britz-Cunningham SH, Shah MM, Zuppan CW, Fletcher WH 1995 Mutations of the connexin43 gap-junction gene in patients with heart malformations and defects of laterality. N Engl J Med 332:1323–1329

Bruzzone R, White TW, Scherer SS, Fischbeck KH, Paul DL 1994 Null mutations of connexin32 in patients with X-linked Charcot-Marie-Tooth disease. Neuron 13:1253–1260

Chang M, Werner R, Dahl G 1996 A role for an inhibitory connexin in testis? Dev Biol 175:50–56

Chanson M, Fanjul M, Bosco D et al 1998 Enhanced secretion of amylase from exocrine pancreas in connexin32-deficient mice. J Cell Biol 141:1267–1275
Denoyelle F, Weil D, Maw MA et al 1997 Prelingual deafness: high prevalence of a 30delG mutation in the connexin26 gene. Hum Mol Genet 6:2173–2177
Deschênes SM, Bone LJ, Fischbeck KH, Scherer SS 1996 Connexin32 and X-linked Charcot-Marie-Tooth disease. In: Spray DC, Dermietzel R (eds) Gap junctions in the nervous system. Springer-Verlag, Heidelberg, p 213–227
Duflot-Dancer A, Mesnil M, Yamasaki H 1997 Dominant negative abrogation of connexin-mediated growth control by mutant connexin genes. Oncogene 15:2151–2158
Gabriel H-D, Jung D, Bützler C et al 1998 Transplacental uptake of glucose is decreased in embryonic lethal connexin26 deficient mice. J Cell Biol 140:1453–1461
Gebbia M, Towbin J, Casey B 1996 Failure to detect connexin43 mutations in 38 cases of sporadic and familial heterotaxy. Circulation 94:1909–1912
Gong X, Li E, Kliev G et al 1997 Disruption of $\alpha3$ connexin gene leads to proteolysis and cataractogenesis in mice. Cell 91:833–843
Goodenough DA, Goliger JA, Paul DC 1996 Connexins, connexons and intercellular communication. Annu Rev Biochem 65:475–502
Guerrero PA, Schuessler RB, Davies LM et al 1997 Slow ventricular conduction in mice heterozygous for a connexin43 null mutation. J Clin Invest 99:1991–1998
Houghton ED, Thönnissen E, Kidder GM et al 1999 Doubly mutant mice, deficient in connexin32 and 43, show normal prenatal development of organs where the two gap junction proteins are expressed in the same cells. Dev Gen, in press
Kelsell DP, Dunlop J, Stevens HP et al 1997 Connexin26 mutations in hereditary non-syndromic sensorineural deafness. Nature 387:80–83
Kikuchi T, Kimar R, Paul DL, Adams J 1995 Gap junctions in rat cochlea: immunochemical and ultrastructural analysis. Anat Embryol 91:101–118
Kirchhoff S, Nelles E, Hagendorff A, Krüger O, Traub O, Willecke K 1998 Reduced cardiac conduction velocity and predisposition to arrhythmias in connexin40-deficient mice. Curr Biol 8:299–302
Kumar NM, Gilula NB 1996 The gap junction communication channel. Cell 84:381–388
Lo CW 1996 The role of gap junction membrane channels in development. J Bioenerg Biomembr 28:379–385
Mackay D, Fonides A, Berry V, Moore A, Bhattacharya S, Shiels A 1997 A new locus for dominant 'Zonular pulverulent' cataract on chromosome 13. Am J Hum Genet 60:1474–1478
Martini R 1997 Animal models for inherited peripheral neuropathies. J Anat 191:321–336
Meda P, Bruzzone R, Bosco D, Orci L 1987 Junctional coupling modulates secretion of exocrine pancreas. Proc Natl Acad Sci USA 84:4901–4904
Nelles E, Bützler C, Jung D et al 1996 Defective propagation of signals generated by sympathetic nerve stimulation in the liver of connexin32-deficient mice. Proc Natl Acad Sci USA 93:9565–9570
Omori Y, Mesnil M, Yamasaki H 1996 Connexin32 mutations from X-linked Charcot-Marie-Tooth disease patients: functional defects and dominant negative effects. Mol Biol Cell 7:907–916
Patel IP, Lupski JR 1994 Charcot-Marie-Tooth disease: a new paradigm for the mechanism of inherited disease. Trends Genet 10:128–133
Penman-Splitt M, Tsai M, Burn J, Goodship JA 1997 Absence of mutations in the regulatory domain of the gap junction protein connexin43 in patients with visceroatrial heterotaxy. Heart 77:369–370
Plum A, Rosentreter A, Traub O, Willecke K 1998 Connexin31 deficient mice can be viable and develop normally. Eur J Cell Biol 75:69

Reaume AG, de Sousa AP, Kulkarni S et al 1995 Cardiac malformation in neonatal mice lacking connexin43. Science 267:1831–1834

Scherer SS, Xu Y-T, Nelles E, Fischbeck K, Willecke K, Bone LJ 1998 Connexin32-null mice develop demyelinating peripheral neuropathy. Glia 24:8–20

Shin BC, Suzuki T, Matsuzaki T et al 1996 Immunolocalization of GLUT1 and connexin26 in the rat placenta. Cell Tissue Res 285:83–89

Simon AM, Goodenough DA, Li E, Paul DL 1997 Female infertility in mice lacking connexin37. Nature 385:525–529

Simon AM, Goodenough DA, Paul DL 1998 Mice lacking connexin40 have cardiac conduction abnormalities characteristic of first-degree atriaventricular block and bundle branch block. Curr Biol 8:295–298

Temme A, Buchmann A, Gabriel H-D, Nelles E, Schwarz M, Willecke K 1997 High incidence of spontaneous and chemically induced liver tumors in mice deficient for connexin32. Curr Biol 7:713–716

Temme A, Ott T, Dombrowski F, Willecke K 1998 Synchronous initiation and termination of DNA synthesis in regenerating mouse liver is dependent on connexin32-expressing gap junctions, in prep

Valiunas V, Niessen H, Willecke K, Weingart R 1998 Electrophysiological properties of gap junction channels in hepatocytes isolated from connexin32 deficient and wild type mice, submitted

White TW, Goodenough DA, Paul DL 1997 Ocular abnormalities in connexin50 knockout mice. Mol Biol Cell 8:93a

Zelante L, Gasparini P, Estivill X et al 1997 Connexin26 mutations associated with the most common form of non-syndromic neurosensory autosomal recessive deafness (DFNB1) in Mediterraneans. Hum Mol Genet 6:1605–1609

## DISCUSSION

*Beyer:* It is gratifying to see that connexin40 (Cx40; $\alpha_5$) knockout animals have a phenotype and a conduction block. This makes us all feel more comfortable about the importance of Cx40 in the heart. All of the connexins expressed in the heart are probably adequate to allow current to pass between cells, but we also have to contemplate that connexins likely have additional roles in the heart besides the conduction of current, and that this may be one of the reasons for connexin diversity in the heart. The diversity allows not just from variations in expression patterns but also differences in permeability properties of channels formed of different connexins.

*Gilula:* I understand that Dan Goodenough is also publishing paper on heart-related connexin knockouts, so I would like to ask him to comment on this work.

*Goodenough:* The only discrepancy that Klaus Willecke and David Paul and I have is that so far we have not been able to see any of the arrhythmias that they report. We see elongated PR intervals and splits in the QRS complexes as though the right and left ventricles are contracting asynchronously, which is consistent with a delayed and non-coordinated excitation going down the right and left and right bundle branches. But why we don't see arrhythmias is puzzling. It may be because the conditions for doing the measurements are stressing the mice, but we haven't yet done any stress tests to see if we can induce arrhythmia.

*Beyer:* What is the nature of the arrhythmia?

*Willecke:* Six out of 31 Cx40-deficient mice had arrhythmias that were of different types; for example, sinus arrhythmia, atrial ectopia, total atrioventricular block and intra-atrial re-entrant tachycardia, although they all appeared to originate in the atrium.

*Yeager:* The metabolic environment may influence whether arrhythmias are observed. For instance, significant differences in the serum electrolytes between the transgenic mice generated in the Goodenough and Willecke laboratories could account for differences in the incidence of arrhythmias.

*Goodenough:* All our measurements were carried out on animals lightly anaethesized with nembutol.

*Willecke:* I have suggested to Dan Goodenough that we should exchange the Cx40-deficient mice between our laboratories, in order to confirm the occurrence of arrhythmias.

*Severs:* You both had one common observation, which was a slowed conduction through the ventricular conduction system. This is entirely consistent with an absence of Cx40, which is reassuring.

*Goodenough:* Your paper on Cx45 ($\alpha_6$) was also satisfying (Coppen et al 1998), in that we were wondering which connexin was rescuing the Cx40 knockout mice from total heart block. Our mice show only a first degree atrioventricular block, indicating that a conduction pathway still exists through the His/Purkinje system.

*Sanderson:* It must be difficult to study arrhythmias in the mouse because it is so small. If a particular cell type has Cx43 ($\alpha_1$) but not Cx40, for example, surely the electrical coupling between those the abnormal cells will change relative to normal cells, so can you estimate the changes in propagation velocities from cells taken from these animals and put in culture?

*Beyer:* There are difficulties with this approach. We were not successful in recording epicardial surface conduction in the homozygous Cx43 knockout mice. The reason for this is probably that the hearts were just too fragile. However, we do have data from the heterozygous Cx43 knockouts that suggest ventricular conduction is slowed. This surprised me because I expected it to be normal. Our interpretation of this is that there is a gene dosage effect, i.e. half as much protein results in half the slower conduction rate. However, we don't yet know whether there are half as many open channels between two cells.

*Lo:* We sent some of our heterozygous Cx43 knockout mice to Mario Delmar and Jose Jalife, but they could not find any differences in conduction velocity. We don't know whether this is because of the different strain backgrounds or differences in the methodologies.

*Gilula:* In the interest of relating some of the conduction observations to the potential movements of second messengers, Michael Sanderson should be encouraged to obtain some of these animals to apply his imaging procedures.

*Sanderson:* We have developed a culture system so that we can look at calcium waves, not in heart cells but in mice airway epithelial cells. These cells have different connexins, so it will be exciting to determine whether cells from knockout mice missing certain connexins have calcium waves.

*Lo:* Is it possible to do these experiments in cultures of cardiac myocytes?

*Beyer:* We've been trying to do these experiments in collaboration with André Kléber, who has been growing strands of neonatal cardiac myocytes and using optical mapping to quantitate conduction in ordered cellular arrays. Murine cultures of cardiac myocytes from the knockout animals are not yet available. It should be easier to do this in the Cx40 knockout animals because for the Cx43 knockout mice we have to generate each culture from individual animals.

*Green:* One problem with this approach is that once you culture neonatal cardiac myocytes, you change the entire regulation of gap junction expression. The cells in culture can range from those with next to no communication to complete 'zipping up' of adjacent cells with Cx43.

*Willecke:* As a general rule, primary cells taken into culture often lose Cx40 and Cx45 expression, although expression of Cx43 is maintained or even turned on in established cell lines.

*Yeager:* Do you have any evidence for Cx43 up-regulation in Cx40 knockouts?

*Goodenough:* We have no evidence for Cx43 up-regulation. And Cx45 is probably also not up-regulated, given the slowing down of conduction. However, these experiments should be repeated with better antibodies.

*Willecke:* We have not observed any changes in Cx45 or Cx43 expression in Cx40 knockouts.

*Beyer:* I'm not aware that anyone has evidence in any of the connexin knockouts of up-regulation of the remaining connexins.

*Musil:* Have you seen any increase in the incidence of tumours in any of the connexin knockouts apart from the Cx32 ($\beta_1$) knockout?

*Willecke:* These experiments take over a year to complete. We have looked at the Cx32 knockout and we now intend to do similar experiments with the Cx31 ($\beta_3$) knockout, but it will be another year before the results are known.

*Lo:* I have a question about the Cx26 ($\beta_2$) knockout. The presumption is that the defect has something to do with placental function. Have you looked at aggregation chimeras, comprising homozygous Cx26 null mutant embryos and wild-type embryos? Because if a placental defect is involved in the lethality, you should be able to achieve rescue of the knockout lethality.

*Willecke:* We have thought of doing that, and this experiment will probably be carried out in another laboratory.

*Dermietzel:* I have a question about the expression of Cx31 in the embryonic brain. Have you done any further analysis of the nervous system? Because 3 and 5 rhombomeres form most parts of the hindbrain and you also detected expression in neuroblasts. When we studied the differentiation of immortalized human neuroblasts, we found that at a certain step in differentiation Cx31 was expressed in these cells.

*Willecke:* We haven't yet looked at this. Achim Plum in my laboratory found a metameric expression pattern near the spinal cord in late-stage embryos, and with the help of the reporter gene (instead of the Cx31 gene), we could also look at expression in neural tissues of adult mice. At this stage this is all preliminary work, so there is much to be done.

*Gilula:* Rolf Dermietzel, were your Cx31 analyses carried out using antibody detection of the protein product?

*Dermietzel:* No, we used reverse transcriptase (RT)-PCR.

*Gilula:* And Klaus Willecke, have you used the reporter gene as a way of defining where you should use antibodies to look for the production of Cx31 protein in the wild-type mouse?

*Willecke:* Again, these results are too preliminary. We are now trying to verify expression of the Cx31 protein in those areas in which we have detected expression of *lacZ*.

*Gilula:* I asked both you and Rolf Dermietzel this question because at the level of transcript detection, different observations have been made by different people, and obtaining evidence that those transcripts are actually making a detectable protein product seems to be a different challenge.

*Beyer:* Your question raises an issue we struggled with for years, i.e. is there translational regulation of connexins? There are a number of cases where mRNA can be detected but gap junction proteins cannot be identified by immunofluorescence. This would suggest that the proteins are there but we have failed to detect them, or that there is translational regulation.

*Goodenough:* Jean Jiang has looked at RCAS (a replication-competent chicken retrovirus)-driven expression of connexins in the chick lens (Jiang & Goodenough 1998). These connexins are Cx43, Cx56 and Cx45.6, the latter two corresponding to Cx46 and Cx50 ($\alpha_8$), respectively. The connexins were expressed with epitope tags so that they could be distinguished from the endogenous proteins. Jean found that all virally encoded connexins were expressed and that the proteins were found only in their appropriate locations. In contrast, RCAS-driven alkaline phosphatase was non-selectively expressed in all lens cells. These experiments do not distinguish between regulated expression of the connexins in certain cells vs. unregulated expression followed by selective degradation in the appropriate cells.

*Werner:* I would like to give a short progress report on the transcriptional regulation of the Cx43 gene. It is expressed constitutively in the heart, whereas its expression is induced in the uterus by oestrogen and repressed by progesterone. At the International Gap Junction meeting in Key Largo, USA (July 12–17 1997), my student Elisa Oltra, presented evidence for the isolation of a novel transcription factor with a molecular mass of 11 kDa that is possibly involved in this process (Oltra & Werner 1998). We don't know much yet about this transcription factor. We have found a similar sequence in yeast and we are trying to knock out this gene. We also attached the piece of DNA that was used to isolate the transcription factor to a column to isolate three more potential transcription factors. We don't yet have the sequences for these. It's likely that the 11 kDa transcription factor is ubiquitous. We found the mRNA in many cell types. We have done some co-transfections with the promoter construct, and we find that there is at most a twofold increase in transcription in the presence of the transcription factor cDNA. We haven't yet done the opposite experiment, i.e. put the antisense transcription factor cDNA into the cells.

There's one more point I would like to make. It is often the case that cells in culture behave differently from those *in vivo*. We made a Cx43 transgenic mouse that had the 6.1 kb regulatory region and 8.1 kb intron of the Cx43 gene attached to the luciferase reporter gene. My student was somewhat depressed when he looked at luciferase expression in this mouse. Luciferase activity was not found in the heart and uterus as expected, but in the testes and lung. It is possible that our construct was inserted into the genome at a position that is regulated in a different way.

*Lo:* We have also made some transgenic mice using the 6.5 kb Cx43 promoter sequence attached to the *lacZ* reporter (Lo et al 1997). We observed strong expression only in the dorsal neural tube and in all neural crest cell lineages, but not in the myocardium. We first thought that this was because the element involved in driving expression in the myocardium is absent. From various *in vitro* studies we also have evidence that much of the 6.5 kb sequence is involved in repression, and we have now found that when only 800 bp of the upstream sequence is present in transgenic mice, reporter gene expression is observed in the myocardium. In such transgenic mice, we don't see expression in all neural crest cell lineages, so the regulation of Cx43 expression is complicated and involves upstream silencing sequences as well as activating sequences.

*Beyer:* What happens to the expression of the construct in the uterus?

*Lo:* We haven't looked at this in detail. In the 6.5 kb transgenic the construct is expressed in the gonads during late embryogenesis, but I don't know what happens in adult animals.

*Werner:* Is the intron present in your construct?

*Lo:* No.

*Werner:* Then I am suspicious of your results, because regulatory regions are probably present in the intron. The mouse and human downstream regions of the intron are about 98% identical at the nucleotide level, which suggests to me that they are involved in regulation.

*Lo:* I agree. I'm not saying that we have seen expression in all the places where you see Cx43, but where we have seen expression, the pattern is reproducible in multiple independently derived transgenic lines so I'm sure that it is not a position effect based on site of insertion.

*Gilula:* I would like to hear what Andy Forge has to say about Klaus Willecke's comments on connexin expression in the ear.

*Forge:* I can think of at least two things that gap junctions do in the ear—one involves the transfer of ions, and the other involves the formation of new junctions when cells are lost and the tissue is repaired—yet it is difficult to relate these to how the different connexins are expressed in the ear. I wondered whether it is possible that following a traumatic event a different subpopulation of connexins are used to enable the transport of a different signal during the repair process.

*Severs:* With regard to altered connexin expression after trauma, when arteries are damaged, a massive increase in Cx43 gap junctions between the smooth muscle cells follows, which is therefore an example of increased gap junction expression during a repair process after injury.

*Sanderson:* What is the time-scale of that process?

*Severs:* Following balloon catheter injury of an artery, a new intima starts to be formed after about nine days, so that increased gap junction expression is evident from nine days as part of the repair mechanism.

*Nicholson:* Have any hepatectomies been performed on the Cx32 knockout mice? This might be interesting given the apparent role of the dramatic drop in, and subsequent reformation of, gap junctions during liver regeneration in normal mice (Yee & Revel 1978), and the tendency of the knockout mice to develop hepatic tumours. One might anticipate a bizarre product.

*Willecke:* We have performed hepatectomies and analysed the consequences by standard techniques, such as bromodeoxyuridine incorporation and measurements of liver weight. We did not observe any differences compared to wild-type mice, with the possible exception of changes in the levels of apoptosis.

*Yamasaki:* We created transgenic mice in which one of the X-linked Charcot-Marie-Tooth syndrome Cx32 mutants was driven by the albumin promoter and expressed specifically in the liver. We didn't see any deficiencies in communication, but after partial hepatectomy we saw a high mortality rate amongst these transgenic mice. When we looked at bromodeoxyuridine incorporation following partial hepatectomy, we found that regeneration occurred more slowly than in the wild-type.

*Willecke:* Valiunas et al (1998) have measured conductivity in primary hepatocyte cultures taken from wild-type versus Cx32 knockout mice, and have found that the Cx32-deficient hepatocytes have only 15% of wild-type conductance. Thus, the cells are not uncoupled: coupling can be demonstrated either by dye transfer of Lucifer yellow or by electrical conductivity. Originally, I did not expect to see an effect on tumorigenesis following a reduction in gap junctional communication by 90%. In order to explain our findings, one could argue that Cx32 channels are qualitatively different from Cx26 channels, and this may cause the effect on liver tumorigenesis in Cx32-deficient mice.

*Gilula:* In the homozygous Cx43 knockout, which dies at birth, why don't the other connexins rescue the phenotype? Why does this animal not have the ability to recruit and utilize other connexins?

*Lo:* The best explanation is probably that in this instance Cx43 may have a unique role. The model we have is that the perturbation of neural crest cells is related to the defect. We presume that Cx43 has a unique role in the modulation of cardiac neural crest cells, and that there are no connexins present which can replace the function of Cx43 in these cells.

*Beyer:* We have to speculate that there is a unique function there. There are cases of relative functional redundancy, when different connexins are coexpressed, but I'm not sure that there are any examples of compensatory alterations.

*Gilula:* Another example is that in the lens Cx50 cannot rescue a Cx46 knockout, in the context of the specific cells that are solely using them. The questions are, have we learned anything about the unique utilization of certain connexins in certain regions developmentally, and if so, can we develop strategies to overcome it?

*Yamasaki:* I'm not sure whether knockouts that involve the removal of the gene from the entire animal can be used to address these questions. It may be better to develop tissue-specific knockouts.

*Goodenough:* It is possible that the phenotype observed in a knockout animal will derive from a cell that has lost the ability to express another connexin, i.e. a cell that may be too specialized. For example, in the case of Cx37 ($\alpha_4$) the oocyte, which is locked in meiotic prophase, may not have access to certain parts of its genome to enable it to express another connexin gene.

*Kistler:* One aspect that we have not yet considered is how many channels are required to maintain a function. It may not be a question of one connexin replacing the role of another, rather that a certain number of channels are required.

*Gilula:* That was a fundamental question 25 years ago, which indicates simply how little we still understand about the quantity versus function issue.

Could you expand on the double Cx43/Cx32 knockout, where you saw no detectable effect on the limb buds.

*Willecke:* This analysis was carried out by Eva Thönnissen in my laboratory (Houghton et al 1999). The genotype of the double-deficient mouse embryos has been verified by PCR and Southern blot analysis. Our co-workers have looked at the thyroid gland, the developing teeth and the limbs. In all three cases, and in agreement with the results in other laboratories, they found coexpression of Cx32 and Cx43 in the same cells of these organs. Yet when they looked by immunostaining and hormone production in the thyroid gland and cross-sections of the developing teeth or limbs of double-deficient mice, they did not see any difference. The double-deficient embryos died just after birth, around the same time as the Cx43 knockout mice die. We could not distinguish one phenotype from the other. Since the double-deficient mice die shortly after birth, we can't tell if any abnormalities show up later on. Also, the number of double-deficient embryos was in the range of 10–20, so we would not have picked up any defects with low penetrance. However, given these precautions, the experimental result was quite clear. We carried out the experiment expecting to see additional abnormalities but we did not see any.

*Warner:* Have you looked at whether any other connexins are expressed in any of these situations? We find Cx43 and Cx32 coexpressed in the developing limb bud, but we have not done an exhaustive search to see if other connexins are present. All we can say is that they're both expressed. We don't know whether they are the only ones that are expressed, or if they are the only ones capable of being expressed.

*Willecke:* At the age of 10.5 days post coitum we observed strong expression of Cx31 in the limb buds.

*Warner:* It is becoming clear that in many situations multiple connexins can be expressed, but we don't know what determines which ones are expressed.

*Willecke:* One experiment that might help to clarify the situation is the so-called 'knock-in' experiment, where one connexin is replaced by another. We are in the middle of doing these experiments, but we don't yet have any results.

*Gilula:* You showed us a slide of the different connexin genes that have been knocked out. Is that inclusive? Are there any other genes that people have not yet knocked out?

*Willecke:* It is all the information I have.

*Goodenough:* The Cx33 ($\alpha_7$) knockout experiments are in progress.

*Gilula:* How soon will it be before they are all knocked out?

*Willecke:* We don't yet know how many there are, and there is also the question of whether we will be any smarter once we have knocked them all out.

## References

Coppen SR, Dupont E, Rothery S, Severs NJ 1998 Connexin45 expression is preferentially associated with the ventricular conduction system in mouse and rat heart. Circ Res 82:232–243

Houghton ED, Thönnissen E, Kidder GM et al 1999 Doubly mutant mice, deficient in connexin32 and 43, show normal prenatal development of organs where the two gap junction proteins are expressed in the same cells. Dev Gen, in press

Jiang JX, Goodenough DA 1998 Retroviral expression of connexins in embryonic chick lens. Invest Ophthalmol Vis Sci 39:537–543

Lo CW, Cohen MF, Huang GY et al 1997 Cx43 gap junction gene expression and gap junctional communication in mouse neural crest cells. Dev Genet 20:119–132

Oltra EJ, Werner R 1998 Cloning of a potential regulatory protein involved in connexin43 gene expression. In: Werner R (ed) Gap junctions. IOS press, Amsterdam, p 321–325

Valiunas V, Niessen H, Willecke K, Weingart R 1998 Electrophysiological properties of gap junction channels in hepatocytes isolated from connexin32 deficient and wild type mice, submitted

Yee AG, Revel J-P 1978 Loss and reappearance of gap junctions in regenerating liver. J Cell Biol 78:554–564

# Connexins in the lens: are they to blame in diabetic cataractogenesis?

Joerg Kistler, Jun Sheng Lin*, Jacqui Bond, Colin Green*, Reiner Eckert, Rachelle Merriman, Mark Tunstall and Paul Donaldson†

*School of Biological Sciences, *Department of Anatomy with Radiology, School of Medicine, and †Department of Physiology, School of Medicine, University of Auckland, Auckland, New Zealand*

*Abstract.* The pathohistology of the diabetic lens is an enigma. Under normal conditions the lens behaves as a functional syncitium, whereas the diabetic lens exhibits a localized zone of fibre cell swelling and rupture that is confined to the lens outer cortex. Because the lens fibre cells are extensively coupled by gap junction channels, it is believed that the abnormal closure of these channels is responsible for this phenomenon. New evidence concerning regional differences in gap junction gating supports this contention, and it is used to propose a new hypothesis that may explain the cellular changes observed in the diabetic lens.

*1999 Gap junction-mediated intercellular signalling in health and disease. Wiley, Chichester (Novartis Foundation Symposium 219) p 97–112*

## The pathohistology of the diabetic cataract

Experimentally induced diabetic and galactosaemic cataracts in rats have been widely used as models to study the mechanisms underlying lens opacification. Discrete opacities form early during cataractogenesis as a result of fluid accumulation and tissue breakdown in the lens cortex. It is well documented that at least initially in sugar cataractogenesis, only a limited tissue segment in the lens cortex is affected (Kuwabara et al 1969, Datiles et al 1982, Robison et al 1990). More recently, we have utilized confocal laser scanning microscopy of lens equatorial sections labelled with fluorescent membrane markers to visualize the time course of cellular changes that are initiated by the swelling of individual fibre cells (Bond et al 1996). In rats made diabetic with a single injection of streptozotocin (Mitton et al 1993), such swollen cells become apparent after two weeks and are concentrated in a discrete zone in the lens cortex (Fig. 1a). After three weeks, swollen cells are so numerous that they form a coherent band of tissue disorder in this region (Fig. 1b). Four weeks after the onset of diabetes, extensive cell breakdown and formation of fluid lakes, a process also referred to as tissue liquefaction, becomes evident (Fig. 1c).

We still do not understand why cell swelling and tissue liquefaction are restricted to a narrow band in the lens cortex despite all fibre cells across the lens being extensively connected via gap junctions (Fig. 1d). The gap junction channels are normally open and provide direct links for cell–cell communication. This has been demonstrated by patch clamp analysis of isolated fibre cell pairs and with tracer

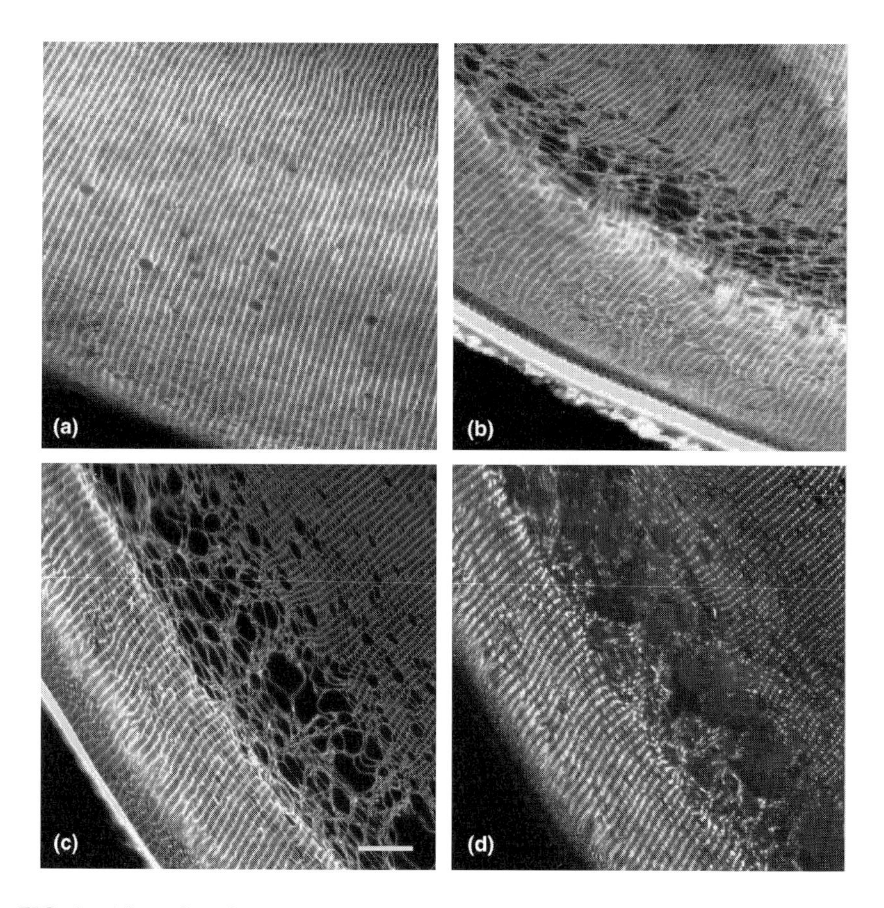


FIG. 1. Tissue liquefaction in the cortex of the diabetic rat lens. Confocal laser scanning microscopy of lens equatorial sections showing cellular changes two weeks (a), three weeks (b) and four weeks (c) after injection with streptozotocin. Labelling with fluorescent wheat germ agglutinin visualizes cell membranes, and reveals progressive cell swelling and breakdown in a narrowly defined cortical zone. (d) Same section as in (c), dual labelled with anti-connexin50 (Cx50; $\alpha_8$) showing that cell swelling starts in a lens region where cells are connected with abundant gap junctions. Bar=50 $\mu$m.

studies in whole lenses (Donaldson et al 1995, Rae et al 1996). Hence, we would have expected that changes in electrolyte levels and osmolyte concentrations associated with diabetes average out over large tissue areas, resulting in a more uniform pattern of cell swelling. The images show that this is clearly not the case, and raise the possibility that cell–cell communication is impaired in the diabetic lens.

In this chapter we will review the role gap junction channels are believed to play in the normal lens, and present data which show that their gating properties are not uniform across the entire lens. We will make a case supporting an aberrant closure of gap junction channels in the diabetic lens, and discuss how this might be responsible for the tissue breakdown in the lens cortex. Finally, we will highlight other mechanisms that could accelerate this destructive process.

## Transport properties of the normal lens

Without vascularization the survival of fibre cells in the lens interior depends on their ability to communicate with cells at the lens surface, thereby ensuring ionic homeostasis and delivery of nutrients. This notion of a freely communicating lens is supported by a number of observations including metabolic coupling (Goodenough et al 1980), tracer spreading (Rae et al 1996) and electrophysiological recording in whole lenses (Mathias et al 1997). However, passive diffusion alone is not sufficient to deliver nutrients to fibre cells situated more than 150 $\mu$m from the lens surface. The lens has overcome this difficulty by developing an internal circulation that drives mass transport (Mathias et al 1997). This system utilizes $Na^+/K^+$ pumps in the epithelium to provide the driving force for a circulating current that consists of a potassium outflux at the epithelium and a sodium and chloride influx in the fibre cells (Fig. 2). In the fibre cells this influx of solutes would create a large osmotic gradient and is balanced by water uptake into the cells. The asymmetric distribution of conductances and gap junctions in the lens, therefore, results in a circular solute and water flow directed inward towards the lens centre via the extracellular space between the fibre cells, and outward towards the lens surface via a cytoplasmic route mediated by gap junction channels. Since the extracellular space between the fibre cells is narrow and tortuous, this fluid flow also leads to convectional transport of larger molecules and nutrients into the lens far beyond the distance expected from diffusion alone.

This model of circulation is supported by a number of electrophysiological observations in whole lenses (Mathias et al 1997), and by an extensive molecular, structural and functional characterization of the lens gap junction proteins. Connexin46 (Cx46; $\alpha_3$) and Cx50 ($\alpha_8$) co-localize in the gap junction plaques of lens fibre cells. These gap junctional plaques are initially localized to the broad faces of cortical fibre cells, but then the gap junction channels disperse and

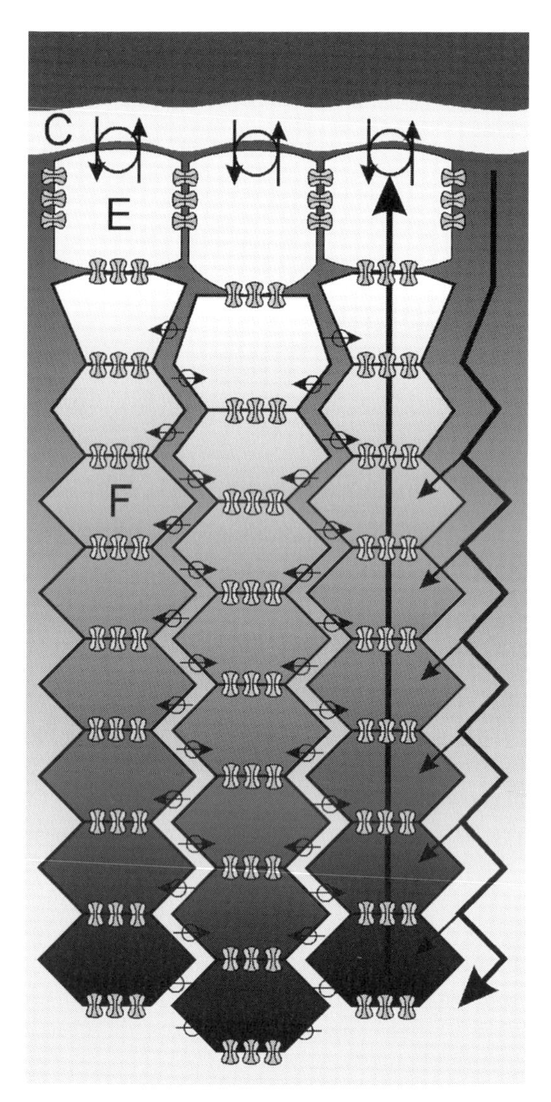


FIG. 2. The lens circulation model in the normal lens. Fluid carrying ions and nutrients moves inward via the extracellular space between fibre cells, traverses the plasma membranes and moves outward via the cytoplasmic route through gap junction channels. This is driven by active transport processes in the lens epithelium and by passive uptake in the fibre cells. C, capsule; E, epithelium; F, fibre cells.

become more uniformly distributed as the fibre cells mature (Grujters et al 1987). When Cx46 and Cx50 are expressed individually in the *Xenopus* oocyte system, they have the ability to form communicating channels (Paul et al 1991, White et al 1992). Patch clamp records of isolated lens fibre cell pairs demonstrate that these connexins also form functional channels in the 'native source' cells (Donaldson et al 1995). Their gating properties differ somewhat from those observed with oocytes, consistent with biochemical evidence that the fibre cell channels are heteromeric (Jiang & Goodenough 1996). The single epithelial cell layer that covers the anterior face of the lens is coupled by gap junction channels composed of Cx43 ($\alpha_1$; Beyer et al 1989). Dye tracer studies have recently established that the epithelial cells and underlying fibre cells are also communicating with each other (Rae et al 1996).

## Spatial differences in gap junction channel gating

Although fibre cells are communicating with each other throughout the normal rat lens, there is evidence for spatial variations in channel gating (Baldo & Mathias 1992). Acidification of the lens with $CO_2$ reversibly uncouples cells only in the outer 20% ($\sim$400 $\mu$m) of the lens radius. The remaining inner 80% of the lens gap junctions are insensitive to acidification and do not uncouple. Spatial differences also exist at the molecular level: both Cx46 and Cx50 have their carboxyl tail segments cleaved when the fibre cells become fully elongated and shed their cell organelles. Lin et al (1997) identified calpain as the lens endogenous protease that removes the carboxyl tails of both Cx46 and Cx50, leaving behind a 38 kDa N-terminal fragment that remains embedded in the membrane. Although classical gap junctions cannot be detected in the lens core by freeze-fracture electron microscopy (Prescott et al 1994), gap junction channels can still be isolated by detergent solubilization of lens core membranes (Lin et al 1997). In addition, tracer coupling experiments on isolated lens cores have confirmed that fibre cells from the lens core region do indeed communicate with each other (Lin 1997). Hence, it appears that although gap junction channels disperse and their connexins become cleaved, the individual channels remain functional in the lens core.

In the rat lens, which has an equatorial diameter of about 4 mm, the transition between cleaved and uncleaved connexin, and the transition between pH-sensitive and pH-insensitive gap junction channels, both occur at a depth of approximately 400 $\mu$m from the equator. The question therefore arises whether the two phenomena are related.

To answer this question we need to determine the functional consequences of connexin cleavage. This has recently been achieved with ovine lens Cx50. The cleaved carboxyl tail segment for Cx50 was purified, and protein sequencing was

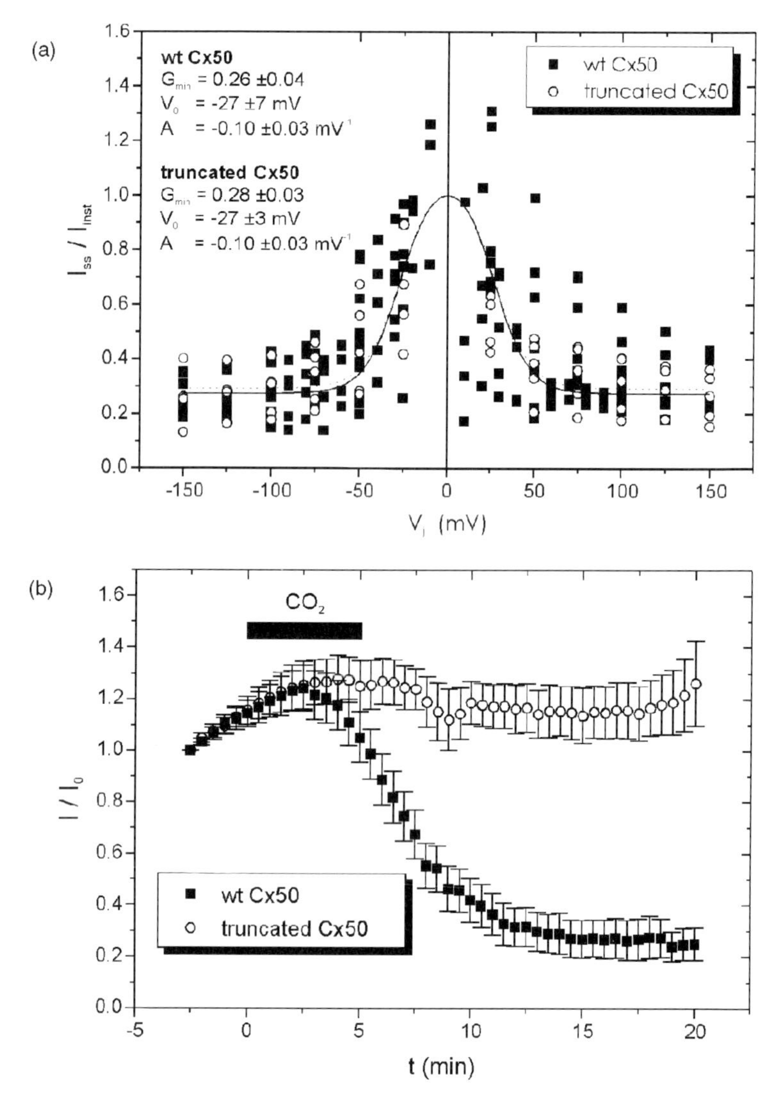


FIG. 3. The functional consequences of connexin50 (Cx50; $\alpha_8$) truncation. Dual voltage clamp analysis of full-length and truncated Cx50 in *Xenopus* oocyte pairs reveals no significant difference in voltage gating (a), but greatly reduced pH-sensitive gating in the truncated variant (b). A, voltage sensitivity; $G_{min}$, voltage-independent residual conductance; I, junctional current; $I_{inst}$, instantaneous current; $I_o$, junctional current at time zero; $I_{ss}$, steady-state current; $V_j$, transjunctional voltage difference; $V_o$, voltage for half maximal inactivation; wt, wild-type.

used to identify the calpain cleavage site in the Cx50 molecule. This knowledge was used to construct a truncation mutant of Cx50 for the expression in *Xenopus* oocyte pairs (Lin 1997). The truncated channels are indeed communicating. They exhibit the same voltage sensitivity as full-length Cx50 (Fig. 3a). However, they react differently to acidification: whereas channels made of full length Cx50 close upon acidification, the truncated form has a greatly reduced pH sensitivity (Fig. 3b).

The outcome of these experiments with Cx50 was similar to previous studies of truncated forms of Cx43, which had also lost pH sensitivity (Liu et al 1993). It is fair to assume that channels made of Cx46, which is closely related to Cx43 and Cx50, and which is also cleaved by calpain, undergo similar changes in gating. This would directly link the two phenomena in the lens, the spatial differences in channel gating and the connexin cleavage.

## A new concept for tissue liquefaction

The loss of pH sensitivity of gap junction channels in the lens core is probably essential for those cells to remain in contact with the more peripheral fibre cells. The lens core is acidic, and there is a pH gradient that gradually rises to neutral pH in the peripheral tissue. This gradient is the result of the anaerobic metabolism of glucose, producing lactate which accumulates in the lens core. The estimated $pK_a$ value for the closure of uncleaved Cx50 channels in the oocyte system is about 6.4. It is likely that in the normal lens this pH value is reached well within the inner lens portion where gap junction channels are cleaved and are thus pH insensitive (Fig. 4a). In the normal lens, therefore, fibre cells across all regions are well coupled.

The situation is likely to be different in the diabetic lens. Increased amounts of glucose are metabolized, resulting in higher lactate levels and further acidification of the lens interior. Hence, the pH gradient in the lens is expected to be steeper, and the intracellular pH becomes lower in the more peripheral fibre cells that contain uncleaved connexins and, therefore, pH-sensitive channels. In some of these cells the intracellular pH will drop below the threshold for gap junction channel closure and these cells will uncouple from their neighbours (Fig. 4b). Many layers of fibre cells in the deeper cortex probably have a mixture of cleaved and uncleaved channels, and these cells may well remain coupled. However, if the pH reaches threshold levels in the young fibre cells containing exclusively uncleaved channels, total cell uncoupling would occur. This will generate a discrete zone of uncoupled cells with a relatively sharp outer boundary.

The damaging effects of irregular uncoupling of fibre cells in the diabetic lens cortex can be understood in the context of the lens circulation model (Fig. 5). The fluid flow via extracellular routes into the lens is not expected to be impaired in the diabetic lens. The outward flow via the cytoplasmic route, however, would be stopped in the cortical zone where cells uncouple in response to the decreased

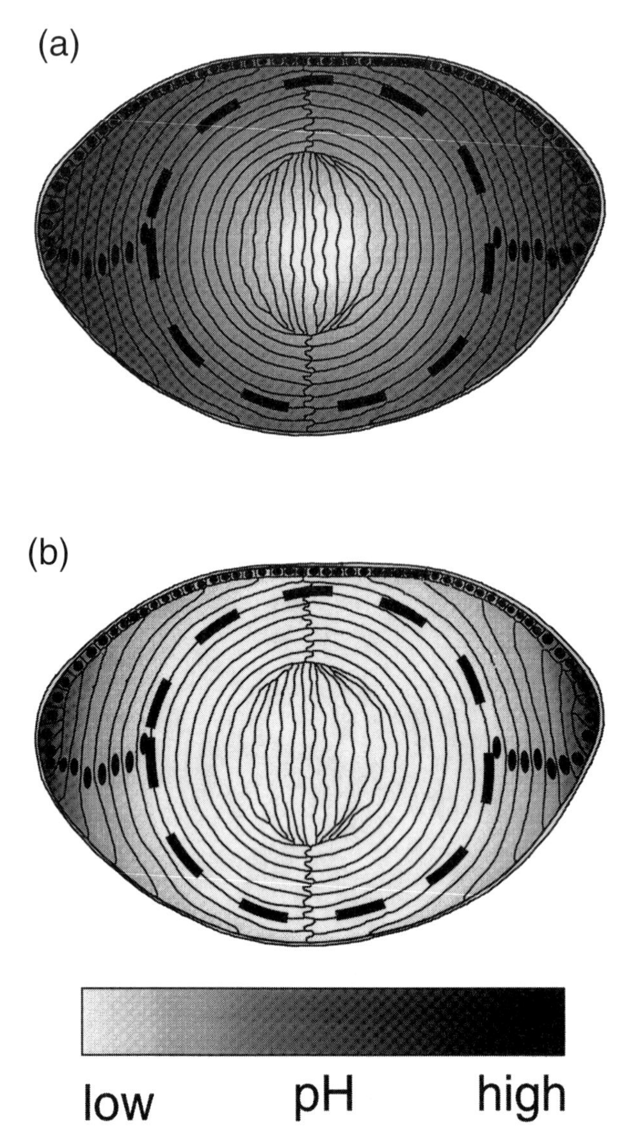


FIG. 4. Irregular closure of gap junction channels in the diabetic lens. The broken line indicates the cleavage zone for connexins. Channels inside this region consist of cleaved connexin46 (Cx46; $\alpha_3$)/Cx50 ($\alpha_8$) and are pH insensitive. (a) The situation in the normal lens: the pH outside the cleavage zone is still above the threshold for channel closure, and all the cells are coupled via gap junctions. (b) The situation in the diabetic lens: increased lactate levels cause the pH in the cortical region to drop below the threshold for channel closure for gap junctions composed of uncleaved connexins. Channels are postulated to close in this zone, effectively uncoupling the lens core from peripheral tissue.

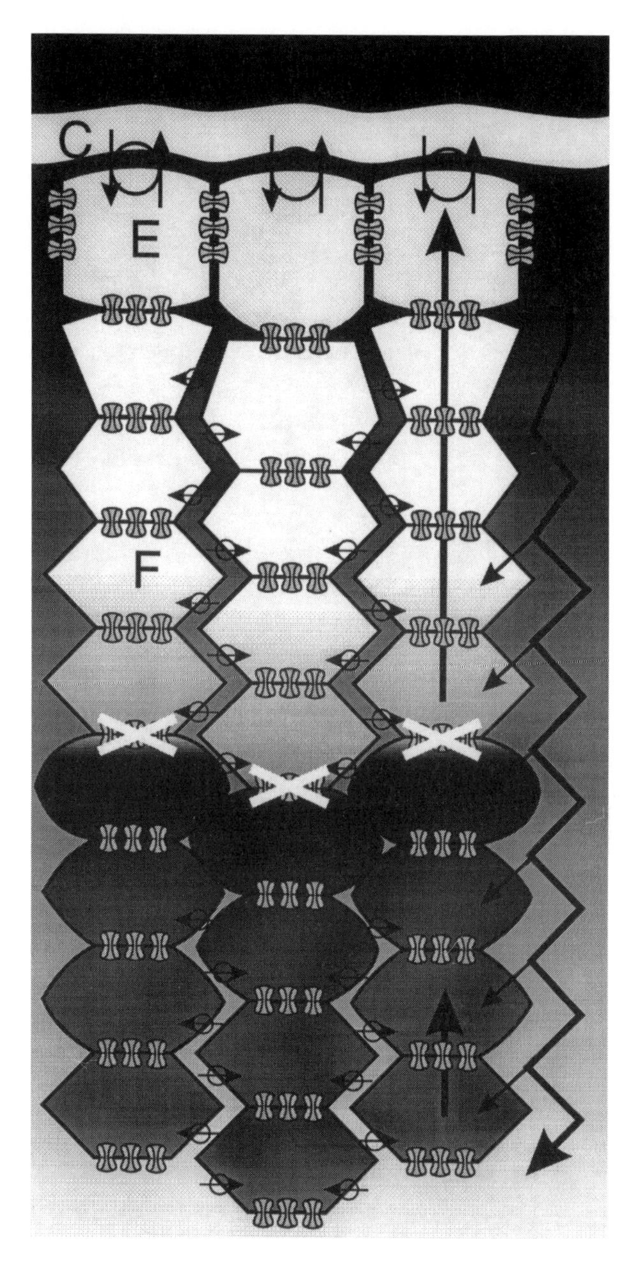


FIG. 5. The lens circulation model in diabetic cataract. Closure of gap junction channels due to lens acidification blocks the outflow pathway of the circulation current in a discrete zone of the lens cortex and leads to fluid accumulation, cell swelling and finally to cell rupture within that zone. C, capsule; E, epithelium; F, fibre cells.

intracellular pH. A backstop would be created, causing the fibre cells in this region to swell uncontrollably. As more cells swell and burst, the tissue disintegrates and liquefies. This possible scenario of events fits well with the histology of the diabetic lens represented in Fig. 1.

## Outlook

We have mainly focused on the contribution gap junction channels could make to the early events in diabetic cataractogenesis. However, gap junctions are only one part of the circulatory pathway. It is likely that any changes in the balance of fluid and ion fluxes may result in a similar phenotype. In addition, the manifestation of severe tissue breakdown may also require further aggravating factors. For example, increased levels of the osmotically active sugar alcohols sorbitol or galactitol may accelerate fluid uptake in the lens and, hence, make the cells at the backstop swell faster. Inhibition of aldose reductase, the enzyme responsible for the production of sugar alcohols, is known to delay diabetic lens opacification in rats (Datiles et al 1982), and its overexpression in transgenic mice accelerates it (Lee et al 1995). P2Y purinoceptors, recently identified in lens fibre cells, could also accelerate cortical tissue breakdown (Merriman-Smith et al 1998). This is because exposure of the isolated rat lens to extracellular ATP induces additional membrane currents in the fibre cells, resulting in increased water uptake. The initial rupture of fibre cells would release cellular ATP which would in turn activate additional P2Y purinoceptors in neighbouring cells thus increasing water uptake, cell swelling, rupture and further ATP release. Thus, both the accumulation of sugar alcohols and the activation of P2Y purinoceptors would accelerate cortical tissue breakdown initially started by the irregular closure of the gap junction channels.

Of course, it is goes without saying that this new concept requires thorough testing. An obvious starting point is to measure lactate levels and intracellular pH in the diabetic lens, and to demonstrate cortical uncoupling. It is expected that our hypothesis of diabetic cataract will stimulate relevant new experiments and, in the longer term, help to identify possible drug targets with the aim to prevent diabetic opacification.

*Acknowledgements*

This work was supported by grants from the New Zealand Health Research Council, the New Zealand Lottery Grants Board, the Marsden Fund, and the Auckland University Research Committee.

## References

Baldo GJ, Mathias RT 1992 Spatial variations in membrane properties in the intact rat lens. Biophys J 63:518–529

Beyer EC, Kistler J, Paul DL, Goodenough DA 1989 Antisera directed against connexin43 peptides react with a 43-kD protein localized to gap junctions in myocardium and other tissues. J Cell Biol 108:595–605

Bond J, Green C, Donaldson P, Kistler J 1996 Liquefaction of cortical tissue in diabetic and galactosemic rat lenses defined by confocal laser scanning microscopy. Invest Ophthalmol Vis Sci 37:1557–1565

Datiles M, Fukui H, Kuwabara T, Kinoshita JH 1982 Galactose cataract prevention with sorbinil, an aldose reductase inhibitor: a light microscopic study. Invest Ophthalmol Vis Sci 22:174–179

Donaldson PJ, Dong Y, Roos M, Green C, Goodenough DA, Kistler J 1995 Changes in lens connexin expression lead to increased gap junctional voltage dependence and conductance. Am J Physiol 269:C590–C600

Goodenough DA, Dick JSB, Lyons JE 1980 Lens metabolic cooperation: a study of mouse lens transport and permeability visualized with freeze-substitution autoradiography and electron microscopy. J Cell Biol 86:576–589

Grujters WTM, Kistler J, Bullivant S 1987 Formation, distribution and dissociation of intercellular junctions in the lens. J Cell Sci 88:351–359

Jiang JX, Goodenough DA 1996 Heteromeric connexons in lens gap junction channels. Proc Natl Acad Sci USA 93:1287–1291

Kuwabara T, Kinoshita JH, Cogan DG 1969 Electron microscopic study of galactose induce cataract. Invest Ophthalmol Vis Sci 8:133–149

Lee AYW, Ching SK, Chung SSM 1995 Demonstration that polyol accumulation is responsible for diabetic cataract by use of transgenic mice expressing the aldose reductase gene in the lens. Proc Natl Acad Sci USA 92:2780–2784

Lin JS 1997 Post-translational cleavage of lens connexin50 and its functional implications. PhD thesis, University of Auckland, Auckland, New Zealand

Lin JS, Fitzgerald S, Dong Y, Knight C, Donaldson P, Kistler J 1997 Processing of the gap junction protein connexin50 in the ocular lens is accomplished by calpain. Eur J Cell Biol 73:141–149

Liu SG, Taffet S, Stoner L, Delmar M, Vallano ML, Jalife J 1993 A structural basis for the unequal sensitivity of the major cardiac and liver gap junctions to intracellular acidification. Biophys J 64:1422–1433

Mathias RT, Rae JL, Baldo GJ 1997 Physiological properties of the normal lens. Physiol Rev 77:21–50

Merriman-Smith R, Tunstall M, Kistler J, Donaldson P, Housley G, Eckert R 1998 Expression profiles of P2 receptor isoforms P2Y1 and P2Y2 in the rat lens. Invest Ophthalmol Vis Sci 39: in press

Mitton KP, Dean PAW, Dzialosynski T, Xiong H, Sanford SE, Trevithick JR 1993 Modelling cortical cataractogenesis. 13. Early effects on lens ATP/ADP and glutathione in the streptozotocin rat model of the diabetic cataract. Exp Eye Res 56:187–198

Paul DL, Ebihara L, Takemoto LJ, Swenson KI, Goodenough DA 1991 Connexin46, a novel lens gap junction protein, induces voltage gated currents in nonjunctional plasma membrane of *Xenopus* oocytes. J Cell Biol 115:1077–1089

Prescott A, Duncan G, Van Marle J, Vrensen G 1994 A correlated study of metabolic cell communication and gap junction distribution in the adult frog lens. Exp Eye Res 58:737–746

Rae JL, Bartling C, Rae J, Mathias RT 1996 Dye transfer between cells of the lens. J Membr Biol 150:89–103

Robison WG, Houlder N, Kinoshita JH 1990 The role of lens epithelium in sugar cataract formation. Exp Eye Res 50:641–646
White TW, Bruzzone R, Goodenough DA, Paul DL 1992 Mouse Cx50, a functional member of the connexin family of gap junction proteins, is the lens fiber protein MP70. Mol Biol Cell 3:711–720

## DISCUSSION

*Beyer:* Is there a way that you could block proteolysis? Because if you could prevent cleavage in your model you would predict that you would see swelling at a different level.

*Kistler:* We thought about making a transgenic mouse that overexpresses calpastatin, for example.

*Beyer:* Or you could make a connexin that is resistant to cleavage.

*Kumar:* Can sorbitol go through lens channels?

*Kistler:* Lens channels have a large current conductance, which may imply that they are large. I would imagine that sorbitol is small enough to pass through but we haven't done the experiment.

*Musil:* Dan Goodenough and co-workers have shown that in the embryonic chick there is also a gradient of gap junction sensitivity to pH such that fibre cells, but not epithelial cells, remain coupled after cytoplasmic acidification (Miller & Goodenough 1986, Schuetze & Goodenough 1982). This would imply that there must also be a mechanism to regulate pH sensitivity that is independent of removal of the connexin cytoplasmic tail because as far as I know there isn't any cleavage of connexins in the chick lens.

*Beyer:* The only time we saw cleavage, at least of connexin56 (Cx56; $\alpha_3$) in that system was if we were sloppy in preparing the protein extracts and failed to chelate divalent cations. If we isolated the protein with enough EDTA or EGTA we didn't see cleavage, so we believe it is normally intact.

*Gilula:* Alternatively, calcium may be required to activate the enzyme, either immediately or during the isolation procedure. The regulation of protease activity in the lens is not a trivial issue. It is a pressing concern. If calpain does what you're proposing *in vivo*, what kind of regulatory scheme do you have in mind for making certain that calpain isn't active as a protease when it's not supposed to be? It could be detrimental if it is activated in the lens cortex.

*Kistler:* It is striking that cleavage occurs in the same region where the organelles are degraded. Therefore, one could argue that a release of calcium in that region creates a calcium spike that is sufficient to activate calpain. However, the story is probably not quite as simplistic because the lens also contains its own calpain inhibitor, namely calpastatin. We don't yet know where calpastatin is expressed, but we will be doing *in situ* PCR hybridization experiments to show where both the calpains and calpastatin are localized.

*Yeager:* Using point mutagenesis and deletions of Cx43 ($\alpha_1$), Delmar and his colleagues (Morley et al 1996) have shown that the C-terminus of Cx43, even when detached, can act as a pH-induced gating element, analogous to the 'ball and chain' mechanism for inactivation of voltage-gated channels (Armstrong & Bezanilla 1977, Hoshi et al 1990, Zagotta et al 1990). A titratable histidine (His95) in the cytoplasmic loop near M2 of Cx43 has been implicated as part of the 'receptor' for the 'ball' (Ek et al 1994). Have you compared the amino acid sequences of the C-terminal domains of Cx46, Cx50 and Cx43 to try and localize the 'ball' region in each?

*Kistler:* There is an area close to the C-terminus which is homologous between the various connexins, and charged amino acids are present in that area. We will have to look in greater detail in this region, and we haven't yet done any further truncation experiments. We hope to mass produce the piece that is cleaved off, crystallize it and do X-ray crystallography.

*Lau:* In collaboration with Mario Delmar, we have been trying to localize the region in Cx43 that seems to be involved in creating the particle region involved in pH-induced gating. It seems to correspond to amino acids 274–284, which is a proline-rich region. It would be good to know whether that region is present in Cx50.

*Fletcher:* That protein is loaded with charges. As Alan Lau said, there is a negatively charged proline-rich domain, which is a site for proline-directed protein kinases. There are also many lysine/arginine positive charges, which need to be investigated because they're intrinsically charged.

*Lau:* In the model that we're entertaining, and it may be true for pH-induced gating as well as phosphorylation-induced gating, phosphorylation of the tail may create a 'particle' that binds to the loop of the connexin or some other region, and in some way induce channel closure. We're also investigating the possibility that interacting proteins may bind to the proline-rich region and contribute to the gating process.

*Fletcher:* We have transgenic mice in which the cytoplasmic loop lysine residues that are in $\alpha$-helical configuration have been replaced with uncharged residues. This is some work we did with Mark Yeager a couple of years ago (K. Balli, M. Yeager & W. H. Fletcher, unpublished observations 1996). We found that this manipulation destroyed coupling in transfected cells, although we don't know how this contributes to malformations in the animal.

*Nicholson:* We have some data which suggest that interacting factors play a role in pH gating. When we perfuse the inside of one oocyte of a pair, we eliminate all pH and calcium gating of Cx43 on that side. We know that the gap junctions are still capable of normal responses, because the pH and calcium gating of the intact oocyte can be shown to be normal by changing pH and calcium levels through addition of extracellular factors that cross the membrane. Also, if we add oocyte extract to the perfused oocyte we can reconstitute closure of the channels at low pH, suggesting that a soluble factor is required for pH gating. Mario Delmar has found a small peptide that has an inhibitory effect on pH gating (Calero et al 1998), so it is possible that this is binding to the soluble factor and inactivating it. It might be possible to introduce this, or an analogous peptide, into your system to eliminate the pH gating and perhaps alleviate your problems in the lens. At least this is a smaller fragment that may be easier to work with.

*Kistler:* We have some plans with Andreas Engel at the M. Müller Institute in Basel, to look for conformational changes in the full-length channels at pH 6 and pH 7. We are going to use atomic force microscopy to look at *in vitro* reconstituted arrays, which we can make from detergent-solubilized lens hemichannels. Using the liquid cell option of the microscope, we should be able to detect any movement or conformational changes in the carboxyl tail when we change the pH.

*Nicholson:* It is crucial to search for the cleaved tail to make sure it has been removed, because if it hasn't then there is a good chance that pH gating will occur.

*Kistler:* We have purified the 32 kDa cleaved piece from Cx50 in the lens cortex, but we have not been able to purify it from the lens core, suggesting that it is degraded further in the lens core. Cx46 is even more elusive: we have never found any cleaved peptides of Cx46, although we know that the carboxyl tail is cleaved because we see the amino-terminal portion in the 38 kDa band. We are going to make the entire carboxyl tail of Cx46 as a His-tag protein, purify it with an affinity column and cleave it with calpain while it is attached to the column. The cleaved fragment(s) can then be eluted and sequenced, and the cleavage site in the Cx46 molecule determined. We will do this in parallel with Cx50, because we know where the *in vivo* cleavage site is, and if we obtain the same Cx50 cleavage site on the column, then we will be confident that the Cx46 cleavage site will also be correct.

*Beyer:* What is the relationship between this protease activity and that observed in the cataract formed in the Cx50 knockout?

*Gilula:* The impact of the knockout is profound, particularly in the context of trying to understand how these two connexins contribute to normal lens function. It's clear that Cx46 is required in the lens region. This is probably the most poignant observation of Xiaohua Gong's knockouts. Dan Goodenough has done some work on the Cx50 knockout. Do you want to say anything about that now?

*Goodenough:* I don't have any information on the points we have been discussing, except to say that the timing of the cataract in the Cx50 knockout is different to that for the Cx46 knockout, and there are also some differences in the development of the lens in terms of growth rate.

*Gilula:* Gong et al (1997) have found that there is a specific proteolytic cleavage event, although the protease is not known. Xiaohua Gong has found that calpain doesn't cleave crystallin, and in the Cx43 knockout mouse, we don't have any evidence that the calcium-dependent calpain identified by Joerg Kistler and other investigators plays a direct role in cleaving the $\gamma$ crystallin. One of the reasons why protease activity in the lens is interesting is that proteases can have devastating effects once they are activated. In that context, moving back into the clinical realm, if this diabetic model is relevant for humans, do these cleavages take place in humans? Has anyone observed the cleavage of these two connexins, and if so, and if calpain is involved, is it possible to come up with a therapy to deliver an inhibitor that would at least delay the onset of human diabetic caractogenesis?

*Kistler:* The answer to your first question is that, with a sample size of $n$=1, we have looked at the human lens, and at least for Cx50 the cleavage patterns are the same as those in sheep, mice and rats. Therefore, it is likely to be a general phenomenon among mammals.

*Beyer:* How did you do that experiment?

*Kistler:* By immunostaining with anti Cx50 carboxyl tail-specific antibodies.

The other question, i.e. the possibility of inhibiting protease activity, has been addressed by Thomas Shearer and colleagues. They have been able to show convincingly that the cysteine protease inhibitor E64 reduced the rate of cataractogenesis in cultured lenses and in the whole animal (Shearer et al 1991, Azuma et al 1991).

*Gilula:* In the context of human complications any kind of suppression that delays either the onset or the progression could be extremely valuable. I propose that even though you might have activated the cleavage of some of the crystallins that contribute to opacification, if you could delay that process, it would still be of benefit.

*Kistler:* Our thinking does not follow the thinking behind the current clinical trials, which have all focused on aldose reductase inhibitors. Aldose reductase inhibitors delay the formation of cataracts in the rat, but in humans the results are mixed. Our scheme is a new one, in which we look for drugs that control lens hydration and could induce the diabetic lens to release tissue water.

*Gilula:* The simplest hypothesis is that the delivery of a calpain inhibitor at the appropriate time should have a significant impact on connexin cleavage, and that alone should have some influence on the progression of diabetic cataractogenesis. The inhibitor is involved directly in inhibiting calpain activity in the normal situation.

*Goodenough:* Except that the rates of cleavage and lens growth are relatively slow compared to the onset of diabetic cataractogenesis.

*Gilula:* But that could work to your advantage, because one could deliver the inhibitor for long periods of time using an ointment, for instance.

*Nicholson:* That would make the situation worse, because some amount of cleavage is required.

*Kistler:* And also connexins are not the only proteins that are cleaved by calpain.

*Gilula:* Connexins in the lens: are they to blame in diabetic cataractogenesis? One could take the position that they themselves may not be to blame for cataractogenesis, but they are one of the players, and other components may be just as important, if not more important.

*Kistler:* I agree. They take some of the blame.

*Goodenough:* Your hypothesis states that the original problem is a change in the pH gradient, so what you could do is disrupt the pH gradient.

*Gilula:* Which could be done independently of the connexins.

*Kistler:* Yes. The approach we will take will be independent of the connexins. We will zoom in on the regulation of chloride channels and other channels to try to shrink the lens. We have to take a broader view because there's a lot more to the lens membranes than just the gap junction channels.

## References

Armstrong CM, Bezanilla F 1977 Inactivation of the sodium channel. II. Gating current experiments. J Gen Physiol 70:567–590

Azuma M, David LL, Shearer TR 1991 Cysteine protease inhibitor E64 reduces the rate of formation of selenite cataract in the whole animal. Curr Eye Res 10:657–666

Calero G, Kanemitsu M, Taffet SM, Lau A, Delmar MA 1998 A 17mer peptide interferes with acidification-induced uncoupling of connexin 43. Circ Res 82:929–935

Ek JF, Delmar M, Perzova R, Taffet SM 1994 Role of His95 in pH gating of cardiac gap junction protein, connexin 43. Circ Res 74:1058–1064

Gong X, Li E, Klier G et al 1997 Disruption of $\alpha 3$ connexin gene leads to proteolysis and cataractogenesis in mice. Cell 91:833–843

Hoshi T, Zagotta WN, Aldrich RW 1990 Biophysical and molecular mechanisms of Shaker potassium channel inactivation. Science 250:533–538

Miller TM, Goodenough DA 1986 Evidence for two physiologically distinct gap junctions expressed by the chick lens epithelial cell. J Cell Biol 102:194–199

Morley GE, Taffet SM, Delmar M 1996 Intramolecular interactions mediate pH regulation of connexin43 channels. Biophys J 70:1294–1302

Schuetze SM, Goodenough DA 1982 Dye transfer between cells of the embryonic chick lens becomes less sensitive to $CO_2$ treatment with development. J Cell Biol 92:694–705

Shearer TR, Azuma M, David LL, Murachi T 1991 Amelioration of cataracts and proteolysis in cultured lenses by cysteine protease inhibitor E64. Invest Ophthalmol Vis Sci 32:533–540

Zagotta WN, Hoshi T, Aldrich RW 1990 Restoration of inactivation in mutants of Shaker potassium channels by a peptide derived from ShB. Science 250:568–571

# Neuronal coupling in the central nervous system: lessons from the retina

David I. Vaney

*Vision, Touch and Hearing Research Centre, Department of Physiology and Pharmacology, University of Queensland, Brisbane 4072, Queensland, Australia*

*Abstract.* The retina is a model system for studying gap junctional intercellular communication in the CNS. The cellular coupling can be graphically visualized in retinal whole mounts by injecting small cationic tracers into microscopically identified neurons; the pattern of tracer coupling shown by each type of retinal neuron is highly stereotyped, with many types of amacrine cells and ganglion cells showing complex patterns of both homologous and heterologous coupling. Parallel physiological studies have demonstrated that the gap junctions can be modulated dynamically by neurotransmitters and by the level of ambient illumination. Taken together, the numerous structural and functional studies on gap junctions in the retina provide powerful support for the concept that electrical synapses are complex components of neuronal circuits, having many of the attributes normally ascribed to chemical synapses.

*1999 Gap junction-mediated intercellular signalling in health and disease. Wiley, Chichester (Novartis Foundation Symposium 219) p 113–133*

The business of the brain is intercellular communication, and one might expect that evolution would have dealt gap junctions their most extravagant hand in the CNS, comparable to the extraordinary diversification of ligand-gated and voltage-regulated ion channels. Although there is mounting evidence that gap junctional intercellular communication in the CNS is both more complex and perhaps more subtle than in other tissues, it is not clear to what extent this is a reflection of heterogeneity in the connexons. While there is unequivocal evidence that the supporting cells in the CNS express a variety of connexin types, identification of the connexins in the neurons themselves has often been inconclusive and sometimes contradictory (Nadarajah & Parnavelas 1999, this volume). Notwithstanding the evidence from cell expression studies that connexins alone are necessary for the generation of gap junctions, albeit in the presence of generic cell adhesion molecules, it seems probable that complex neuronal interactions will underlie the specificity of all synaptic communication, whether mediated by chemical synapses or gap junctions (electrical synapses). Thus, it is not necessary to invoke a broad repertoire of connexin types to account for the stereotyped patterns of neuronal coupling observed in the CNS. Nevertheless, the cellular localization of connexin26 (Cx26; $\beta_2$), Cx32 ($\beta_1$) or Cx43 ($\alpha_1$) in some CNS neurons potentially allows for a diverse combination of homotypic, heterotypic and heteromeric pairings, which may contribute both to the selectivity of the gap junctional connections and to their functional heterogeneity.

The vertebrate retina has many features that make it an outstanding model for studying both neuronal coupling and glial coupling in the CNS. The retina is a particularly familiar part of the CNS because its structure and function have been elucidated in remarkable detail. Consequently, new observations can be related seamlessly to countless previous studies and their significance assessed informatively. The intact retina can be isolated readily from the rest of the CNS and also from the underlying pigment epithelium and sclera. The isolated retina, which has been described as 'nature's brain slice', retains many of the important advantages of *in vivo* preparations (intact neuronal circuitry, responsiveness to natural stimuli), while gaining the significant benefits of *in vitro* preparations (recording stability, microscopic identification of cells, controlled application of drugs).

The study of the retina is facilitated by the highly ordered neuronal architecture, both in the plane of the retina (horizontal) and through its depth (vertical; reviewed by Wässle & Boycott 1991). Identification of the major classes of retinal neurons is guided by the retina's laminar structure, which comprises three somatic layers separated by two synaptic layers (Fig. 1). The rod and cone photoreceptors are located in the outer nuclear layer; most of the retinal interneurons are located medially in the inner nuclear layer; the projection neurons of the retina as well as some interneurons are located in the ganglion cell layer. A direct 'vertical' pathway connects photoreceptors to bipolar cells to ganglion cells, with the axons of photoreceptors synapsing on the dendrites of bipolar cells in the outer plexiform layer, and the axons of bipolar cells synapsing on the dendrites of ganglion cells in the inner plexiform layer. Two classes of interneurons provide 'lateral' connections in the retina: the horizontal cells are located at the scleral margin of the inner nuclear layer and branch in the outer plexiform layer; whereas most of the amacrine cells are located at the vitread margin and branch in the inner plexiform layer. The five major classes of retinal neurons together comprise as many as 100 'natural types', each of which is defined by the regular association of particular morphological, synaptic, neurochemical and physiological properties, which may vary systematically between species. In general terms, the visual information transduced by the photoreceptors undergoes simple spatial processing in the outer retina and then more complex temporal processing in the inner retina.

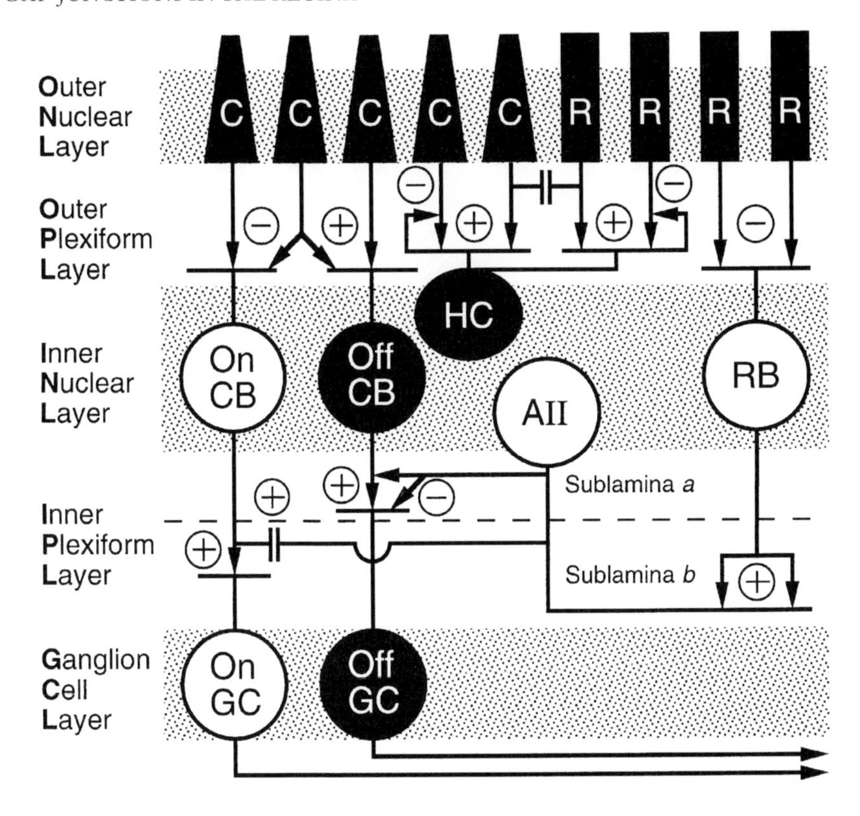


FIG. 1. Schematic circuit of the mammalian retina showing the direct neuronal pathways from the cone (C) and rod (R) photoreceptors to the ganglion cells (GC), whose axons form the optic nerve. Excitatory (+) and inhibitory (–) chemical synapses are marked with arrows; electrical synapses between different types of neurons are marked with parallel lines, representing gap junctions. The white neurons depolarize at light On, whereas the black neurons depolarize at light Off. AII, rod amacrine cell; CB, cone bipolar cell; HC, horizontal cell; RB, rod bipolar cell.

## Patterns of neuronal coupling in the retina

It is now apparent that direct intercellular communication between retinal neurons is both more common and more diverse than previously thought. Although gap junctions between horizontal cells and between photoreceptors have long been known to play important roles in visual processing in the outer retina (Yamada & Ishikawa 1965, Raviola & Gilula 1973), the widespread occurrence of neuronal coupling in the inner retina had not been appreciated until recently. The development of a sensitive technique for visualizing the coupling, by injecting small biotinylated tracers into microscopically identified cells (Vaney 1991), has revealed remarkable patterns of cellular coupling throughout the

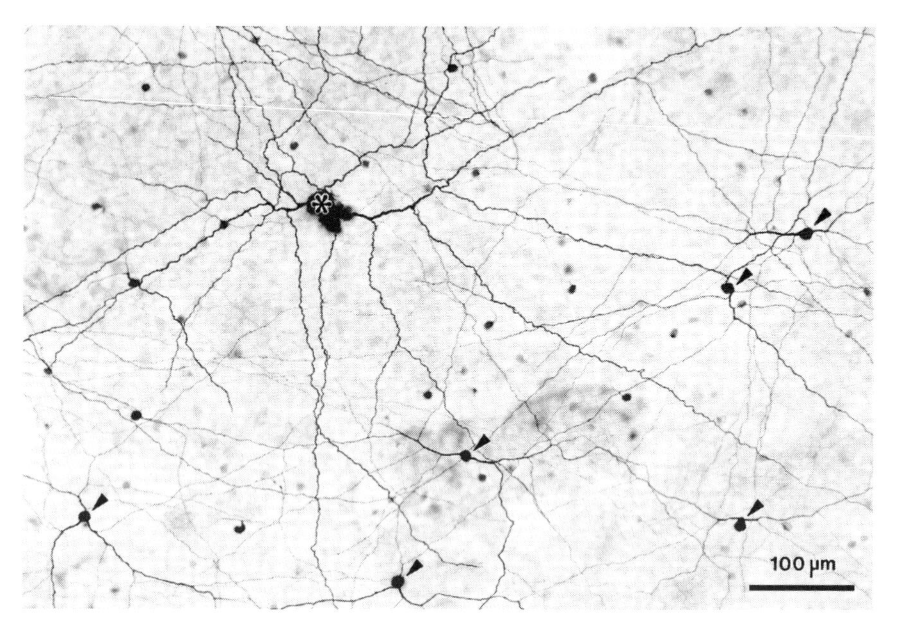


FIG. 2. Tracer coupling pattern of the type 1 nitrergic (ND1) amacrine cells in the rabbit retina. The ND1 cell marked with an asterisk was injected with Neurobiotin, and it shows both strong homologous coupling to neighbouring ND1 cells (arrowheads) and weaker heterologous coupling to other amacrine cells with smaller somata. Given that each ND1 cell accounts for only one amacrine cell in 1000, it is apparent that the ND1 cells are heterologously coupled to less than 1% of the amacrine cells in the rabbit retina.

retina, involving dozens of neuronal types (reviewed by Vaney 1994a). As expected from electron microscopic studies, most cases of tracer coupling occurred between cells of the same type (homologous coupling). Rather surprisingly, many cases of tracer coupling were also observed between cells of different types (heterologous coupling), including amacrine and ganglion cells, different types of amacrine cells, and different types of glial cells, in addition to the well-known example of heterologous coupling between rod amacrine cells and cone bipolar cells (see below). The heterologous coupling is highly stereotyped, with each cell showing tracer coupling to only a few other types of retinal neurons, and these patterns appear to be conserved in diverse mammals (Fig. 2).

The finding that many types of retinal ganglion cells show homologous and/or heterologous tracer coupling (Vaney 1991, Dacey & Brace 1992, Vaney 1994b, Xin & Bloomfield 1997) was unexpected because there was no ultrastructural evidence that ganglion cell dendrites make gap junctions (cf. Jacoby et al 1996). However, a freeze-fracture study on monkey and rabbit retinas (Raviola & Raviola

1982) did reveal that many of the gap junctions in the inner plexiform layer are quite small or comprise loosely arranged connexons enclosing islands of particle-free membrane, and thus they might not be recognized in ultra-thin sections. In freeze-fractures it is often difficult to identify the class of neuron, let alone the particular type, that gives rise to a particular gap junction. Given that the input resistance of neurons may be high, a junction of minimal size may still provide effective electrical coupling, at least under steady-state conditions or near threshold. Indeed, elegant electrophysiological experiments on both the cat retina (Mastronarde 1983) and the salamander retina (Meister et al 1995) provide functional evidence for electrical coupling of retinal ganglion cells.

It has been suggested that some of the intercellular connections revealed by tracer coupling in the mature retina are simply vestiges of the more promiscuous patterns of neuronal coupling observed in the developing retina (Becker et al 1996, Vaney 1996). In particular, the homologous coupling shown by all types of horizontal cells, many types of amacrine cells and bipolar cells, and some types of ganglion cells could mediate the cellular interactions required to produce a regular two-dimensional array of neurons (Cook & Becker 1995). However, there are clear examples of retinal neurons that do not show any tracer coupling, either in the mature retina or prior to eye opening: the uncoupled cell types differ markedly in their dendritic field coverage of the retina, ranging from neurons with small overlap, such as the beta ganglion cells (Dacey & Brace 1992, Penn et al 1994, Vaney 1994a), to neurons with extensive overlap, such as the cholinergic amacrine cells (Vaney 1991) and the radiate amacrine cells in fish retina (Hitchcock 1997). Taken together, these findings indicate that neuronal coupling is unlikely to be essential for the development or maintenance of the spatial organization of neuronal arrays in general, although such a role cannot be discounted where coupling is present at an appropriate stage of development.

Important exceptions may be provided by those retinal neurons whose spatial extent appears to be delimited by dendritic contact with neighbouring cells: such territorial organization achieves seamless coverage of the retina with minimal overlap of the dendritic trees. The best characterized example is provided by the On-Off transient amacrine cells in fish retina, which are electrically coupled by large gap junctions that are permeable to Lucifer yellow. The coupled amacrine cells make dendritic contacts either from tip-to-shaft or from tip-to-tip, forming a remarkable pattern of closed loops in which it is difficult to tell where the coupled cells begin and end (Negishi & Teranishi 1990). Such exclusive territoriality is exhibited by diverse types of bipolar cells, amacrine cells and ganglion cells, most of which show strong homologous tracer coupling (Vaney 1996). The gap junctions on the distal dendrites that mediate this coupling are appropriately placed to transmit intercellular signals that would inhibit further dendritic growth, as proposed for the anterior pagoda neurons in the leech embryo (Wolszon et al 1994).

The recognition that gap junctions are sophisticated components of neuronal circuits is due in good measure to the demonstration that retinal gap junctions are modulated dynamically by neurotransmitters and by the level of ambient illumination (reviewed by Baldridge et al 1998). This modulation has been characterized most extensively in the horizontal cells of lower vertebrates, which were the subject of the pioneering studies in this field (Piccolino et al 1984, Teranishi et al 1984, Lasater & Dowling 1985). This chapter focuses on the more recent studies on the rod signal circuit (reviewed by Vaney 1997), which provides some of the clearest evidence for both the synaptic specificity and the functional heterogeneity of gap junctional communication in the retina.

## Gap junctions in the rod signal circuit

In the mammalian retina the rod photoreceptors synapse on a single morphological type of rod bipolar cell, which terminates deep in the inner plexiform layer. The rod bipolar cells do not contact the ganglion cells directly but synapse on specialized interneurons, the rod (AII) amacrine cells. These narrow-field bistratified neurons depolarize in response to illumination and this On response is then fed into the cone circuits through either sign-inverting glycinergic synapses with Off-centre cone bipolar cells in sublamina *a*, or sign-conserving electrical synapses with On-centre cone bipolar (CB) cells in sublamina *b*. The cone bipolar cells 'piggy-back' the rod signal (Strettoi et al 1992), carrying it both to the ganglion cells and to the interposed amacrine cell circuitry that underlies complex visual processing in the inner retina.

The AII amacrine cells make both homologous gap junctions with each other and heterologous gap junctions with three of the five types of CB cells that terminate in sublamina *b* of the inner plexiform layer (Kolb & Famiglietti 1974, Strettoi et al 1992). In the cat retina, the AII cells make large gap junctions with the CB$b_1$ cells and small gap junctions with the CB$b_2$ and CB$b_4$ cells (Cohen & Sterling 1990). Although the AII cells do not appear to make gap junctions with the CB$b_3$ and CB$b_5$ cells, they may provide indirect input to the CB$b_3$ cells through the small heterologous gap junctions between the CB$b_4$ and CB$b_3$ cells; in addition, the CB$b_4$ cells make homologous gap junctions with each other. This complex pattern of gap junctional connectivity can be visualized graphically by injecting an AII amacrine cell with Neurobiotin (Vaney 1991, 1997): the somata of the tracer-coupled bipolar cells differ in their size, depth, labelling intensity and calbindin immunoreactivity, indicating that they comprise several types of neurons (Hampson et al 1992, Massey & Mills 1996). Thus, the pattern of tracer coupling qualitatively reflects the pattern of gap junctional connectivity in this well-studied retinal circuit.

The AII amacrine cells traverse the full depth of the inner plexiform layer and, potentially, they could be heterologously coupled to any of the dozens of different types of amacrine and ganglion cells that appear to make gap junctions. In fact, both the electrical synapses and the chemical synapses of the AII cells are restricted to only a few types of neurons, suggesting that common mechanisms underlie the generation of synaptic specificity. For example, the formation and regulation of gap junctions may depend on the expression of specific cell adhesion molecules in the coupled neurons ('the precedence hypothesis'; Edelman 1988). However, it cannot be excluded that the selectivity of the gap junctional coupling in the retina arises, at least in part, from diversity in the connexins or their post-translational modification. Despite tantalizing preliminary reports, the retinal connexins have yet to be securely identified, which suggests that their expression is labile and may involve connexin types that have not been characterized in other tissues.

However, there is clear evidence from studies on the rabbit retina that the homologous and heterologous gap junctions of the AII amacrine cells differ significantly in their functional properties, although the molecular basis is presently unknown. Both types of gap junctions appear to be impermeable to conventional anionic probes, such as Lucifer yellow and carboxyfluorescein, but quite permeable to comparable-sized cationic probes, such as biocytin and Neurobiotin (Vaney 1991). However, the AII→AII junctions are more permeable than the AII→CB junctions to the larger cation, biotin-X cadaverine, suggesting that the heterologous junctions have a smaller pore size than the homologous junctions (Mills & Massey 1995).

## Modulation of homologous and heterologous coupling

The homologous tracer coupling between AII cells is dramatically reduced by low concentrations of dopamine, and this effect is mediated by $D_1$ dopamine receptors through an increase in intracellular cAMP (Hampson et al 1992), comparable to that described for horizontal cells in many retinas (Fig. 3). The AII amacrine cells receive substantial synaptic input from another type of amacrine cell that contains both dopamine and $\gamma$-aminobutyric acid (GABA; Pourcho 1982), although studies on the rat retina have failed to localize $D_1$ receptors on the AII cells (Veruki & Wässle 1996). Recordings from AII amacrine cells in the rabbit retina revealed that both the extent of homologous tracer coupling and the extent of electrical coupling, as measured by receptive field size, varied systematically with the level of ambient illumination (Bloomfield et al 1997). Weak illumination caused a fivefold increase in coupling compared with the dark-adapted retina, but the network uncoupled when the illumination was further raised to a level that saturated the responses of the rod photoreceptors and the rod bipolar cells.

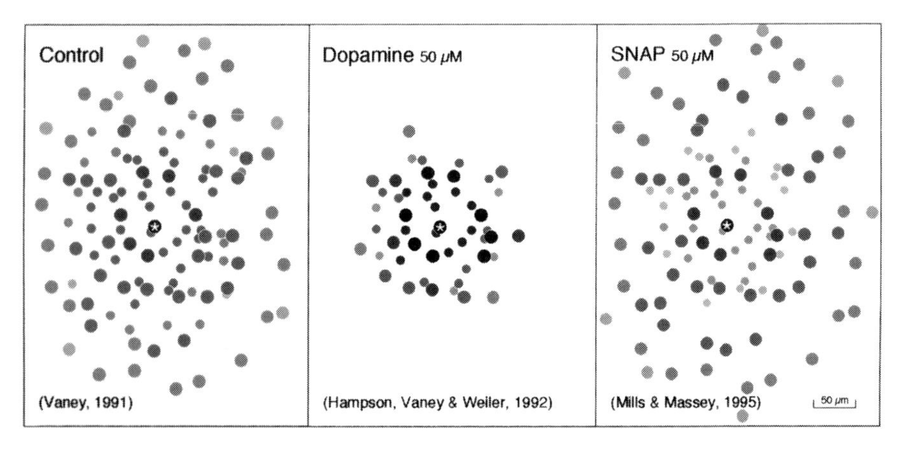


FIG. 3. Neuromodulation of amacrine cell coupling. Diagrammatic representation of the patterns of tracer coupling after the injection of Neurobiotin into a single AII amacrine cell (asterisk), as seen in rabbit retinal whole mounts under control conditions, with dopamine and with the nitric oxide donor, S-nitroso-N-acetylpenicillamine (SNAP). The injected cell shows homologous tracer coupling to surrounding AII amacrine cells (large somata) and heterologous tracer coupling to underlying cone bipolar cells (small somata). Exogenous dopamine selectively reduces the homologous coupling, whereas SNAP selectively reduces the heterologous coupling.

However, it has yet to be demonstrated that the uncoupling observed in either the light-adapted or dark-adapted retina is mediated endogenously by dopamine.

The heterologous tracer coupling between the AII amacrine cells and the CB cells appeared to be unaffected by dopamine and cAMP, but it was reduced by nitric oxide and cGMP agonists (Mills & Massey 1995). Bright illumination may increase the cGMP levels in the On-centre cone bipolar cells, either directly through deactivation of the metabotropic glutamate receptors or indirectly through the light-evoked release of nitric oxide. Although it might be expected that the cone circuit would be uncoupled from the rod circuit in the light-adapted retina, so as to preserve the spatial acuity of the system, there is no direct evidence that the heterologous coupling is modulated endogenously by the ambient illumination. In fact, the tracer-coupling pattern of the AII amacrine cells under conditions of rod saturation appeared indistinguishable from that in the dark-adapted rabbit retina (Bloomfield et al 1997).

The differences in the pharmacological regulation and the permeability of the homologous and heterologous gap junctions of the AII cells might reflect differences in either the subconductance state or the connexin composition of the hemichannels, whether arising from the expression of different types of connexins or from post-translational modification. There is intriguing structural and functional evidence that the heterologous gap junctions are asymmetric, which would support the hypothesis that they are heterotypic, notwithstanding the lack of direct molecular evidence. Electron microscopy shows that these gap junctions have an asymmetric morphology, with 'fluffy material' lining the amacrine side; if this is an ultrastructural correlate of connexin diversity, the expression pattern might be complex because the homologous gap junctions formed by the AII cells appear to lack the fluffy material.

There have been no reports that cone bipolar cells injected with Neurobiotin show tracer coupling to AII amacrine cells, suggesting that the heterologous gap junctions are asymmetrically permeable to small metabolites. Such 'chemical rectification' appears to be a common property of heterologous gap junctions in the retina, including ganglion cell→amacrine cell coupling (Vaney 1996, Xin & Bloomfield 1997) and astrocyte→oligodendrocyte coupling (Robinson et al 1993). There is good evidence from other parts of the CNS that the heterologous coupling between different classes of neuroglial cells is mediated by different types of connexins (e.g. Ochalski et al 1997). Although heterotypic gap junctions may indicate a capacity for asymmetric coupling, it needs to be emphasized that chemical rectification is not synonymous with electrical rectification, which is dependent on a potential difference across the gap junction. In fact, recent intracellular recordings from AII amacrine cells in both rabbit and monkey retinas revealed robust cone driven responses, which appeared to be transmitted from CB cells through the heterologous gap junctions. However, none of these experiments directly address the question of whether the heterologous gap junctions are asymmetrically permeable to current, which is largely carried by hydrated potassium ions.

In contrast to the findings on the AII amacrine cells in mammalian retina, previous studies on the horizontal cells in fish retina had shown that the gap junction permeability is regulated independently by cAMP and cGMP, with both second messengers acting on the same conductance (deVries & Schwartz 1989). However, the cAMP pathway affected both the open duration and open frequency of the gap junction channels, whereas the cGMP pathway appeared to affect only the open frequency without significantly affecting the open duration (Lu & McMahon 1997).

## Neurotransmitter coupling through gap junctions

There is now wide appreciation that gap junctions may be permeable to second-messengers such as calcium ions, cyclic nucleotides (cAMP, cGMP) and inositol-1,4,5-trisphosphate ($InsP_3$), and such metabolic coupling could play important roles in the development of the CNS (Peinado et al 1993, Penn et al 1994, Becker et al 1996). However, there has been little recognition that small amino acid transmitters should pass readily through neuronal gap junctions, thus mediating

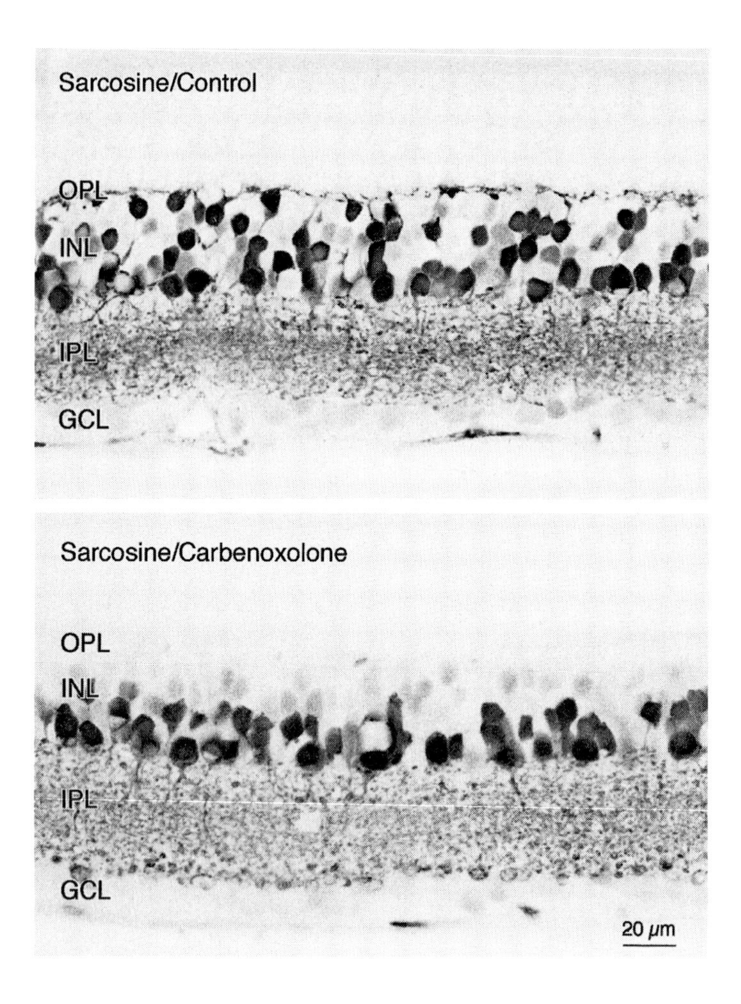


FIG. 4. Neurotransmitter coupling through gap junctions. Confocal micrographs of transverse sections of rat retina labelled by glycine immunocytochemistry. Both preparations were initially incubated in 0.5 mM sarcosine for 4 h to deplete the endogenous glycine in the amacrine cells and the cone bipolar cells (Pow 1998). Subsequent incubation of the retina in control medium for 2 h led to replenishment of the glycine stores, which showed the normal pattern of glycine immunoreactivity (top). When the replenishment was undertaken in the presence of 50 $\mu$M carbenoxolone, a potent gap junction blocker, only the amacrine cells showed strong glycine uptake (bottom). This supports the hypothesis that the bipolar cells obtain their glycine through heterologous gap junctions with glycinergic amacrine cells. GCL, ganglion cell layer; INL, inner nuclear layer; IPL, inner plexiform layer; OPL, outer plexiform layer (Vaney et al 1998).

neurotransmitter coupling. Although it appears that all retinal bipolar cells are glutamatergic, the CB cells arborizing in sublamina *b* of the inner plexiform layer also contain and accumulate glycine, leading to the hypothesis that these bipolar cells obtain their glycine through the heterologous gap junctions with AII amacrine cells (Cohen & Sterling 1986, Marc 1989).

Strong support for this hypothesis is provided by three independent lines of evidence obtained from recent experiments on rabbit and rat retinas (Vaney et al 1998). First, the glycine transporter, GLYT1, is expressed by the glycine-containing amacrine cells but not by the glycine-containing bipolar cells, suggesting that only the amacrine cells are functionally glycinergic. Second, lipophilic gap junction blockers such as carbenoxolone greatly reduce exogenous glycine accumulation into the bipolar cells but not the amacrine cells. Moreover, when the endogenous glycine stores in both cell classes are depleted by incubating the retina with a potent glycine-uptake inhibitor (Pow 1998), carbenoxolone blocks the subsequent glycine replenishment of the bipolar cells but not the amacrine cells (Fig. 4). Third, combined tracer coupling and immunocytochemical experiments indicate that all of the bipolar cells coupled to AII amacrine cells show glycine immunoreactivity. Taken together, these findings indicate that the glycine in On-centre CB cells is not derived by either high affinity uptake or *de novo* synthesis, but is obtained by neurotransmitter coupling through gap junctions with glycinergic amacrine cells. Thus, transmitter phenotype may be an unreliable indicator of transmitter function for neurons that make heterologous gap junctions.

*Acknowledgements*

D.I.V. is a Principal Research Fellow of the National Health and Medical Research Council (Australia).

## References

Baldridge WH, Vaney DI, Weiler R 1998 The modulation of intercellular coupling in the retina. Semin Cell Dev Biol 9:311–318

Becker D, Bonness V, Catsicas M, Mobbs P 1996 Determination of the pattern of gap-junctional communication between neurons of the embryonic chick retina by whole-cell patch clamping and confocal microscopy. J Physiol (Lond) 494:11P

Bloomfield SA, Xin D, Osborne T 1997 Light-induced modulation of coupling between AII amacrine cells in the rabbit retina. Vis Neurosci 14:565–576

Cohen E, Sterling P 1986 Accumulation of (3H)glycine by cone bipolar neurons in the cat retina. J Comp Neurol 250:1–7

Cohen E, Sterling P 1990 Demonstration of cell types among cone bipolar neurons of cat retina. Philos Trans R Soc Lond B Biol Sci 330:305–321

Cook JE, Becker DL 1995 Gap junctions in the vertebrate retina. Microsc Res Tech 31:408–419

Dacey DM, Brace S 1992 A coupled network for parasol but not midget ganglion cells in the primate retina. Vis Neurosci 9:279–290 deVries SH, Schwartz EA 1989 Modulation of an electrical synapse between solitary pairs of catfish horizontal cells by dopamine and second messengers. J Physiol (Lond) 414:351–375

Edelman GM 1988 Morphoregulatory molecules. Biochemistry 17:3533–3543

Hampson EC, Vaney DI, Weiler R 1992 Dopaminergic modulation of gap junction permeability between amacrine cells in mammalian retina. J Neurosci 12:4911–4922

Hitchcock PF 1997 Tracer coupling among regenerated amacrine cells in the retina of the goldfish. Vis Neurosci 14:463–472

Jacoby R, Stafford D, Kouyama N, Marshak D 1996 Synaptic inputs to ON parasol ganglion cells in the primate retina. J Neurosci 16:8041–8056

Kolb H, Famiglietti EV 1974 Rod and cone pathways in the inner plexiform layer of cat retina. Science 186:47–49

Lasater EM, Dowling JE 1985 Dopamine decreases conductance of the electrical junctions between cultured retinal horizontal cells. Proc Natl Acad Sci USA 82:3025–3029

Lu C, McMahon DG 1997 Modulation of hybrid bass retinal gap junctional channel gating by nitric oxide. J Physiol (Lond) 382:689–699

Marc ER 1989 The role of glycine in the mammalian retina. In: Osborne N, Chader J (eds) Progress in retinal research. Pergamon Press, Oxford, p 67–107

Massey SC, Mills SL 1996 A calbindin-immunoreactive cone bipolar cell type in the rabbit retina. J Comp Neurol 366:15–33

Mastronarde DN 1983 Interactions between ganglion cells in cat retina. J Neurophysiol 49:350–365

Meister M, Lagnado L, Baylor DA 1995 Concerted signaling by retinal ganglion cells. Science 270:1207–1210

Mills SL, Massey SC 1995 Differential properties of two gap junctional pathways made by AII amacrine cells. Nature 377:734–737

Nadarajah B, Parnavelas JG 1999 Gap junctions and connexin expression in the adult and developing cerebral cortex. In: Gap junction-mediated intercellular signalling in health and disease. Wiley, Chichester (Novartis Found Symp 219) p 157–174

Negishi K, Teranishi T 1990 Close tip-to-tip contacts between dendrites of transient amacrine cells in carp retina. Neurosci Lett 115:1–6

Ochalski PA, Frankenstein UN, Hertzberg EL, Nagy JI 1997 Connexin-43 in rat spinal cord: localization in astrocytes and identification of heterotypic astro-oligodendrocytic gap junctions. Neuroscience 76:931–945

Peinado A, Yuste R, Katz LC 1993 Gap junctional communication and the development of local circuits in neocortex. Cereb Cortex 488–498

Penn AA, Wong RO, Shatz CJ 1994 Neuronal coupling in the developing mammalian retina. J Neurosci 14:3805–3815

Piccolino M, Neyton J, Gerschenfeld HM 1984 Decrease of gap junction permeability induced by dopamine and cyclic adenosine 3′:5′-monophosphate in horizontal cells of turtle retina. J Neurosci 4:2477–2488

Pourcho RG 1982 Dopaminergic amacrine cells in the cat retina. Brain Res 252:101–109

Pow DV 1998 Transport is the primary determinant of glycine content in retinal neurons. J Neurochem 70:2628–2636

Raviola E, Gilula NB 1973 Gap junctions between photoreceptor cells in the vertebrate retina. Proc Natl Acad Sci USA 70:1677–1681

Raviola E, Raviola G 1982 Structure of the synaptic membranes in the inner plexiform layer of the retina: a freeze-fracture study in monkeys and rabbits. J Comp Neurol 209:233–248

Robinson SR, Hampson EC, Munro MN, Vaney DI 1993 Unidirectional coupling of gap junctions between neuroglia. Science 262:1072–1074

Strettoi E, Raviola E, Dacheux RF 1992 Synaptic connections of the narrow-field, bistratified rod amacrine cell (AII) in the rabbit retina. J Comp Neurol 325:152–168

Teranishi T, Negishi K, Kato S 1984 Regulatory effect of dopamine on spatial properties of horizontal cells in carp retina. J Neurosci 4:1271–1280
Vaney DI 1991 Many diverse types of retinal neurons show tracer coupling when injected with biocytin or Neurobiotin. Neurosci Lett 125:187–190
Vaney DI 1994a Patterns of neuronal coupling in the retina. In: Osborne N, Chader J (eds) Progress in retinal and eye research. Pergamon Press, Oxford, p 301–355
Vaney DI 1994b Territorial organization of direction-selective ganglion cells in rabbit retina. J Neurosci 14:6301–6316
Vaney DI 1996 Cell coupling in the retina. In: Spray DC, Dermietzel R (eds) Gap junctions in the nervous system. RG Landes, Austin, p 79–102
Vaney DI 1997 Neuronal coupling in rod-signal pathways of the retina. Investig Ophthalmol Vis Sci 38:267–273
Vaney DI, Nelson JC, Pow DV 1998 Neurotransmitter coupling through gap junctions in the retina. J Neurosci: in press
Veruki ML, Wässle H 1996 Immunohistochemical localization of dopamine $D_1$ receptors in rat retina. Eur J Neurosci 8:2286–2297
Wässle H, Boycott BB 1991 Functional architecture of the mammalian retina. Physiol Rev 71:447–480
Wolszon LR, Gao WQ, Passani MB, Macagno ER 1994 Growth cone 'collapse' *in vivo*: are inhibitory interactions mediated by gap junctions? J Neurosci 31:999–1010
Xin D, Bloomfield SA 1997 Tracer coupling pattern of amacrine and ganglion cells in the rabbit retina. J Comp Neurol 383:512–528
Yamada E, Ishikawa T 1965 The fine structure of the horizontal cells in some vertebrate retinae. Cold Spring Harb Symp Quant Biol 30:383–392

## DISCUSSION

*Gilula:* Is there electrophysiological cell activity in these preparations?

*Vaney:* We haven't recorded from these preparations, but electrophysiological experiments on brain slices indicated that 30–100 $\mu$M carbenoxolone had no effect on the cell conductance or the kinetics of stepped currents (Osborne & Williams 1996). The effects of sarcosine are less predictable, but this glycine-uptake blocker did not affect the amino acid content of the Müller glial cells, which are known to be sensitive to metabolic perturbations.

*Gilula:* An issue of biological relevance under these conditions is whether or not the changes you are observing reflect the pathway available to these cells under normal conditions.

*Vaney:* The cone bipolar (CB) cells that show high affinity glycine uptake do not express high affinity glycine transporters; thus, the heterologous gap junctions with the rod (AII) amacrine cells appear to provide the only pathway for these bipolar cells to acquire extracellular glycine selectively. The glycinergic amacrine cells are effective controls because they show normal uptake of glycine in the presence of gap junction blockers, whereas the bipolar cells do not.

*Goodenough:* How do you look at glycine immunoreactivity?

*Vaney:* We use antibodies raised against glycine coupled to bovine serum albumin with glutaraldehyde or paraformaldehyde. The latter antibodies are effective on lightly fixed retinal whole mounts.

*Goodenough:* If glycine is being used as a transmitter, presumably it is contained within vesicles so one wouldn't expect it to go through gap junctions.

*Vaney:* That's certainly true. However, the high affinity glycine transporter, GLYT1, concentrates the glycine intracellularly before it is taken up by the vesicular glycine transporter. Because GLYT1 is distributed throughout the plasma membrane of the AII amacrine cells, it lies in close proximity to the gap junctions, which are remote from the chemical synapses and the associated glycine-containing vesicles.

*Goodenough:* In the images you showed it is not possible to tell whether we were looking at glycine that will be used as a transmitter or glycine that will be used metabolically.

*Vaney:* The glycine levels in the glycine-immunopositive amacrine cells and bipolar cells are significantly higher than the metabolic pools present in other retinal neurons, including the $\gamma$-aminobutyric acid-immunopositive amacrine cells (Wright et al 1997). Further evidence that glycine is used as a transmitter by these amacrine cells is provided by the localization of both the glycine and GLYT1 in the cells. By contrast, there is no clear evidence that the glycine in the CB cells is used as a neurotransmitter.

*Gilula:* Your use of Neurobiotin, rather than Lucifer yellow, provokes some consideration by this audience, who have mostly worked with Lucifer yellow over the years. Could you comment on why the experiments with Lucifer yellow have not been successful in the retina.

*Vaney:* Retinal networks that are permeable to both Neurobiotin and Lucifer yellow, such as the axonless horizontal cells, show more extensive tracer coupling than dye coupling, and thus Neurobiotin is more sensitive than Lucifer yellow for detecting gap junctional coupling. However, other neuronal networks, such as the axon-bearing horizontal cells or the AII amacrine cells, appear to be totally impermeable to Lucifer yellow but are quite permeable to biotinylated tracers of comparable molecular weight. Thus, the fundamental difference appears to be the net charge, with most retinal gap junctions selectively passing cationic tracers and excluding anionic dyes.

*Gilula:* Can you use Neurobiotin to define the coupling association between heterologous cells, and then demonstrate viable electrical coupling between those cells?

*Vaney:* That's a difficult experiment to do in intact tissue because non-junctional membrane conductances mask weak electrical coupling through small gap junctions. However, we have been able to do the reverse experiment. Prior to using Neurobiotin there was no morphological evidence that retinal ganglion cells make gap junctions, although there was indirect physiological evidence that the $\alpha$ ganglion cells are electrically coupled (Mastronarde 1983). Subsequently, tracer coupling experiments demonstrated conclusively that the $\alpha$ ganglion cells show both homologous and heterologous coupling (Vaney 1991). The tracer coupling technique appears to be more sensitive than other methods for demonstrating cellular coupling, including electron microscopy and connexin immunocytochemistry. Ironically, this has led to questions about the validity of the technique, because it has often been difficult to obtain independent confirmation of the gap junctional pathway using other morphological and physiological criteria.

*Warner:* In many systems, and particularly in developmental systems, people often rely entirely on Lucifer yellow as an indicator of coupling. In the amphibian embryo we know that the selectivity of gap junctions is different in different regions. Therefore, when looking at systems where gap junction properties are regulated it is important to use more than one tracer. One may obtain misleading results from a single tracer, especially as Lucifer yellow does seem, for reasons that are not entirely clear, to behave in a selective way. For instance, in early amphibian embryos Lucifer yellow transfer occurs more frequently between cells in dorsal regions than between cells in ventral regions. By contrast, EDTA, which is almost the same size as Lucifer yellow, is transferred between cells with equivalent frequency in both regions, as is the smaller molecule silver dicyanoargentate (Guthrie et al 1988).

*Goodenough:* The same is true for Neurobiotin.

*Nicholson:* Gary Goldberg in our lab has developed a capture technique, in which he pre-loads 'donor' cells with radioactivity and an immobile fluorescent dye, allows the radioactive label to pass to unlabelled 'recipient' cells, and then captures the recipient cells through a fluorescence cell sorter (Goldberg et al 1998). He has found that a connexin32 (Cx32; $\beta_1$)-transfected clone of C6 cells transfers Lucifer yellow about 10–20-fold more effectively than a Cx43 ($\alpha_1$)-transfected clone C6 cell clone. However, when he did the same comparison using his radioactive capture technique, he found that radioactive transfer is almost completely inverted: the Cx43-transfected clone transfers radioactive metabolites about 10-fold more effectively than the Cx32-transfected clone.

*Warner:* These are radioactively labelled normal metabolites, which highlights another problem in that we generally monitor transfer through gap junctions with molecules that are not normally found in cells.

*Goodenough:* What is known about the distribution of different connexins in the retina?

*Becker:* Cx43, Cx32 and Cx26 ($\beta_2$) are expressed in the retinas of rabbit, ferret and chick. We were unable to detect Cx40 ($\alpha_5$) and Cx37 ($\alpha_4$), by immunostaining, in the neural retina, although they could be detected in the blood vessels of vascularized retinas. The distribution of these connexins between the different laminae is complicated. It changes from central to peripheral retina and according to the ambient lighting conditions.

*Goodenough:* Mills & Massey (1995) showed that AII amacrine cells are connected to other AII cells and to bipolar cells. Neurobiotin (286 kDa) passed through both types of gap junctional pathways, whereas biotin-X cadaverine (442 kDa) passed easily through AII/AII junctions but only weakly through AII/bipolar gap junctions. Also, the two pathways showed different regulation. Dopamine and cAMP agonists, known to reduce AII/AII communication, did not change AII/bipolar communication. Nitric oxide and cGMP agonists selectively diminished bipolar/AII communication. These data suggest that there are different connexin genes expressed by these different cell types; indeed, the AII cell may use one connexin in homologous intercellular pathways and another in heterologous intercellular pathways. Have you found this possibility to be true in your immunofluorescence studies of the retina?

*Becker:* At this point there is no definitive correlation. However, there are distinct sublaminar distributions of the different connexin types within the inner plexiform layer. The inner plexiform layer has five distinct sub laminae with differential expression of Cx32 and Cx26. Expression of large and small plaques of these connexins within the sublaminae correlate well with electron microscope evidence concerning the distribution of large and small gap junctions on the AII amacrine cell and cone bipolar subtypes.

*Dermietzel:* We tried to clone individual connexins in different fish cells by single-cell reverse transcriptase (RT)-PCR analysis. We found that Cx43 was present in horizontal cells, and we also found some new types of connexins. We are only starting to understand the diversity of neuronal connexins, and my guess is that there are more connexins in the retina that have not yet been found by using conventional techniques, i.e. immunocytochemistry.

I would also like to stress another important point. The old story that there is a dichotomy between the electrical and chemical synapse is incorrect because we know that there is an interference between chemical and electrical transmission, either by providing neurotransmitters such as glycine through gap junctions or direct effects of neurotransmitters on electrical coupling, as David Vaney showed by the effect of dopamine on horizontal cells. I would like David Vaney to expand on this. How does dopamine affect horizontal cells? And what are the effects on gap junction coupling? Is your idea that a receptor mediates the effect from the cell membrane to the gap junctions, or does dopamine have a direct effect on the channels?

*Vaney:* John Dowling and his colleagues have done a lot of work on this over the years. The uncoupling effects of dopamine are mediated through $D_1$ receptors: the increased cAMP activates protein kinase A, which may phosphorylate the gap junction protein directly (Lasater 1987, McMahon et al 1994). Retinal gap junctions may also be uncoupled by increased cGMP, but the molecular mechanism isn't clear. Patch-clamp recordings from horizontal cells in fish retina show that cGMP and cAMP close the same gap junctions (deVries & Schwartz 1989), whereas tracer-coupling experiments on AII amacrine cells in the mammalian retina indicate that the two second messengers act on different gap junctions (Mills & Massey 1995).

*Willecke:* Is there any evidence for rectifying junctions between interneurons in the retina?

*Vaney:* Our studies on retinal neuroglial cells showed that Lucifer yellow and Neurobiotin passed more efficiently from the astrocytes to the oligodendrocytes than vice versa (Robinson et al 1993). Such asymmetric permeability seems to be a common property of heterologous gap junctions in the retina, including ganglion cell→amacrine cell coupling and AII amacrine cell→CB cell coupling. If the tracer coupling results largely from passive diffusion through the gap junctions rather than iontophoretic movement of the tracer in the injection current, then the asymmetric permeability cannot be attributed to electrical rectification, which is dependent on a potential difference across the cells. Loewenstein (1981) made a clear distinction between chemical rectification and electrical rectification, but it is not clear how chemical rectification could be implemented in molecular terms without violating the Second Law of Thermodynamics.

*Giaume:* I would like to make a few comments about the regulation of gap junction permeability and the control of propagation of intercellular calcium waves by bioactive lipids in glial cells. This is related to our studies with anandamide (N-arachidonoyl-ethanaolamine), an endogenous arachidonic acid derivative that activates cannabinoid receptors (Devane et al 1992). We have shown that anandamide blocks electrical and dye coupling between astrocytes cultured from rat or mouse striata (Venance et al 1995). As shown in Fig. 1b, this inhibitory effect is reversible and can be observed using low concentrations of this compound, since 50% uncoupling is obtained with 1 $\mu$M of anandamide and inhibition is total for 5 $\mu$M. Since the chemical structure of anandamide includes the polysaturated fatty acyl moiety of arachidonic acid, we have also tested the effect of this fatty acid. Compared to anandamide the full inhibitory effect on astrocytic gap junctions was observed with a 10-fold higher concentration of arachidonic acid, suggesting that the action of the endogenous cannabinoid is not a fatty acid-like effect. Further indications were obtained by treating the cells with pertussis toxin (PTX), which ADP-ribosylates and inhibits Gi/o $\alpha$-proteins. Such treatment prevents the anandamide (5 $\mu$M)-induced uncoupling, whereas it has no effect on that evoked by arachidonic acid (50 $\mu$M). This observation indicates that in striatal astrocytes the block of gap junction permeability by anandamide is mediated through the stimulation of a membrane receptor. This

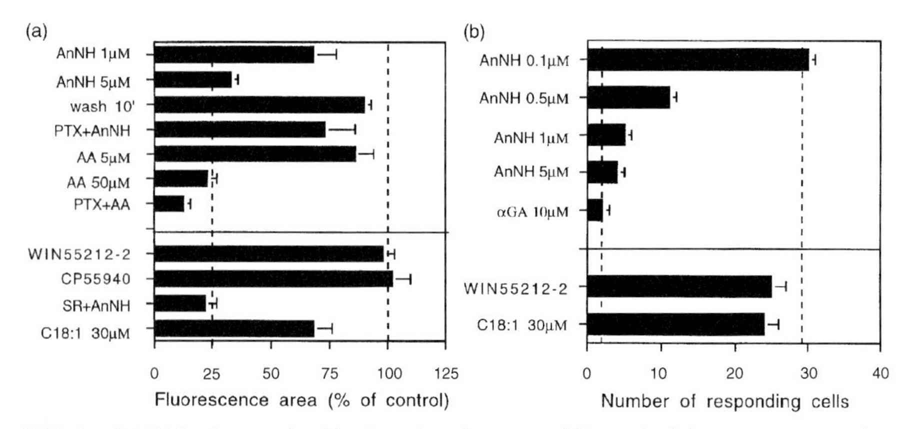


FIG. 1. Inhibition by anandamide of gap junction permeability and calcium wave propagation in striatal astrocytes. (a) Scrape-loading dye transfer experiments showing the properties of inhibitory effect of anandamide on junctional permeability in mouse striatal astrocytes. The level of Lucifer yellow diffusion was quantified by measuring the averaged values (mean $\pm$S.E.M., number of independent experiments ranged from 4 to 34) of the surface area stained by the dye. The vertical dashed lines indicates the control level (100%) of dye spread for each internal controls (right) and the restricted dye spread observed in the presence of the uncoupling agent αGA (in a). When not mentioned, the concentrations used for anandamide and arachidonic acid were 5 $\mu$M and 50 $\mu$M, respectively. The cannabinoid agonists WIN55212-2 and CP55940 and the antagonist SR141716A were used at 5 $\mu$M, 1 $\mu$M and 0.5 $\mu$M, respectively. Pertussis toxin (1 $\mu$g/ml) treatments were performed 24 h before the experiments. AA, arachidonic acid; AnNH, anandamide; PTX, pertussis toxin; SR, SR141716A. (b) Calcium imaging experiments showing the blockage by anandamide of propagation of calcium waves in primary cultures of rat striatal astrocytes (mean $\pm$S.E.M., number of independent experiments ranged from 5 to 25). Mechanical stimulation of a single astrocyte resulted in the propagation of increase in intracellular calcium concentration in 28 $\pm$ 1 neighbouring cells (left vertical dashed line). Perfusion of the astrocytes with increasing concentrations of anandamide shown a dose-dependent inhibition of the propagation. At 5 $\mu$M this endogenous cannabinoid exerts a potent inhibitory effect comparable to that observe in the presence of the uncoupling agent αGA (left vertical dashed line).

statement was strengthened by additional data obtained with N-ethylmaleimide, an alkylating agent that inhibits G proteins, and 4-bromophenacyl bromide, an histidine alkylating agent. These two compounds prevented the uncoupling effect of anandamide (5 $\mu$M). Selectivity of anandamide was supported by experiments showing that structural analogues which are co-released with anandamide by striatal neurons (Di Marzo et al 1994) were less effective, and that the uncoupling effect of anandamide was much more pronounced in striatal astrocytes compared to other brain regions. Altogether, these observations indicate that in striatal astrocytes anandamide inhibits gap junction permeability through a mechanism which involves the activation of a membrane receptor. These observations have to be distinguished from a recent study (Guan et al

1997) which reports that high concentrations (50 μM) of oleamide or anandamide inhibit gap junctions in astrocytes cultured from whole brain. In this former study, the inhibitory effect of these two compounds is similar to that observed in our study with high doses of arachidonic acid (25–50 μM) or of the co-released analogue C18:1 (30 μM), i.e. a non-receptor-mediated effect.

A consequence of the inhibitory effect of low doses of anandamide is the block of intercellular calcium signalling in astrocytes cultured from rat striatum. We have recently shown that functional gap junction channels were necessary for the propagation of these calcium waves (Venance et al 1997a). As illustrated in Fig. 1a, anandamide at low concentration (0.5 μM) reduces the extent of calcium waves and blocks their propagation at 1–5 μM. This receptor-mediated inhibition contrasts with a recent report showing that anandamide and oleamide used at a concentration of 50 μM, which inhibits gap junctional permeability, did not block the propagation of intercellular calcium waves (Guan et al 1997).

In conclusion, from our observations and those reported by others, it is likely that active biolipids may have different mechanisms of action, depending on the concentration used and the type of glial cells investigated. In our study, anandamide inhibits astrocytic gap junction permeability and blocks calcium waves through the activation of a receptor coupled to a pertussis-sensitive G protein. This glial receptor is likely distinct from the central CB1 cannabinoid receptor present in neurons. Indeed, the pharmacological profile of the anandamide inhibition of gap junction and calcium waves propagation were neither reproduced by classical synthetic cannabinoids (WIN55212-2, CP55940) nor prevented by a CB1 antagonist (SR141716A). This statement has now been strengthened by biochemical assays of cAMP formation, binding assays and immunocytochemistry with anti-CB1 antibody performed and pure cultures of striatal neurons or astrocytes (Sagan et al 1998).

*Sanderson:* One of the criticisms that people put forward when calcium waves are inhibited is whether anandamide affects the calcium release mechanism. Have you controlled for that?

*Giaume:* We have controlled the filling of the internal calcium stores. Indeed, R-methanandamide, a chiral analogue of anandamide, used at a concentration of 5 μM induces a depletion of internal pools tested with ionomycin in the absence of external calcium (Venance et al 1997b). This emptying is sensitive to PTX treatment, which again suggests the involvement of a membrane receptor. Such effects on the calcium release mechanism combined with the inhibition of gap junction channels may explain why the block of calcium wave propagation occurred at low concentration of anandamide (see Fig. 1).

*Goodenough:* Have you looked at any of the known blockers of intracellular signalling pathways, such as Wortmanin, to see if PI3 kinase is involved in the effect of anandamide?

*Giaume:* We have performed several attempts in order to identify which intracellular signalling pathway is involved in the effect of anandamide on gap junction permeability. When cells are pre-treated with of the phosphodiesterase inhibitor 3-isobutylmethylxanthine (IBMX) and agents known to increase cAMP concentration (forskoline or isoproterenol), the effectiveness of the anandamide uncoupling is not altered, suggesting that cAMP level is not involved in this process. In addition, protein kinase inhibitors that have a large spectre of action such as H7 or staurosporine have no effect, and the phosphatase inhibitor okadaic acid does not affect the efficiency of anandamide to uncouple striatal astrocytes. So far, only two compounds that are known to interact with second messenger pathways have been found to interfere with the anandamide-induced uncoupling. 4-bromophenacyl bromide, a PLA2 inhibitor, and U73122, a PLC inhibitor, both prevent the inhibitory effect of 5 $\mu$M anandamide. However, the mechanism involved in the receptor-mediated uncoupling of astrocytic gap junctions by anandamide is not yet fully understood.

*Gilula:* These compounds that affect junction pathways have several different possible mechanisms. This is interesting in terms of the pathways that regulate transduction responses in different cell types. Christian Giaume and I have talked about the possibility that there are several different mechanistic approaches for understanding their effects. The one he's talked about today might be receptor mediated, but there are others that may not be receptor mediated.

*Fletcher:* But the reason he may be getting cGMP effects is that he may be activating a cGMP-activated cAMP phosphodiesterase, so the two would merge.

*Becker:* But the actions of cAMP and cGMP on the gating of horizontal cells are quite distinctive. One affects the frequency of opening and the other affects the duration of opening, indicating a separate pathway.

## References


Devane WA, Hanuš L, Breuer A et al 1992 Isolation and structure of a brain constituent that binds to the cannabinoid receptor. Science 258:1946–1949

deVries SH, Schwartz EA 1989 Modulation of an electrical synapse between solitary pairs of catfish horizontal cells by dopamine and second messengers. J Physiol (Lond) 414:351–375

Di Marzo V, Fontana A, Cadas H et al 1994 Formation and inactivation of endogenous cannabinoid anandamide in central neurons. Nature 372:686–691

Goldberg GS, Lampe PD, Sheedy D, Stewart CC, Nicholson BJ, Naus CCG 1998 Direct identification and analysis of transjunctional ADP from Cx43 transfected C6 glioma cells. Exp Cell Res 239:82–92

Guan X, Cravatt BF, Ehring GR et al 1997 The sleep-inducing lipid oleamide deconvolutes gap junction communication and calcium wave transmission in glial cells. J Cell Biol 139: 1785–1792

Guthrie S, Turin L, Warner AE 1988 Patterns of junctional communication during development of the early amphibian embryo. Development 103:769–783

Lasater EM 1987 Retinal horizontal cell gap junctional conductance is modulated by dopamine through a cyclic AMP-dependent protein kinase. Proc Natl Acad Sci USA 84:7319–7323

Loewenstein WR 1981 Junctional intercellular communication: the cell-to-cell membrane channel. Physiol Rev 61:829–913

Mastronarde DN 1983 Interactions between ganglion cells in cat retina. J Neurophysiol 49:350–365

McMahon DG, Rischert JC, Dowling JE 1994 Protein content and cAMP-dependent phosphorylation of fractionated white perch retina. Brain Res 659:110–116

Mills SL, Massey SC 1995 Differential properties of two gap junctional pathways made by AII amacrine cells. Nature 377:734–737

Osborne PB, Williams JT 1996 Forskolin enhancement of opioid currents in rat locus coeruleus neurons. J Neurophysiol 76:1559–1565

Robinson SR, Hampson EC, Munro MN, Vaney DI 1993 Unidirectional coupling of gap junctions between neuroglia. Science 262:1072–1074

Sagan S, Venance L, Torrens Y, Cordier J, Glowinski J, Giaume C 1998 Anandamide and WIN 55212-2 inhibit cyclic AMP formation through G protein-coupled receptors distinct from CB1 cannabinoid receptors in cultured astrocyes. Eur J Neurosci, in press

Vaney DI 1991 Many diverse types of retinal neurons show tracer coupling when injected with biocytin or Neurobiotin. Neurosci Lett 125:187–190

Venance L, Piomelli D, Glowinski J, Giaume C 1995 Inhibition by anandamide of gap junctions and intercellular calcium signalling in striatal astrocytes. Nature 376:590–594

Venance L, Stella N, Glowinski J, Giaume C 1997a Mechanisms involved in initiation and propagation of receptor-induced intercellular calcium signaling in cultured rat astrocytes. J Neurosci 17:1981–1992

Venance L, Sagan S, Giaume C 1997b R-methandamide inhibits receptor-induced calcium responses by depleting internal calcium stores in cultured astrocytes. Pflueg Arch Eur J Physiol 434:147–149

Wright LL, Macqueen CL, Elston GN, Young HM, Pow DV, Vaney DI 1997 The DAPI-3 amacrine cells of the rabbit retina. Vis Neurosci 14:473–492

# Gap junctions and connexin expression in the inner ear

A. Forge, D. Becker*, S. Casalotti, J. Edwards, W.H. Evans†, N. Lench‡ and M. Souter

*Institute of Laryngology and Otology, University College London, 330–332 Gray's Inn Road, London WC1X 8EE, *Department of Anatomy and Developmental Biology, University College London, Gower Street, London WC1E 6BT, †Department of Medical Biochemistry, University of Wales College of Medicine, Heath Park, Cardiff CF4 4XN, and ‡Molecular Medicine Unit, St. James' University Hospital, Leeds LS9 7TF, UK*

*Abstract.* Several different recessive mutations in the connexin26 (Cx26; $\beta_2$) gene have been associated with non-syndromic hereditary deafness. This suggests gap junctions are important to cochlear function. Numerous large gap junctions are present between adjacent supporting cells in both the vestibular and auditory sensory epithelia of the mature inner ear. In vestibular organs, Cx26 is highly expressed, but antibodies to Cx32 ($\beta_1$) also label the supporting cells. In the organ of Corti of the cochlea, Cx26 is the predominant connexin isoform; neither Cx32 nor Cx43 ($\alpha_1$) can be detected by immunohistochemistry. One role for gap junctions between supporting cells may be to provide a pathway for the rapid removal of ions away from the region of the sensory cells during transduction in order to maintain sensitivity. In the cochlea gap junctions are also associated with the basal cells of the stria vascularis, an ion-transporting epithelium that maintains a positive electrical potential in the potassium-rich endolymph fluid which bathes the apical surfaces of the sensory 'hair' cells and which is crucial for auditory transduction. Gap junctions are present between fibrocytes in the spiral ligament that underlies the stria vascularis, and between these fibrocytes and strial basal cells. During cochlear development, the initial formation and subsequent increase in size and number of gap junctions in the stria vascularis coincides with the initial generation and rise of the endocochlear potential. This and other evidence suggests that one role of gap junctions in the cochlea is to provide a pathway for passage of ions to maintain endolymph and, thus, auditory acuity. Mutations to Cx26 could, therefore, disrupt this ion circulation, resulting in deafness.

*1999 Gap junction-mediated intercellular signalling in health and disease. Wiley, Chichester (Novartis Foundation Symposium 219) p 134–156*

Congenital deafness affects about 1 in 1000 babies born in the UK. In about half of these cases, the cause is an inherited disorder. Autosomal recessive mutations account for about one-third of the inherited disorders associated with non-syndromic deafness, that is where an effect on inner ear function is the only obvious phenotypic consequence. Recently, mutations that map to the chromosomal locus of the gene for connexin26 (Cx26; $\beta_2$) have been shown to be a major cause of hereditary, non-syndromic, pre-lingual deafness (Denoyelle et al 1997, Kelsell et al 1997, Zelante et al 1997). At the time of writing, 22 different mutations in the Cx26 gene that are associated with such deafness in humans have been described (Fig. 1). The mutations are distributed along the gene, resulting in defects at various locations on the Cx26 protein (Fig. 1). The most common, now identified in at least 14 different families, is the 30delG mutation, where a deletion of a guanine in a region of six successive guanines at positions 30–35, results in a premature stop codon and termination of the peptide after 13 amino acids from the N-terminal (Fig. 1). Effectively, such a mutation would seem to result in a Cx26 knockout. That mutations in a connexin gene result in deafness indicates an important role for gap junctions in the inner ear.

We have been using freeze-fracture to examine the distribution of gap junctions in the adult (Forge 1984, 1986, McDowell et al 1989) and developing (Souter &

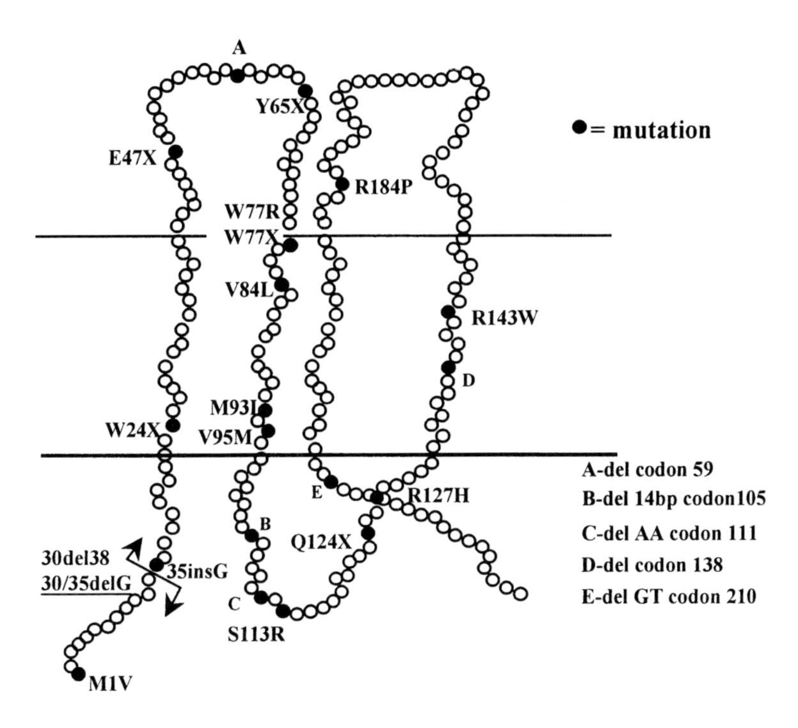


FIG. 1. Diagrammatic representation of the connexin26 (Cx26; $\beta_2$) protein showing the sites and nature of those mutations associated with non-syndromic hereditary deafness that have been described to date.

Forge 1998) inner ear. Recently, we have also applied a panel of antibodies to three different connexin isoforms, Cx26, Cx32 ($\beta_1$) and Cx43 ($\alpha_1$), to examine the pattern of connexin expression in the inner ears of guinea-pigs, gerbils and mice (Forge et al 1997). The expression of Cx26 in the inner ear of rats has also been examined (Kikuchi et al 1994, 1995). The results provide a possible explanation for the effect of the Cx26 mutations on auditory function.

## General anatomy and physiology of the inner ear

The inner ear consists of the hearing organ, the cochlea and the end organs of balance in the vestibular system. It is formed from a labyrinth of membranous ducts enclosed in bony channels at the base of the skull (Fig. 2a). The fluid surrounding the membranous ducts, and enclosed by the bony channels, is perilymph, which has a composition similar to that of most extracellular fluids, with a high $Na^+$ and low $K^+$ content. The fluid enclosed by the membranous ducts is endolymph and is unusual as it has a high $K^+$ concentration and low $Na^+$. In the mammalian cochlea, but not the vestibular system, endolymph also has a high positive electrical potential of *c*. +80 mV, the endocochlear potential (Wangemann & Schacht 1996).

The inner lining of the membranous labyrinth is formed from: the sensory epithelia; the ion-transporting epithelia that are responsible for the maintenance

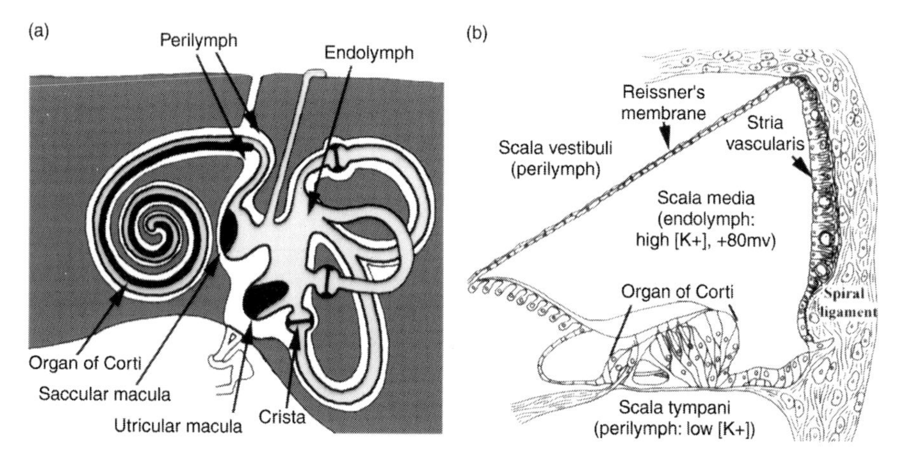


FIG. 2. (a) Diagram of the inner ear depicting the membranous labyrinth, which encloses endolymph, within bony channels that are filled with perilymph. The cochlea is to the left, the vestibular system to the right. Sensory regions are shown in black. (b) Diagram of a cross-section of a single turn of the cochlea to show the fluid spaces and general anatomical relationships of the different tissues described in the text.

of endolymph; and relatively unspecialized epithelia involved in maintaining barriers between endolymph and perilymph. The organ of Corti, a long, narrow strip of cells coiled in a spiral, is the auditory sensory epithelium in the cochlea (Fig. 2a, b). In the vestibular system there are five sensory patches: three cristae, one in each semi-circular canal; and two flat sheets, the utricular and saccular maculae (Fig. 2b). The sensory elements of these epithelia are the 'hair' cells, so called because each one bears at its apical end an organized bundle of rigid, erect projections, the stereocilia, which are modified microvilli. These extend into (or just beneath) an extracellular matrix mass that overlies the sensory epithelium. Each hair cell is surrounded by and separated from its neighbours by supporting cells, so that the apical surface of the epithelium appears like a mosaic. The bodies of the supporting cells also intervene between the base of the hair cell and the underlying basement membrane. In the vestibular organs (Fig. 3A) the supporting cells closely surround the hair cells and their nerve endings, but in the organ of Corti (Fig. 4A) differentiation of supporting cells occurring in late stages of cochlear maturation results in the separation of the cells along their lateral borders to leave large extracellular spaces in the body of the epithelium. The supporting cells contact hair cells only at the tight-adherens junctional complex at the cells' apices, and contact each other only at their apices and in the cell body region below the level of the hair cells. Those supporting cells on the outer and inner sides of the sensory region contact directly other epithelial cells lining the endolymphatic space (Fig. 2b).

The apical surface of a sensory epithelium is bathed in $K^+$-rich endolymph, whereas the bodies of the hair and supporting cells are bathed in perilymph. Hair cells have an intracellular negative resting potential of $-40$ to $-70$ mV (depending on hair cell type) and a standing current through the hair cell is generated as $K^+$ moves along electrochemical gradients. Sound vibrations (in the cochlea) or head movements (in the vestibular organs) produce motion of the surface of the epithelium relative to the overlying extracellular matrix mass that deflects the hair bundle. This causes opening and closing of ion channels, modulating the standing current and generating receptor potentials. The high positive potential of endolymph in the cochlea provides not only a driving force for current flow that increases receptor cell sensitivity, but also the energy that powers active motile responses of cochlear outer hair cells to sound stimulation (Dallos 1996). These motile responses are thought to be the basis of the 'cochlear amplifier' that underlies the unique frequency discrimination capabilities of the mammalian hearing organ. The absence of cochlear amplification results in severe loss of hearing acuity. Endocochlear potential is therefore essential for cochlear function. It is generated and maintained by the stria vascularis, a specialized ion-transporting epithelium that lines the lateral wall of the cochlear duct (Fig. 2b).

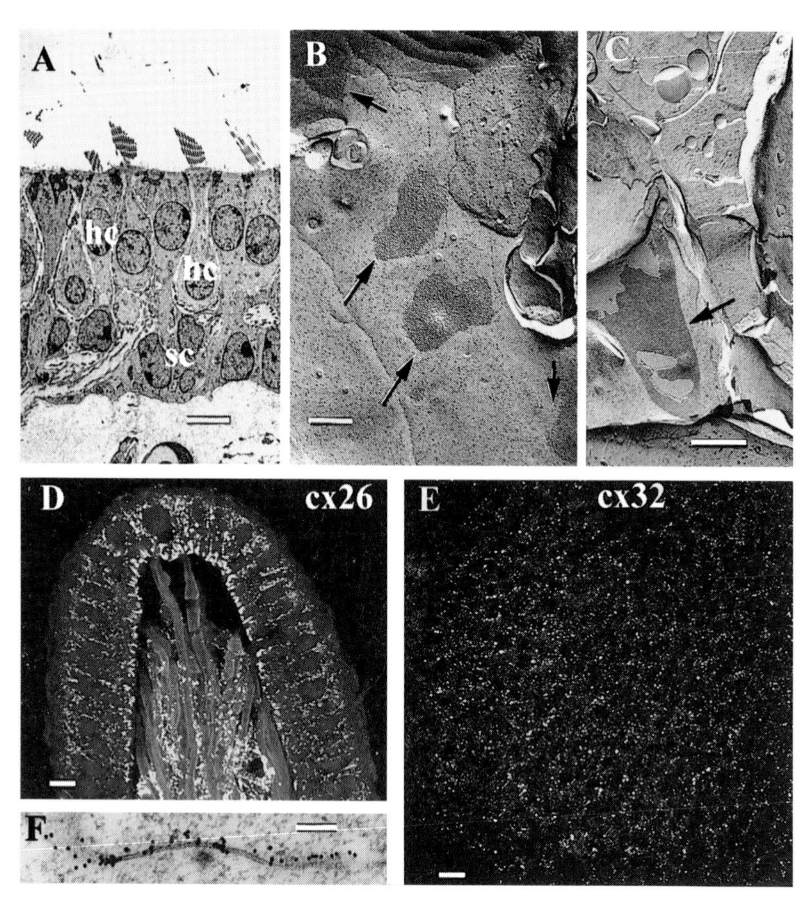


FIG. 3. The vestibular sensory epithelia. (A) Thin section of gerbil utricular macula. The sensory epithelium is formed of hair cells (hc) and supporting cells (sc), which separate each hair cell from its neighbours and intervene between the base of the hair cell and the underlying basement membrane. Bar = 10 $\mu$m. (B) Freeze-fracture revealing numerous large gap junctions (arrows) on the lateral membranes between two adjacent supporting cells at the apical end of the cell (as indicated by the strands of the tight junctional complex at the top of the figure). Bar = 500 nm. (C) Freeze-fracture showing large gap junction (arrow) in the cell body region at the basal end of the supporting cell. This junction is about 1 $\mu$m across. Bar = 500 nm. (D) Frozen section of guinea-pig crista labelled with antibodies to connexin26 (Cx26; $\beta_2$). Intense labelling is present in large puncta at the base of the epithelium (equivalent to junction shown in C) and all down the lateral membranes between supporting cells. Unlabelled 'islands' represent the sites of the hair cells and their calcyeal nerve endings. Bar = 20 $\mu$m. (E) Whole-mount preparation of gerbil utricular macula labelled with antibodies to Cx32 ($\beta_1$). The unlabelled holes represent the sites of the hair cells, and labelled puncta appear to define the borders of the supporting cells. Bar = 20 $\mu$m. (F) Gap junction in thin section of gerbil utricular macula, immunogold labelled using an antibody to Cx26. Gold particles are numerous on one side of the junction, but absent from the opposite side. Bar = 250 nm.

## Gap junctions in the sensory epithelia

Early studies of gap junctions in the sensory epithelia of the inner ear (Ginzberg & Gilula 1979) showed that in the developing vestibular organs of chicks, just prior to the differentiation of hair cells, all the cells in the presumptive sensory epithelium are connected by gap junctions. The cells that begin to differentiate as hair cells internalize and lose their gap junctional plaques, but gap junctions are retained between those cells that differentiate as supporting cells. Consequently, in the mature sensory epithelia, gap junctions are present between adjacent supporting cells, but they do not appear to be associated with hair cells. Freeze-fracture of mammalian vestibular organs (Fig. 3B,C) shows that gap junctional plaques are numerous and present all the way down the lateral membranes between adjacent supporting cells from the level of the tight junctions at the cells' apices (Fig. 3B) to the cell body region at the base (Fig. 3C). These junctions are of various sizes and are often quite large, several micrometers across in some cases. The ubiquitous presence of gap junctions between supporting cells and their absence from the lateral membranes of hair cells suggests that the supporting cell population is a functional syncitium, but that hair cells are functionally isolated from the supporting cells.

Immunohistochemical methods applied to whole-mount and to frozen section preparations of the vestibular sensory epithelia show that antibodies to Cx26 label numerous puncta of various sizes at the borders between supporting cells in both the cristae and the utricular maculae (Fig. 3D). The cells in the connective tissue underlying the sensory epithelium also label intensely with Cx26 antibodies. Some labelling with antibodies to Cx32 has also been obtained (Fig. 3E). Immunogold-labelling procedures applied to thin sections for transmission electron microscopy confirmed the presence of Cx26 in the gap junctions between adjacent supporting cells. It has also suggested the presence of connexins other than Cx26. A few junctions show extensive labelling with Cx26 antibodies on the cytoplasmic side of the junction in one cell, but almost no labelling on the cytoplasmic side of the junction of the adjacent cell (Fig. 3F). This suggests the possible presence of heterotypic junctions, but the result needs to be confirmed, and at present the connexin isoform present in the side of the junction opposite to that which contains Cx26 has not been determined.

As in the vestibular sensory organs, gap junctions are present in the organ of Corti where supporting cells are adjacent. Thus, they are present between supporting cells at their apical contacts, but are most pronounced between the cell body regions of the cells. Cx26 antibodies produce intense labelling of puncta in these regions of the tissue. In sections (Fig. 4B) numerous puncta labelled with Cx26 are present between adjacent Deiters' cells (the supporting cells that surround the outer hair cells) in the same row, i.e. in the longitudinal direction, as well as between adjacent cells in the radial direction, that is between Deiters' cells in one

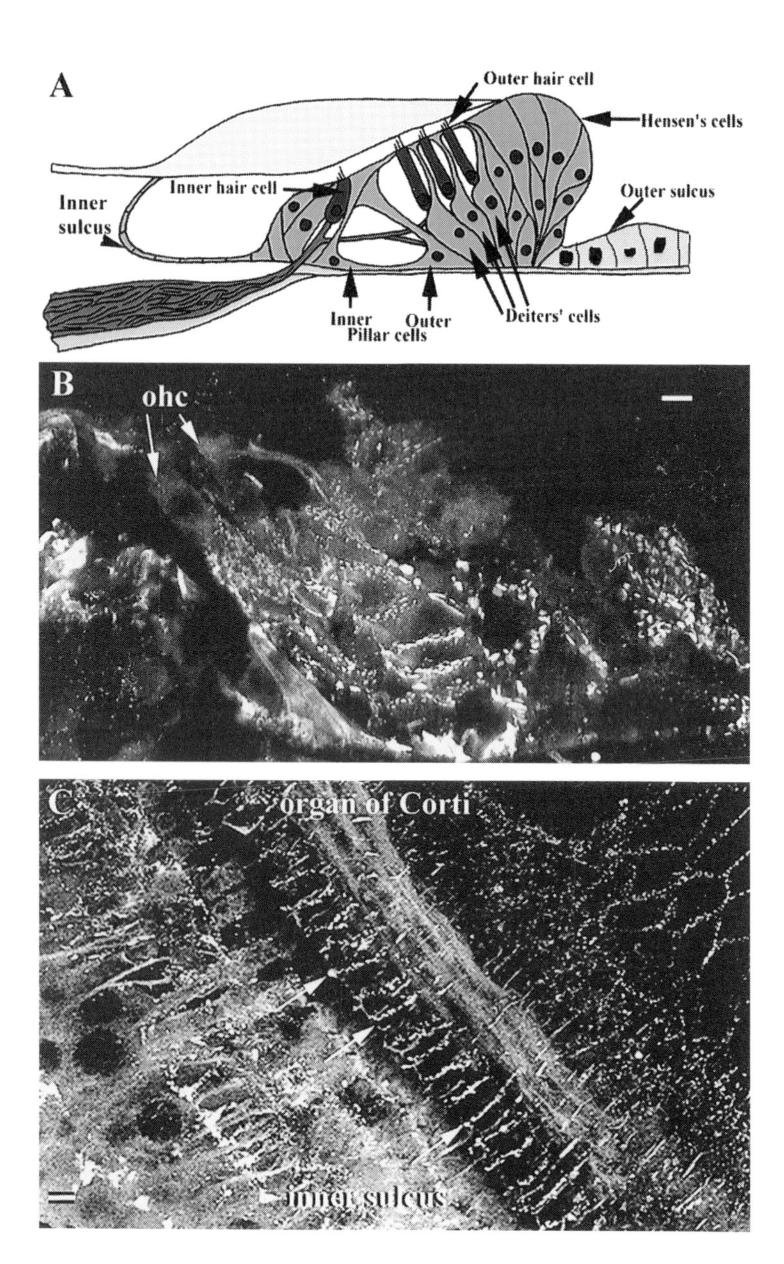
A
Outer hair cell
Hensen's cells
Inner hair cell
Outer sulcus
Inner
sulcus
Inner
Outer
Pillar cells
Deiters' cells
B
ohc
C
organ of Corti
inner sulcus

row with those in the next. Cx26-labelled junctions are also numerous between the Hensen's cells on the outer side of the organ of Corti, and between the cells that comprise the outer sulcus outwards from the Hensen's cells. In whole-mount preparations (Fig. 4C) Cx26 labelling occurs in lines between pillar cells (the cells forming the arch that separates the inner and outer hair cells) along their borders in the longitudinal direction. Each of these lines may represent single large gap junctions, since freeze-fracture shows that the junctions between the cell bodies of adjacent pillar cells are extremely large, perhaps occupying the greater part of the membrane that covers the cell body (Forge 1986). In the organ of Corti there was no positive labelling for either Cx32 nor Cx43, indicating that Cx26 protein is the only one of these three isoforms that is expressed in the auditory sensory epithelium.

The supporting cells of the organ of Corti are electrically coupled (Santos-Sacchi 1991) and coupling is sensitive to changes in intracellular pH but not to increased intracellular $Ca^{2+}$ levels except where these become abnormally high (Sato & Santos-Sacchi 1994). Dye transfer experiments using isolated segments of the organ of Corti have shown Lucifer yellow and 6-carboxyfluorescein injected into the Hensen's cells on the outer edge of the organ of Corti will spread quite rapidly both longitudinally to adjacent Hensen's cells, and radially, inwards into Deiters' cells (Santos-Sacchi 1986). However, in this study dye did not appear to spread across the region of the pillar cells towards the inner hair cell region. A recent study in which Lucifer yellow was injected *in vivo* into the basal cell body regions of Deiters' cells or pillar cells (Oberoi & Adams 1998) showed that dye injected into Deiters' cells (i.e. below the outer hair cells) spread radially outwards and longitudinally, but rarely spread radially inwards into pillar cells; on the other hand, when dye was injected into pillar cells it predominantly spread in the longitudinal direction and inwards into epithelial cells on the inner side of the organ of Corti but rarely spread outwards into Deiters' cells. This pattern of dye spread is essentially consistent with the observed pattern of Cx26

FIG. 4. The organ of Corti. (A) Diagram of the organ of Corti to show the different cell types. (B) Connexin26 (Cx26; $\beta_2$) labelling in a frozen section of guinea pig organ of Corti. Large puncta surround the bodies of Deiters' cells below the level of the outer hair cells (ohc) and appear to be organized in distinct lines at different levels in this region. Numerous puncta also surround the Hensen's cells (at right). Bar = 10 $\mu$m. (C) Cx26 labelling in a whole-mount preparation of gerbil organ of Corti as seen from a projection series obtained by confocal microscopy. The borders of the Hensen's cells at the right are delineated by labelled puncta. The region of the Deiters' cells (and outer hair cells) also shows puncta surrounding cell borders. At the level of the inner pillar cells, arrows point to puncta in lines at the borders between inner pillar cells in the longitudinal direction. There appear to be few puncta between pillar cells and the Deiters' cell region. The cells of the inner sulcus also show intense labelling for Cx26. Bar = 10 $\mu$m.

labelling described above, i.e. there is relatively little connexin labelling at the borders between the Deiters' and pillar cells, but extensive labelling between adjacent Deiters' cells, Deiters' cells and Hensen's cells, and between Hensen's cells and the cells of the outer sulcus. It suggests that there may be two separate routes for intracellular passage of molecules: one running from the pillar cells inwards and one from the Deiters' cells outwards towards the lateral wall of the cochlea. Supporting cells in the organ of Corti possess inward $K^+$ currents (Santos-Sacchi 1991) and a fast transjunctional conductance between at least isolated supporting cells pairs that is dependent upon membrane voltage (Zhao & Santos-Sacchi 1997). These observations support the notion that supporting cells play a role in $K^+$ buffering, such that during sound-evoked activity, as $K^+$ passes into perilymph through the hair cells, it is rapidly removed from the fluids to maintain homeostasis and auditory activity, with gap junctions providing an intracellular route for dissipation of $K^+$ away from the sensory region and ultimately to the perilymphatic compartment at a more distant site. Although there are no similar studies of the vestibular sensory epithelia, it seems reasonable to conjecture that supporting cells in the vestibular sensory organs may also provide $K^+$ sinks. The syncitial nature of the supporting cell populations would be of importance to this activity.

However, gap junctions between the supporting cells may be involved in other activities. When hair cells are killed — for example, by noise, which affects auditory hair cells, or aminoglycoside antibiotics, which are toxic to hair cells in the cochlea and the vestibular system — the supporting cells expand into the region of the dying hair cell to close the lesion in a seemingly controlled manner, such that there is no functional disruption at the epithelial surface during hair cell loss (Forge 1985, McDowell et al 1989, Li et al 1995). We have found that new gap junctions form between supporting cells where their lateral membranes become adjacent in the region of repair, both in the organ of Corti (McDowell et al 1989) and the vestibular sensory epithelia (A. Forge, unpublished observations 1997). This also occurs in regions of hair cell loss following noise trauma in the avian basilar papilla (auditory epithelium; A. Forge & Y. Raphael, unpublished observations 1997). At present it is not known whether the gap junctions formed at this time are composed of the same connexin type that is normally present in the tissue (Cx26) or whether other isoforms are expressed. Some preliminary data suggest that, in addition to mRNA for Cx26, mRNAs for Cx32 and Cx43 are present at low levels in the normal, undamaged organ of Corti as well as the vestibular sensory epithelia (J. Edwards & S. Casalotti, unpublished observations 1998). Following loss of hair cells there is presumably little $K^+$ flow through the epithelium, so the re-modulation of gap junctions may imply an additional role for intercellular coupling in these sensory tissues, perhaps in the repair and recovery processes that follow hair cell loss. This is of interest because there is good evidence for the regeneration of hair cells from supporting cells in the traumatized avian inner ear (Cotanche et al 1994). We have also found evidence for the production of new hair cells to replace those lost in the mammalian vestibular organs (Forge et al 1993). However, there is no evidence for hair cell recovery following losses in the organ of Corti. The formation of new gap junctions during the repair process following hair cell loss would potentially provide a pathway for the transfer of signalling molecules involved in repair, and it is therefore of interest to determine the changing patterns of connexin expression and cell coupling that may occur at this time.

## The gap junctions in the stria vascularis and cochlear connective tissues

The characteristics of endolymph that bathes the apical surface of the sensory tissues are maintained by the ion-transporting epithelia of the respective portions of the inner ear. In the vestibular system, the dark cell region, this is a single layer of cells with a typical morphology of ion-transporting cells. The basolateral membranes are extensively infolded and contain high levels of $Na^+/K^+$ ATPase and a $Na^+/K^+/Cl^-$ co-transporter (Wangemann & Schacht 1996). They are directly exposed to the perilymph in the extracellular spaces of the underlying connective tissue. It is thought that $K^+$ enters the cells directly from perilymph via the transporters and diffuses into endolymph through $K^+$ channels in the apical membrane. The stria vascularis, the ion-transporting epithelium of the cochlea, is structurally more complex than the vestibular dark cell region. It is formed of three cell types—the marginal, intermediate and basal cells—and encloses a complex capillary network (Fig. 5A). The stria not only maintains endolymphatic $K^+$, but is also responsible for the generation of the endocochlear potential. Acute damage to the stria vascularis by a number of agents, including loop diuretics such as furosemide and ethacrynic acid, correlates with rapid reversible inhibition of endocochlear potential (Pike & Bosher 1980, Brown & Forge 1982).

The marginal cells of the stria line the endolymphatic space and are similar to vestibular dark cells; their extensively infolded basolateral membrane possess $Na^+/K^+$ ATPase and a $Na^+/K^+/Cl^-$ co-transporter. The intermediate cells are enclosed within the body of the stria and are pigmented cells of neural crest origin (Steel & Barkway 1989). The two to three layers of elongated basal cells separate the stria from the underlying spiral ligament and are of mesenchymal origin. The spiral ligament is part of the perilymphatic compartment of the cochlea. A complex network of occluding junctions covers the plasma membranes where adjacent basal cells are in contact (Forge 1984) forming a barrier to paracellular diffusion between the extracellular spaces of the ligament and those of stria vascularis (Forge 1981).

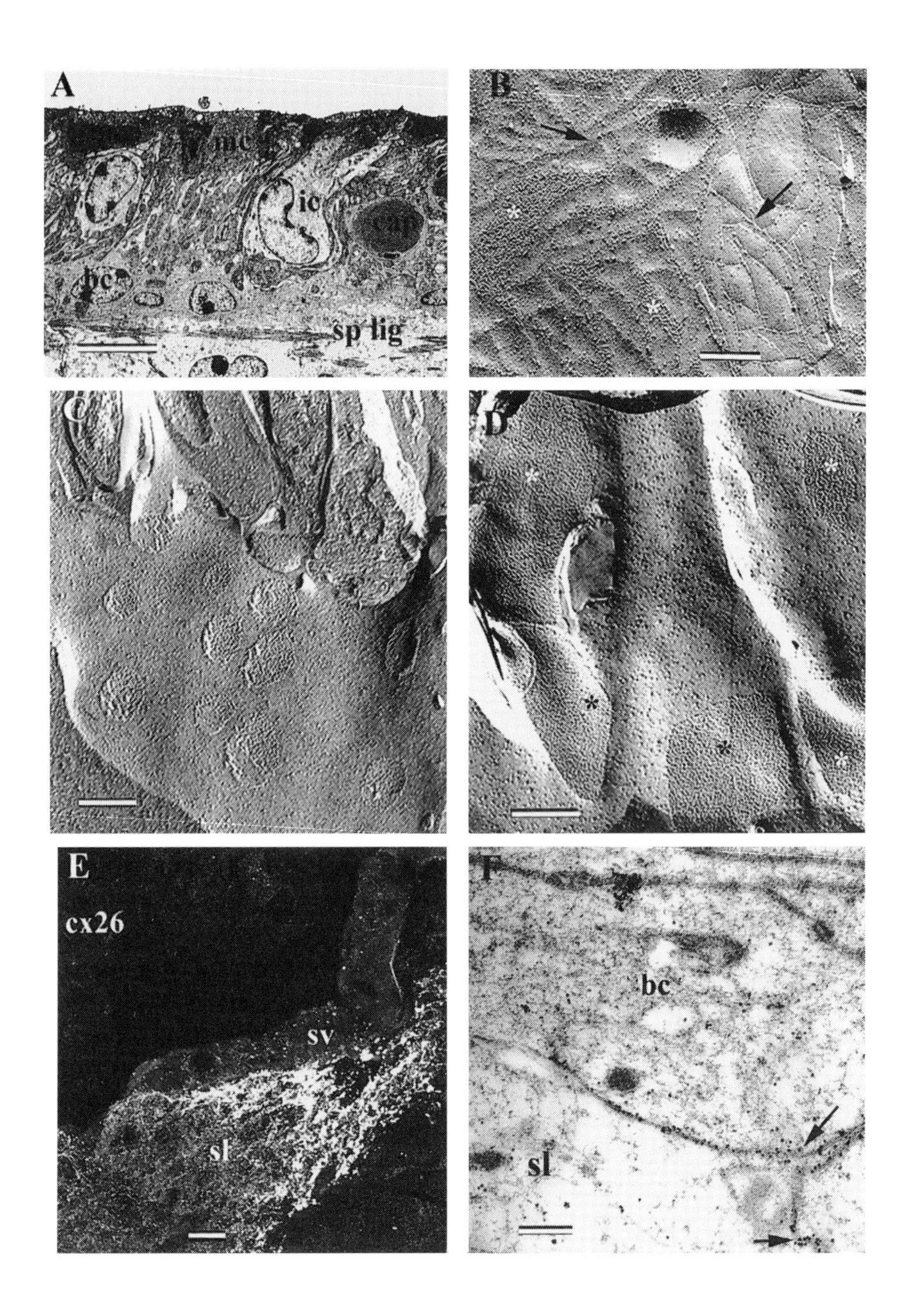
A
mc
ic
cap
bc
sp lig
B
C
D
E
cx26
sv
sl
F
bc
sl

Gap junctions are also associated with the strial basal cells. They are present between adjacent basal cells, localized between the strands that form the occluding junctions (Fig. 5B), both in the longitudinal direction along the length of the cochlea, as well as in the radial direction from the spiral ligament side towards the rest of the stria. Thus, all the basal cells appear to be coupled together so that they may form a functional syncitium. In addition, there are numerous gap junctions on the apical membrane of the basal cell that faces the rest of the stria (Fig. 5C). We have estimated that in the guinea-pig stria vascularis gap junctions occupy about 15% of the total surface area of this inward-facing plasma membrane of the basal cell (Forge 1984), about 16–17% in mice (Carlisle et al 1990) and gerbils (Souter & Forge 1998) and perhaps as much as 20% in bats (A. Forge & M. Vater, unpublished observations 1996). These junctions form predominantly between basal cells and intermediate cells. Careful morphological examinations have failed to reveal gap junctions at the marginal cell–basal cell contact (Kikuchi et al 1995). Furthermore, the gap junctions on the apical membrane of the basal cell are absent in the *viable dominant spotting* mutant mouse strain (Carlisle et al 1990) in which there is a neural crest defect and intermediate cells are absent from the stria. In these animals the marginal cells appear essentially normal morphologically (Carlisle et al 1990) and $Na^+/K^+$ ATPase is present on their basolateral membranes at normal levels (Schulte & Steel 1994). Other than the absence of the apical gap junctions, the basal cells also appear normal, and the gap junctions between adjacent basal cells are apparently unaffected. The absence of a specific population of basal cell gap junctions when intermediate cells are missing but marginal cells are present

---

FIG. 5. The stria vascularis and spiral ligament. (A) Thin section to show the anatomy of the stria vascularis. bc, basal cell; cap, blood capillary; ic, intermediate cell; mc, marginal cell; sp lig, spiral ligament below the stria vascularis. Bar = 10 $\mu$m. (B) Freeze-fracture of lateral membrane of basal cell facing an adjacent basal cell. Membrane face is covered with a network of strands (arrows) of the tight (occluding) junction. Numerous gap junctions (asterisks) are present in the islands between the strands. Bar = 250 nm. (C) Freeze-fracture of apical membrane of the basal cell that faces towards the rest of the stria. The cross-fractured strial cells are seen at the top of the figure. There are numerous gap junctional plaques on this membrane. In this example from the stria vascularis of a bat, the gap junctions occupy about 20% of the total plasma membrane area. Bar = 250 nm. (D) Basal membrane of the basal cell that faces fibrocytes in the spiral ligament. Gap junctions (asterisks) are again large and numerous on this membrane. Bar = 250 nm. (E) Connexin26 (Cx26; $\beta_2$) labelling in the lateral wall of a gerbil. sl, spiral ligament; sv, stria vascularis. Intense labelling amongst cells in the spiral ligament. There is little Cx26 labelling in the stria vascularis except in a fine line at the extreme basal edge of the tissue that is visible where the stria has separated from the underlying ligament during preparation. Bar = 50 $\mu$m. (F) Thin section of basal region of gerbil stria vascularis, immunogold labelled using antibodies to Cx26. A junction between a basal cell (bc) and a fibrocyte in the spiral ligament (sl) is labelled (upper arrow) as is a junction between two cells in the spiral ligament (arrow at extreme bottom right). Other junctions between strial basal cells, at the top of the figure, are not labelled. Bar = 500 nm.

and apparently normal suggests that ordinarily the intermediate and basal cells are coupled but marginal cells do not form gap junctions with basal cells. There are no gap junctions associated with the lateral membranes of the marginal cells either, indicating that marginal cells are coupled neither to each other nor to intermediate nor capillary endothelial cells. It thus appears that the intermediate and basal cells form a coupled unit separate from the marginal cells, which are not directly coupled to other cells. A recent study (Ando & Takeuchi 1998) has suggested that dye injected into strial intermediate cells can spread to basal cells but does not enter marginal cells, in line with this apparent distribution of gap junctions.

Not only are basal cells coupled to each other and to strial intermediate cells, but gap junctions are also present between basal cells and fibrocytes in the spiral ligament (Fig. 5D,F). Fibrocytes in the ligament are themselves also coupled by gap junctions (Fig 5E,F). There is therefore the potential for intracellular communication between the spiral ligament cells and the basal and intermediate cells of the stria. The membranes of the fibrocytes also possess $Na^+/K^+$ ATPase (Schulte & Adams 1989) and, thus, may take up $K^+$ from the perilymphatic environment in which they are bathed. Several lines of evidence suggest that the maintenance of endocochlear potential is dependent upon $K^+$ derived from perilymph (reviewed in Wangemann & Schacht 1996), and that $K^+$ may circulate from endolymph to perilymph and back through the stria. For example, there is a rapid decline in endocochlear potential when the perilymphatic spaces are perfused with a $K^+$-free solution, but it is maintained when $K^+$-free solutions are perfused through the strial vasculature (Marcus et al 1981). It has also been demonstrated that the intracellular potential of strial marginal cells is abnormally high and similar to, or a little higher than, that of cochlear endolymph itself (see Wangemann & Schacht 1996). In current models of endocochlear potential generation (Salt et al 1987, Wangemann & Schacht 1996) it is envisaged that a high positive potential is present in the fluid of the intercellular spaces of the stria which bathes the basolateral membranes of the marginal cells. These cells are proposed to function in a manner essentially identical to that of vestibular dark cells actively taking up $K^+$ and extruding it into endolymph, whereas the endocochlear potential is generated by electrogenic $K^+$ secretion from basal or intermediate cells or the basal/intermediate cell complex into the intercellular spaces of the stria. The gap junctions associated with the basal cells would therefore have a crucial role in providing an intracellular route for entry of $K^+$ from perilymph into basal cells from spiral ligament fibrocytes, a pathway that bypasses the occluding junctions between basal cells.

In support of this model, the *viable dominant spotting* mouse mutant, in which there is no intermediate cell, does not generate an endocochlear potential (Steel et al 1987), although, as described above, marginal cells appear normal, indicating that the intermediate cell or the basal cell/intermediate cell coupled unit is important for endocochlear potential generation. That the gap junctions have an important role in endocochlear potential generation has been suggested from studies of cochlear development (Souter & Forge 1998). In altricial mammals, such as mice, rats and gerbils, which are born deaf, the final stages of cochlear maturation occur postnatally. It is thus possible to follow the course of structural maturation with that of the establishment of physiological activities. In gerbils, in which auditory function is fully mature about 21 days after birth, endocochlear potential can first be recorded on postnatal day (P) 8. It increases to *c.* 15 mV by P12 and then rises at a rate of *c.* 8.5 mV a day to reach the adult level of *c.* 80 mV by P20. Marginal cells have developed basal infoldings and possess $Na^+/K^+$ ATPase activity at close to adult levels by P6, that is before an endocochlear potential can be recorded. It has been shown in rats that $K^+$ levels in endolymph are like those of the adult some days before the onset of endocochlear potential (Bosher & Warren 1971) and, thus, the maturation of marginal cells may be related to the establishment of the ionic composition of endolymph rather than to endocochlear potential generation. Pigmented cells, the presumptive intermediate cells, also become closely associated with the maturing marginal cells by P6 and basal cells can be identified by this time forming in layers below the pigmented cells. Tight junctional elcments first appear on basal cell membranes at P6 and are quite well developed by P10 (Table 1), i.e. they appear before endocochlear potential is recordable and are well developed while endocochlear potential is still relatively low. The first appearance of endocochlear potential coincides temporally with the first appearance of gap junctions on basal

**TABLE 1 Maturation of junctional specializations of basal cells in the gerbil stria vascularis in relation to onset and rise in endocochlear potential**

| | | | *Gap junction maturation* | |
|---|---|---|---|---|
| *Age* | *Endocochlear potential mean level (mV)* | *Tight junction maturation (no. strands/μm²)* | *Mean area (μm²)* | *Area as % plasma membrane* |
| P6 | 0 | 34.7 | | |
| P8 | 2 | 35.4 | 0.02 | 6.2 |
| P10 | 5 | 72.6[a] | 0.04 | 9.2 |
| P12 | 15 | 118.4[a] | 0.09[a] | 12.7 |
| P16 | 60 | 154.3[a] | 0.25[a] | 26.8 |
| Adult | 80 | 167.5 | 0.33 | 16.9 |

[a]Significant difference from previous age point.

cell membranes at P8, and the subsequent rise in endocochlear potential coincides closely with an increase in the size, number and area of membrane occupied by gap junctional plaques on basal cell plasma membranes (Table 1). These observations suggest that the initial generation of endocochlear potential is dependent not upon the marginal cells, but first upon the formation of tight junctional sealing between basal cells which will presumably isolate the intrastrial spaces from the extracellular spaces of the ligament. The subsequent actual onset and rise in endocochlear potential is associated with the formation and development of the gap junctional coupling associated with basal cells.

Endocochlear potential, then, appears to be dependent upon the circulation of $K^+$ via cell coupling mediated by gap junctions in the spiral ligament and basal cells of the stria. We have found that antibodies to Cx26 provide intense labelling of cells in the spiral ligament in gerbils, mice and guinea-pigs (Fig. 5E), as has also been shown in rats (Kikuchi et al 1995), but neither Cx32 nor Cx43 label in this tissue. Thus, Cx26 appears to be the predominant connexin isoform in the cochlear lateral wall as it is in the organ of Corti. Cx26 also appears to be the connexin isoform associated at least with those gap junctions present between the strial basal cells and the fibrocytes of the spiral ligament (Fig. 5F). In our studies, however, Cx26 antibodies have not provided labelling within the basal cell layer of the stria in any of the species we have examined, but antibodies to Cx32 have labelled within this region of the gerbil stria. Cx26 has been shown to be present in the basal cell layer of the stria in rats, and immunocytochemistry applied to thin sections for electron microscopy suggested that in this species the junctions between basal and intermediate cells contain Cx26 (Kikuchi et al 1995).

## Summary: possible functional consequences of Cx26 mutations

Gap junctions are ubiquitous in the tissues of the inner ear, and Cx26 is the predominant connexin isoform in both the auditory and vestibular organs. The hearing loss suffered by patients with mutations in the Cx26 gene is classified as severe–profound (Denoyelle et al 1997, Estivill et al 1998), indicating significant functional impairment along the length of the cochlea. Vestibular function does not seem to have been tested in these patients, but the absence of reports of balance disorders suggests that there are no overt symptoms of vestibular dysfunction. Balance derives from multiple sensory inputs, and disturbances of vestibular organ function can be compensated for by the other elements. Although this may mask a vestibular phenotype for the Cx26 mutation, it may also be that the hearing organ is more severely affected than the vestibular system. In the vestibular organs the expression of additional connexin proteins might allow compensation for defects in Cx26, whereas in the cochlea Cx26 may be the only isoform expressed. However, it needs to be substantiated that additional connexin proteins are normally expressed in the vestibular organs, and that isoforms other than those which so far have been investigated are absent from the cochlea. Perhaps of greater significance is that hearing sensitivity is dependent upon the high extracellular electrical potential that is present uniquely in the cochlea. Endocochlear potential appears to be maintained by a local circulation of $K^+$, from endolymph to perilymph and back to the stria vascularis, that depends upon cell coupling by gap junctions composed of Cx26. It is likely that a predominant effect of Cx26 mutations is to impede that circulation and thus inhibit the generation of endocochlear potential. This proposal would explain any differences between the hearing organ and the vestibular system in sensitivity to the connexin defects, but, in light of the fact that Cx26 is expressed in many tissues besides the inner ear, does not answer the question of why hearing loss is the only clinical manifestation of Cx26 mutations.

*Acknowledgements*

Work reported here was supported by the Wellcome Trust (A.F.), the UK Medical Research Council (A.F., M.S., W.H.E.) and the George John Livanos Charitable Trust (A.F., S.C.). D.B. is a Royal Society University Research Fellow. We are grateful to Stephen Forge for the drawings in Fig. 2 and Fig. 3A.

## References

Ando M, Takeuchi S 1998 Dye-coupling of intermediate cells with endothelial cells, pericytes and basal cells in the stria vascularis of gerbils. Abs Assoc Res Otolaryngol 25:192 (abstr 765)

Bosher SK, Warren RL 1971 A study of the electrochemistry and osmotic relationships of the cochlear fluids in the neonatal rat at the time of the development of the endocochlear potential. J Physiol (Lond) 212:739–761

Brown AM, Forge A 1982 Ultrastructural and electrophysiological studies of the acute ototoxic effects of furosemide. Br J Audiol 16:109–116

Carlisle L, Steel K, Forge A 1990 Endocochlear potential generation is associated with intercellular communication in the stria vascularis: structural analysis in the *viable dominant spotting* mouse mutant. Cell Tiss Res 262:329–337

Cotanche DA, Lee KH, Stone JS, Picard DA 1994 Hair cell regeneration in the bird cochlea following noise damage or ototoxic drug damage. Anat Embryol 189:1–18

Dallos P 1996 Overview: cochlear neurobiology. In: Dallos P, Popper AN, Fay RR (eds) The cochlea. Springer-Verlag, New York, p 1–43

Denoyelle F, Weil D, Maw MA et al 1997 Prelingual deafness: high prevalence of a 30delG mutation in the connexin 26 gene. Hum Mol Genet 6:2173–2177

Estivill X, Fortina P, Surrey S et al 1998 Connexin-26 mutations in sporadic and inherited sensorineural deafness. Lancet 351:394–398

Forge A 1981 Electron microscopy of the stria vascularis and its response to etacrynic acid. A study using electron-dense tracers and extracellular surface markers. Audiol 20:273–289

Forge A 1984 Gap junctions in the stria vascularis and effects of ethacrynic acid. Hear Res 13:189–200

Forge A 1985 Outer hair cell loss and supporting cell expansion following chronic gentamicin treatment. Hear Res 19:171–182

Forge A 1986 The morphology of the normal and pathological cell membrane and junctional complexes of the cochlea. In: Salvi RJ, Henderson D, Hamernik RP, Colletti V (eds) Basic and applied aspects of noise-induced hearing loss. Plenum, New York, p 55–67

Forge A, Li L, Corwin JT, Nevill G 1993 Ultrastructural evidence for hair cell regeneration in the mammalian inner ear. Science 259:1616–1619

Forge A, Becker D, Evans WH 1997 Gap junction connexin isoforms in the inner ear of gerbils and guinea pigs. Br J Audiol 31:76–77

Ginzberg RD, Gilula NB 1979 Modulation of cell junctions during differentiation of the chicken otocyst sensory epithelium. Dev Biol 68:110–129

Kelsell DP, Dunlop J, Steven HP et al 1997 Connexin-26 mutations in hereditary non-syndromic sensorineural deafness. Nature 387:80–83

Kikuchi T, Adams JC, Paul DL, Kimura RS 1994 Gap junctions in the rat vestibular labyrinth: immunohistochemical and ultrastructural analysis. Acta Otolaryngol 114:520–528

Kikuchi T, Kimura RS, Paul DL, Adams JC 1995 Gap junctions in the rat cochlea: immunohistochemical and ultrastructural analysis. Anat Embryol 191:101–118

Li L, Nevill G, Forge A 1995 Two modes of hair cell loss from the vestibular sensory epithelia of the guinea pig. J Comp Neurol 355:405–417

Marcus DC, Marcus NY, Thalmann R 1981 Changes in cation contents of the stria vascularis with ouabain and potassium-free perfusion. Hear Res 4:149–160

McDowell B, Davies S, Forge A 1989 The effect of gentamicin-induced hair cell loss on the tight junctions of the reticular lamina. Hear Res 40:221–232

Oberoi P, Adams JC 1998 *In vivo* measurements of dye coupling among non-sensory cells in the organ of Corti. Abs Assoc Res Otolaryngol 25:191 (abstr 762)

Pike DA, Bosher SK 1980 The time course of strial changes produced by intravenous furosemide. Hear Res 3:79–89

Salt AN, Melichar I, Thalmann R 1987 Mechanisms of endocochlear potential generation by stria vascularis. Laryngoscope 97:984–991

Santos-Sacchi J 1986 Dye coupling in the organ of Corti. Cell Tiss Res 245:525–529

Santos-Sacchi J 1991 Isolated supporting cells from the organ of Corti: some whole-cell electrical characteristics and estimates of gap junctional conductance. Hear Res 52:89–98

Sato Y, Santos-Sacchi J 1994 Cell coupling in the supporting cells of Corti's organ: sensitivity to intracellular $H^+$ and $Ca^{2+}$. Hear Res 80:21–24

Schulte BA, Adams JC 1989 Distribution of immunoreactive $Na^+$, $K^+$-ATPase in gerbil cochlea. J Histochem Cytochem 37:127–134

Schulte BA, Steel KP 1994 Expression of $\alpha$ and $\beta$ subunit isoforms of Na,K-ATPase in the mouse inner ear and changes with mutations at the $W^v$ or $Sl^d$ loci. Hear Res 78:65–76

Souter M, Forge A 1998 Intercellular junctional maturation in the stria vascularis: possible association with onset and rise of EP. Hear Res, in press

Steel KP, Barkway C 1989 Another role for melanocytes: their importance for normal stria vascularis development in the mammalian inner ear. Development 107:453–463

Steel KP, Barkway C, Bock GR 1987 Strial dysfunction in mice with cochleo-saccular abnormalities. Hear Res 27:11–26

Wangemann P, Schacht J 1996 Homeostatic mechanisms in the cochlea. In: Dallos P, Popper AN, Fay RR (eds) The cochlea. Springer-Verlag, New York, p 130–185

Zelante L, Gasparini P, Estivil X et al 1997 Connexin 26 mutations associated with the most common form of non-syndromic neurosensory autosomal recessive deafness (DFNB1) in Mediterraneans. Hum Mol Genet 6:1605–1609

Zhao HB, Santos-Sacchi J 1997 Fast voltage dependence of gap junctions in cochlear supporting cells. Soc Neurosci Abs

## DISCUSSION

*Nicholson:* How much testing has been done on these mutants to determine whether they are functional defects of connexin26 (Cx26; $\beta_2$)?

*Forge:* We haven't done the analyses yet, but what we're hoping to do is to express these mutations in HeLa cells to see if they affect function and in what way. One of the reasons why some of those mutations might cause hearing loss is because an extremely efficient transfer system is required for ion circulation in the inner ear, and if the capacity to transfer decreases by some fraction it has more dramatic effects on the ear than it would in other systems.

*Gilula:* I would like to ask Nick Lench to comment on the clinical relevance of gap junction expression in the ear.

*Lench:* Even though the most common mutation in Cx26 is 35ΔG, which knocks out almost all of the protein apart from a 13 amino acid peptide, there are a large number of cases where individuals are compound heterozygotes, i.e. they have two separate Cx26 mutations. In a proportion of compound heterozygotes we can identify a Cx26 mutation in the coding region on one allele but we haven't been able to identify the second mutation. It's not in the coding region, which begs the question whether there are promoter mutations or mutations in a second connexin gene nearby on chromosome 13, such as Cx46 ($\alpha_3$). We are interested in the transcription factors driving the transcription of these connexin genes. We have some preliminary evidence that there is a deletion of 10 bp in the 5′ region of the Cx26 gene in some patients, but we don't have any data yet to prove that this is a functional promoter sequence. Finally, the clinical phenotypes do vary, so individual sib pairs with the same mutation have different levels of hearing loss.

*Beyer:* Are there dominant pedigrees as well?

*Lench:* The original paper in *Nature* (Kelsell et al 1997) reported that there was a dominant pedigree and three recessive pedigrees, and we have looked principally at recessive pedigrees. In the dominant pedigree an M34T amino acid substitution was identified in the first transmembrane domain that segregated with individuals who had profound sensorineural hearing loss. As I understand, this mutation has also been identified in the Cx32 ($\beta_1$) gene in patients with Charcot-Marie-Tooth disease (Tan et al 1996). However, Scott et al (1998) and Kelley et al (1998) have identified individuals with the same amino acid substitution who had normal hearing, arguing against that substitution being a causative mutation. The dominant pedigree that originally defined the locus on chromosome 13 was identified by Chaib et al (1994), but I'm not confident that they have proven there is a Cx26 mutation in that pedigree, so the issue of whether the M34T mutation is a causative mutation is still open to debate. David Paul has shown that in *Xenopus* if a wild-type Cx26 is mixed with the M34T mutation channel activity is abolished, arguing in favour of it being a dominant negative mutation.

*Scherer:* Although *Xenopus* oocytes may be optimal for looking at the conductances of mutant channels, they are not ideal for looking at protein trafficking, which may be the more relevant issue in terms of pathophysiology. It is crucial to examine the expression of these mutants in mammalian cells, such as HeLa cells, so that you can determine whether their trafficking is affected.

*Goodenough:* Do you have any autopsy material that allowed you to look at the general structure of the organ of Corti to see whether it had undergone normal development in people affected with severe hearing loss?

*Lench:* The problem is that there's no ethical reason why we should obtain material from a deaf patient.

*Gilula:* Can you obtain any biopsy material from these patients to look at the distribution of protein in other tissues.

*Lench:* We have managed to get a skin biopsy from one patient. We are now collaborating with dermatologists to look at Cx26 expression.

*Vaney:* You mentioned that sensorineural deafness is the only clinical manifestation of the Cx26 mutations. Has anyone looked for subtle effects on retinal function by psychophysical testing of visual perception?

*Forge:* I don't think so.

*Lench:* The original pedigree that Kelsell et al (1997) investigated had sensorineural deafness associated with palmoplantar keratoderma, a condition characterized by hyperkeratinization of the palms and soles. They first thought that the two conditions were co-segregating, but it turned out that only the individuals who were profoundly deaf had the M34T mutation.

*Yeager:* Has anyone checked whether individuals who have no hearing loss but who do have the M34T mutation in Cx26 also have a compensatory mutation that restores function?

*Lench:* I don't know. As far as I'm aware in the two studies I know of (Scott et al 1998, Kelley et al 1998) DNA from 292 random normal-hearing individuals was screened with allele-specific oligonucleotides, and four had the M34T amino acid substitution. However, one deaf individual also had the M34T mutation and a second mutation on the other allele, although obviously that is not a compensatory mutation (Kelley et al 1998).

*Dermietzel:* Cx26 has the same pattern of staining in the vestibular organ. Do the patients, therefore, have any defects in the vestibular organ?

*Forge:* Vestibular dysfunction has not been reported. The primary defect may be related to the production of the endocochlear potential because the vestibular system doesn't have a similar high endolymphatic potential.

*Goodenough:* Isn't the endolymph continuous between the organ of Corti and the vestibular system?

*Forge:* Yes it is continuous, but it is locally produced. The endocochlear potential is specific to the cochlea, so the most likely defect is in the pathway that produces the endocochlear potential, which is why there may not be a vestibular problem. There are also more connexin isoforms in the vestibular system than in the cochlea.

*Gilula:* Klaus Willecke has knocked out this gene in the mouse. Does it have any effects on hearing loss?

*Willecke:* The trouble with the Cx26-deficient mice is that they die at around 11 days post coitum, which is probably before hearing defects would be apparent. As far as I know, no one has looked at this.

*Lo:* It may be possible to bypass this by generating aggregation chimeras between the homozygous Cx26 knockout and wild-type embryos. If the defect is placental then the chimeric placenta would be derived from the wild-type, and the rest of the animal would be mixed, depending on the contribution of cells from the homozygous knockout versus the wild-type embryos. However, an easier way to do this would be to take embryonic stem cells and do a second round of selection, so that you generate homozygous null mutant embryonic stem cells. These can be injected into blastocysts, from which embryos of mice can be derived. If you chose the right strain combination between the recipient blastocysts and the embryonic stem cells, most of the animal would be derived from the embryonic stem cell genotype, so that the animals would basically be homozygous knockouts, but they may survive because the placenta has the wild-type genotype derived from the recipient blastocysts. The main worry with this technique would be that the embryonic stem cells could change and lose their ability to give rise to normal development after prolonged culturing for the second round of selection.

*Gilula:* Some of the perspectives we will gain from Steve Scherer's talk will also be relevant to this, because there is much we need to know about these mutations and the behaviour of the products before we invest the time and resources in developing these sorts of experiments.

*Goodenough:* One of the unusual aspects of the stria vascularis is that it has blood vessels within the epithelium. In the *viable dominant spotting* mutant mouse strain did you see any differences in the pattern of the vasculature?

*Forge:* As far as we could tell, because the vascular network is extremely complex in this region, there were no obvious changes between the mutant and wild-type. The only defect was the loss of the intermediate cells.

*Gilula:* I have a question about the progress in this field. Many years ago there was a real challenge to do organ culture, particularly for pharmaceutical reasons because some of these cells are sensitive to antibiotics. Is it now possible to obtain organ of Corti cultures, and to what extent is this a relevant approach for people like you who might be interested in studying gap junction relationships?

*Forge:* We do have organotypic cultures of adult vestibular organs, but the adult organ of Corti is more difficult to grow in culture because the hair epithelium is more sensitive and falls apart. However, we can do a certain amount with early postnatal organs of Corti in mice, rats and gerbils, which can be maintained for at least two weeks in culture. We have to remove them before the intercellular spaces form, so the epithelium is compact. There is evidence that the outer hair cells normally transduce, so we have been using them to study aminoglycosides and ototoxicity. Therefore, we could use those organ culture systems to look at some aspects of gap junction communication, for example during repair. The stria is more difficult to culture. Some people have tried growing it, but it doesn't last for long, so it could only be used for short-term experiments.

*Scherer:* There is a mouse knockout of the potassium channel *Isk* that causes deafness (Vetter et al 1996). Since potassium is involved in the cochlear potential, where does it localize relative to Cx26, and how does it contribute to the cochlear potential?

*Forge:* That's a good question, but I'm not sure I can answer it in any detail. In the *isk*$^{-/-}$ mouse there is certainly a defect in the potassium channel in the apical membrane of the marginal cells of the stria, but it is presumably involved in potassium transport as opposed to endocochlear potential.

*Lench:* The *isk*$^{-/-}$ mouse is deaf (Vetter et al 1996), amongst other things, but mutations in the *ISK* gene in humans cause the Jervell and Lange-Nielsen syndrome (Tyson et al 1997). People with this syndrome have sensorineural deafness and heart defects that are similar to those with the long QT-type syndrome. We have looked at patients from families with recessive deafness for mutations in the *ISK* gene but haven't found any, so it appears that mutations in this particular potassium channel aren't a common cause of deafness.

*Sanderson:* It seems that the Cx26 gap junction channel is highly specialized in terms of potassium transfer, but potassium is one of the smallest things that can go through gap junctions. Do you know anything about Cx26 that makes it selective for this role?

*Nicholson:* Cx26 is one of many connexins that are at the cation selective end of the spectrum, and it's not particularly good at transferring large molecules, although most of the larger molecules that people have tried to put through, such as Lucifer yellow, are anionic, so it's not clear how much of that selectivity is against size or against charge (Cao et al 1998). I would say that there are other connexins which, by the same criteria, would be equally as selective. For example, Cx45 ($\alpha_6$) is highly selective and doesn't let larger molecules through.

*Gilula:* But in general, isn't the Cx26 channel less dye permeable than the other channels?

*Nicholson:* The Cx45 channel is actually less dye permeable, but Cx26 is also poor.

*Gilula:* An important and interesting question that stems from Michael Sanderson's question and is relevant to the ear is that the use of Cx26 may have a direct relationship to the physiological requirement for a channel with a selective ion preference, such as potassium.

*Nicholson:* The Cx32 channels are much better at transferring dye, so one could argue that the pore is larger. However, it has a conductance of 55 pS, whereas Cx26 has a conductance of 120 pS (Bukauskas et al 1995, personal observations 1998), suggesting that there may be something about the Cx26 channel that facilitates the passage of smaller cations.

*Gilula:* Klaus Willecke, is that consistent with your perspective on the characterization of Cx26?

*Willecke:* Yes. As I mentioned in my presentation, homozygous Cx26-deficient mice die from a defect in the placenta (Gabriel et al 1998). In the wild-type mouse there is a transfer of nutrients across the placenta from the maternal circulation to the fetal circulation, and a transfer of waste products in the opposite direction. Our notion is that in the placenta glucose and ions are transferred via Cx26-containing gap junction channels, but not second messengers because their half-lives are too short.

*Forge:* Am I correct in saying that Cx26 doesn't have a regulatory region because it doesn't have a long C-terminal tail?

*Gilula:* There is some evidence that the C-terminal domains of some connexins have a role in regulating the activity of the channel.

*Nicholson:* And most of that domain is absent in Cx26, including the phosphorylation sites that are known to regulate the other connexins. However, the unique regions have been poorly investigated, so it is not known how much they may play a role. It is fair to say that to date we have not identified any regulatory regions in Cx26, but we probably have not looked hard enough.

*Forge:* If in the ear an extremely efficient ion transfer is required, i.e. the channel is locked open so that ions can pass almost as fast as auditory frequency, would Cx26 have properties that would make it more suitable for this function than the other connexins?

*Nicholson:* It's possible. One interesting point is that it is probably the most voltage insensitive channel: the $V_0$ for voltage closure is 90 mV (Barrio et al 1991), and the next most insensitive channel is 60 mV.

*Sanderson:* It is possible that the large plaques of Cx26 in the membrane affect membrane properties. Therefore, the connexin may have to be of a certain form to be able to make these large plaques.

*Gilula:* The Cx26 plaques Andy Forge showed are certainly amongst the largest ever seen outside the plaques that Joerg Kistler has seen in the lens.

*Kistler:* The lens is a contrasting example, because there are huge plaques and they are composed of the two largest known connexins.

## References

Barrio LC, Suchyna T, Bargiello T et al 1991 Gap junctions formed by connexin 26 and 32 along and in combination are differently affected by applied voltage. Proc Natl Acad Sci USA 88:8410–8414 (erratum: 1992 Proc Natl Acad Sci USA 89:4220)

Bukauskas FF, Elfgang C, Willecke K, Weingart R 1995 Heterotypic gap junction channels (connexin26–connexin32) violate the paradigm of unitary conductance. Pflugers Arch Eur J Physiol 429:870–872

Cao FL, Eckert R, Elfgang C et al 1998 A quantitative comparison of connexin-specific permeability differences of gap junctions to dyes of different charge. J Cell Sci 111:31–43

Chaib H, Lina-Granade G, Guilford P et al 1994 A gene responsible for a dominant form of neurosensory non-syndromic deafness maps to the *NSRD1* recessive deafness gene interval. Hum Mol Genet 3:2219–2222

Gabriel HD, Jung D, Bützler C et al 1998 Transplacental uptake of glucose is decreased in embryonic lethal connexin26-deficient mice. J Cell Biol 140:1453–1461

Kelley PM, Harris DJ, Comer BC et al 1998 Novel mutations in the connexin 26 gene (*GJB2*) that cause autosomal recessive (DFNB1) hearing loss. Am J Hum Genet 62:792–799

Kelsell DP, Dunlop J, Stevens HP et al 1997 Connexin 26 mutations in hereditary non-syndromic sensorineural deafness. Nature 387:80–83

Scott DA, Kraft ML, Stone EM, Sheffield VC, Smith RJ 1998 Connexin mutations and hearing loss. Nature 391:32

Tan CC, Ainsworth PJ, Hahn AF, MacLeod PM 1996 Novel mutations in the connexin 32 gene associated with X-linked Charcot-Marie-Tooth disease. Hum Mutat 7:167–171

Tyson J, Tranebjaerg L, Bellman S et al 1997 *Isk* and *KvLQT1*: mutation in either of the two subunits of the slow component of the delayed rectifier potassium channel can cause Jervell and Lange-Nielsen syndrome. Hum Mol Genet 6:2179–2185

Vetter DE, Mann JR, Wangemann P et al 1996 Inner ear defects induced by null mutation of the *isk* gene. Neuron 17:1251–1264

# Gap junction-mediated communication in the developing and adult cerebral cortex

Bagirathy Nadarajah and John G. Parnavelas*[1]

*Department of Anatomy and Neurobiology, Washington University School of Medicine, St. Louis, MO 63110, USA, and *Department of Anatomy and Developmental Biology, University College London, Gower Street, London WC1E 6BT, UK*

*Abstract.* Recent cell biological and electrophysiological studies have shown that gap junctional coupling and the proteins that mediate this form of communication are present in the developing cerebral cortex from early in corticogenesis to the later stage of neuronal circuit formation. We have used electron microscopy to visualize gap junctions in the developing rat cerebral cortex, and studied the expression patterns and cellular localizations of connexin26 (Cx26; $\beta_2$), Cx32 ($\beta_1$) and Cx43 ($\alpha_1$), which take part in their formation. We found that these connexins are expressed differentially during development, and their patterns of expression are correlated with important developmental events such as cell proliferation, migration and formation of cortical neuronal circuits. We also observed that gap junctions and their constituent connexins were abundant in the adult cerebral cortex. Junctions were predominantly between glial cells or between neurons and glia. The frequency and distribution of gap junctions varied in different regions of the adult cortex, possibly reflecting differences in the cellular and functional organization of these cortical areas.

*1999 Gap junction-mediated intercellular signalling in health and disease. Wiley, Chichester (Novartis Foundation Symposium 219) p 157–174*

The mammalian cerebral cortex, although a structure of bewildering complexity, is characterized by a high degree of organization. It is parcelated into functionally different areas that show distinctive cytoarchitectonic features. All areas contain two main classes of neurons, the pyramidal and non-pyramidal cells, in roughly the same proportions (Rockel et al 1980). Two main classes of glial cells, the astrocytes and oligodendrocytes, are the supporting cells. One of the striking features of the cortex, when viewed in Nissl-stained sections, is the segregation

[1]This chapter was presented at the symposium by John G. Parnavelas, to whom correspondence should be addressed.

of its neurons into separate layers. Pyramidal and non-pyramidal cells in each layer show characteristic size, morphology and packing density. These layers differ further in the functional properties of the cells within them, in the inputs they receive and in the sites to which these cells project (Gilbert 1983). The cerebral cortex also shows a modular structure consisting of vertically organized systems of columns that contain cells with common physiological properties (Hubel & Wiesel 1977).

It is generally thought that intercellular communication in the cerebral cortex takes place through chemical transmission. However, a number of studies in the adult and developing cortex since the early 1980s have demonstrated extensive dye coupling between neurons. These electrophysiological investigations proposed roles for gap junctions in the formation of neuronal domains, which eventually lead to the formation of cortical columns (LoTurco & Kriegstein 1991, Peinado et al 1993a, Yuste et al 1995), and in interneuronal communication in the adult cortex (Gutnick & Prince 1981). Gap junctions, the morphological correlates of low resistance intercellular pathways and electrical coupling, are assemblies of membrane channels that provide a direct signalling pathway between contiguous cells (Bennett et al 1991). The structure of these membrane channels has been deduced by employing a number of methods, including thin-section and freeze-fracture electron microscopy, X-ray diffraction and atomic force microscopy (Revel & Karnovsky 1967, Brightman & Reese 1969, Hoh et al 1991, Unger et al 1999, this volume). In thin-section electron micrographs, gap junctions appear as regions of contact between adjacent cells where apposed plasma membranes are separated by a gap of 2–3 nm. Freeze-fracture preparations in combination with rotary shadowing of exposed surfaces have revealed the hexameric structure of the gap junction channel (Kuraoka et al 1993). A gap junction channel is formed by two hemichannels, or connexons, each composed of six connexin proteins arranged around a central pore. Connexins belong to the family of integral proteins and exhibit both highly conserved and divergent regions. Although the divergent regions may confer functional differences, the general topology of this class of proteins is thought to contain intracellular amino and carboxy termini, four transmembrane regions with an intracellular loop and two extracellular loops (Bennett et al 1991). The amino acid sequence of the carboxyl terminus and intracellular loop varies greatly among connexins, whereas the transmembrane regions and extracellular loop are highly conserved. Further, it is now known that gap junction hemichannels may be homomeric or heteromeric in their composition, implying that cells may express more than one connexin protein.

The use of cellular and molecular approaches has made it possible to localize connexin expression in the CNS. It is now known that connexin 26 (Cx26; $\beta_2$), Cx32 ($\beta_1$) and Cx43 ($\alpha_1$) are the major gap junctional proteins expressed in the brain, with Cx40 ($\alpha_5$) and Cx45 ($\alpha_6$) identified as minor forms (Dermietzel et al 1989, Dermietzel & Spray 1993). Cx43, the most ubiquitous gap junction protein in the brain, is expressed initially in undifferentiated ectodermal cells. As the brain vesicles become established following the closure of the neural tube, Cx26 and Cx43 are expressed in abundance (Dermietzel et al 1989, Ruangvoravat & Lo 1992, Yancey et al 1992). In the adult brain, Cx26 is restricted to non-neuronal cells of the ependyma, leptomeninges and pineal gland (Dermietzel & Spray 1993). Cx32 appears relatively late in certain regions of the developing brain, including the cerebral cortex (Dermietzel et al 1989, Micevych & Abelson 1991).

## Gap junctional communication during corticogenesis

Tritiated thymidine studies have shown that all neurons and some of the glial cells of the cortex arise from a seemingly homogeneous population of epithelial cells lining the lateral ventricles of the embryonic brain, the so-called ventricular zone (Rakic 1981). In the rat all neurons are born during the last week of gestation, i.e. embryonic day (E)14–E21 (Uylings et al 1990). The majority of glia are generated in a second proliferative layer further from the ventricular zone, the subventricular zone (Skoff & Knapp 1991). Newborn neurons migrate towards the margin of the cerebral wall (marginal zone; layer I of the adult cortex), using the processes of radial glial cells as guides, to form the cortex (Rakic 1981). They accumulate below the marginal zone to form the cortical plate (cortical layers II–VI) in an 'inside-out' order, that is to say the deepest cellular layers are formed first and the most superficial last (Fig. 1).

Early in corticogenesis, neuroepithelial cells of the ventricular zone are coupled into clusters by Lucifer yellow-permeable gap junctions; these clusters may be analogous to the 'proliferative units' described in primate neocortex (Rakic 1988, LoTurco & Kriegstein 1991). Although the precise role of cell clustering during corticogenesis is unclear, it is thought to spatially restrict the interactions between ventricular zone cells. The striking resemblance between the columnar organization of coupled cells and the observed radially arranged groups of clonally related cells in the rodent cerebral cortex (Luskin et al 1993, Mione et al 1997) has prompted speculation that coupled cells may be clonally related. However, experimental evidence in support of this notion is yet to be obtained. According to the radial unit hypothesis of Rakic (1988), the 'proliferative units' of the ventricular zone provide a proto-map of prospective cytoarchitectonic areas of the cortex. In such a model, the spatial positions of cohorts of cells in the ventricular zone may be specified by the exchange of diffusible signals through gap junctions, thereby enabling migrating cells to maintain their relative positions and contribute to the formation of areas in the cortex. The clusters contain radial glia

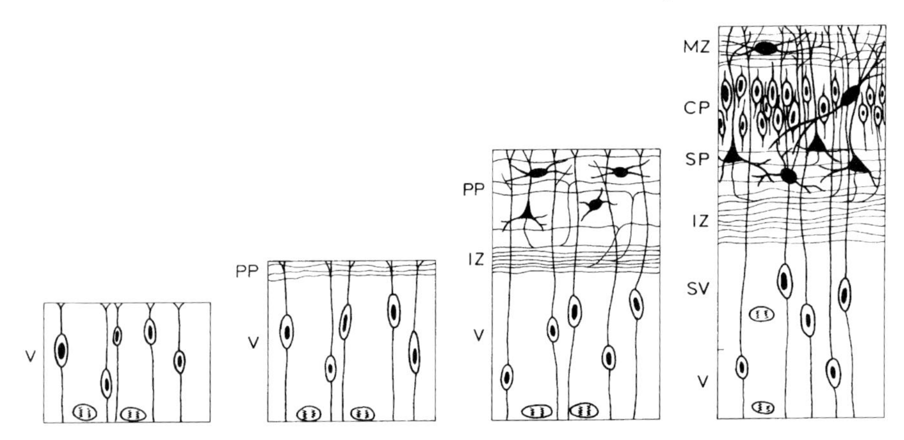


FIG. 1. Illustrations of cross-sections of the cerebral wall at different stages of development. Progenitor cells in the ventricular zone (V) proliferate to give rise to the neurons and some of the glial cells of the cortex. Postmitotic neurons migrate toward the pial surface along the processes of radial glia to form the primordial plexiform zone, or preplate (PP). Between the V and the PP is a layer of transverse fibres, the intermediate zone (IZ; future white matter). At later stages, a second proliferative layer appears, the subventricular zone (SV). Neurons of the cortical plate (CP) take their positions in the cortex in an inside-out manner between the outer part of the PP (the marginal zone, MZ, containing the Cajal-Retzius cells) and the lower part, the subplate (SP), also containing early-generated neurons (from Uylings et al 1990, with permission).

and neuronal precursors, but not differentiating or migrating neurons (Bittman et al 1997). Furthermore, coupling into clusters changes through the course of the cell cycle, suggesting that connexins may regulate cell division.

We have recently characterized the pattern of expression of connexins and the distribution of gap junctions in the proliferative regions of the developing rat cerebral cortex (Fig. 2; Nadarajah et al 1997). In the early stages of

---

FIG. 2. Expression of connexins in the developing dorsal telencephalic wall of rat embryonic brains. (a) Schematic representation of the pattern of distribution of connexin 26 (Cx26; $\beta_2$) and Cx43 ($\alpha_1$) immunoreactivities at various stages of corticogenesis. At embryonic day (E) 12, Cx26 was expressed throughout the neuroepithelium, whereas Cx43 was localized predominantly between cells bordering the ventricle. At E14–E16, both connexins showed increased expression throughout the telencephalic wall. At E19, Cx26 was more concentrated in the proliferative zones — the ventricular zone (VZ) and the subventricular zone (SVZ) — whereas Cx43 showed a more homogeneous expression through the thickness of the expanded telencephalic wall. CP, cortical plate; IZ, intermediate zone; MZ, marginal zone; SP, subplate. Measurements of immunoreactivity were performed in the VZ at E12–E16, and in both the VZ and SVZ at E19; dotted lines represent the upper limits of the areas measured. Bar = 80 $\mu$m. (b) Levels of immunoreactivity (%) of Cx26 and Cx43 measured in the proliferative zones of 12 embryonic brains at each age. The measured levels were corrected taking into account the radial expansion of the developing cortex. Error bars represent S.E.M.

corticogenesis, gap junctions were restricted between neighbouring epithelial cells bordering the ventricle, but at later stages they were observed in other zones of the cerebral wall, away from the ventricular surface. We found that Cx26 and Cx43 were the major gap junction proteins during neurogenesis. Cx26 labelling in the

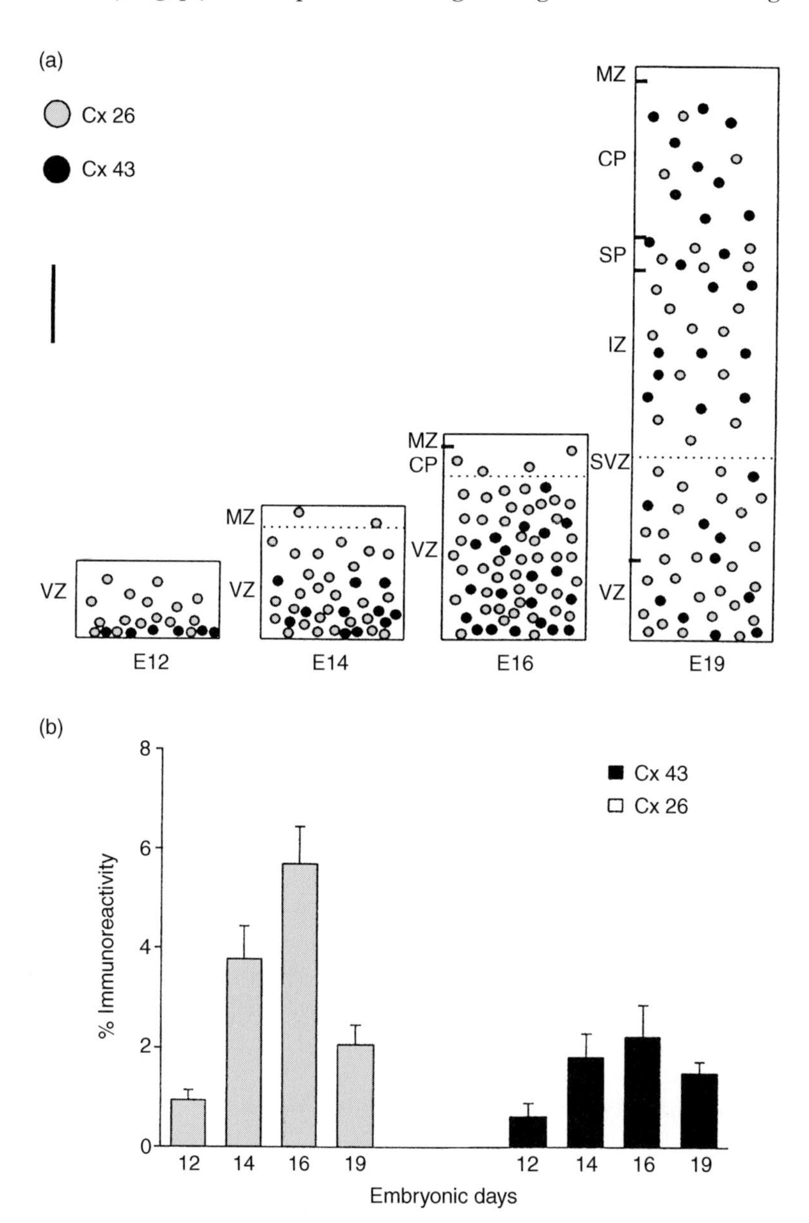

cortical neuroepithelium reached its highest levels at the peak of neurogenesis (E14–E16) and diminished thereafter, at a time when proliferation declined. This observation suggests that Cx26 may be involved in the control of cell generation in the cortex. Evidence that gap junctions are permeant to mitogens and morphogens (Guthrie & Gilula 1989) also supports the notion that they may play a role in cell proliferation (Lowenstein 1981). Immunocytochemistry with cell-specific markers indicated that, during the period of neurogenesis, Cx26 is expressed in cortical progenitor cells and in differentiated neurons, whereas Cx43 appears to be localized in radial glial fibres, often seen in contact with immature neurons migrating toward the cortical plate. The presence of Cx43 in radial glia during migration suggests that neurons establish contact and communication with the scaffolding they use to migrate to their positions in the cortex.

## Gap junctional communication during neocortical circuit formation and in the adult cerebral cortex

The cessation of coupling following the migration of neurons into the cortical plate, and the subsequent reappearance of dye coupling in the cortex during neuronal circuit formation suggests that gap junctions play different roles at different stages of development. Shortly after birth, when the number of synapses is low and synaptic activity sparse, gap junction-mediated intercellular communication could enable developing neurons to interact by a direct, non-synaptic mechanism. This notion has been supported by experiments in which intracellular injections of Lucifer yellow or Neurobiotin showed that 70–80% of rat cortical neurons engage in interneuronal coupling during the first two weeks of postnatal life (Connors et al 1983, Peinado et al 1993b). More recent imaging studies in young rats, using optical recordings of brain slices labelled with a fluorescent calcium indicator, revealed that the cortex is partitioned into distinct neuronal assemblies or 'domains' of spontaneously co-active neurons often arranged in vertical columns spanning several cortical layers (Yuste et al 1995). The persistence of these domains in the presence of the $Na^+$ channel inhibitor tetrodotoxin and their abolition by the gap junction blocker halothane, indicated that cells within a domain may be coupled by gap junctions. The spatial similarities between the Neurobiotin-labelled clusters (Peinado et al 1993b) and the optically recorded domains strongly suggested that both methods reveal the same organization of low resistance pathways between neocortical neurons mediated by gap junctions.

Using immunocytochemical and biochemical techniques in rats, we have demonstrated a differential expression of Cx26, Cx32 and Cx43 during postnatal development (Fig. 3; Nadarajah et al 1997). Cx26 showed high levels of expression during the first two weeks, subsequently declining until it could

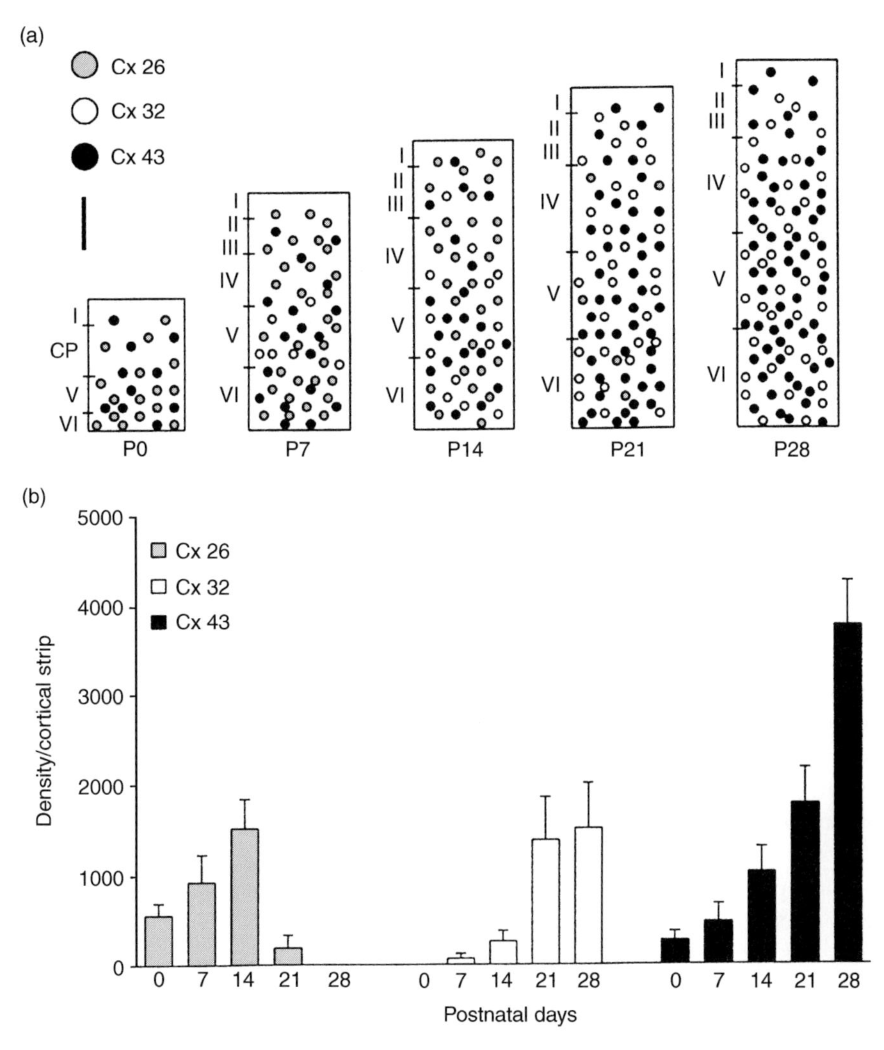


FIG. 3. Expression of connexins in the rat cerebral cortex during postnatal development. (a) Schematic representation of the pattern of distribution of connexin 26 (Cx26; $\beta_2$), Cx32 ($\beta_1$), and Cx43 ($\alpha_1$) in the visual cortex between birth and postnatal day 28. Bar = 200 $\mu$m. (b) Average levels of Cx26, Cx32 and Cx43 immunoreactivities measured from sections of visual cortex of five rats at each of the ages shown. The levels were corrected to take into account the expansion of the cortex during postnatal development. Error bars represent S.E.M.

hardly be detected after the third postnatal week. The observation that the expression of Cx26 matched closely the pattern of development of neuronal coupling (Peinado et al 1993a) suggests that this protein takes part in the establishment of functional coupling during the formation of neuronal circuits in the cortex. While Cx26 is expressed transiently during the first three weeks after birth, the localization and pattern of expression of Cx43 mirrors the maturation of the cortex throughout postnatal development. Further characterization of these gap junction proteins showed that Cx26 is expressed exclusively in cortical neurons, whereas Cx43 is expressed predominantly in astrocytes, but also in a population of neurons. The localization of Cx43 in astrocytes and in some cortical neurons raises the possibility of a homotypic channel system in the cortex, but dye transfer between these two cell types has yet to be demonstrated *in vivo*. However, a recent *in vitro* study has shown that neuroblasts have the potential to express Cx33 ($\alpha_7$), Cx40 and Cx43 (Rozental et al 1998), and since astrocytes contain the mRNA encoding Cx43 and Cx40 (Dermietzel 1996), it is reasonable to assume that homotypic channels composed of Cx40 or Cx43 may provide a direct signalling pathway for these cells to influence neuronal activity as reported previously in cortical cell cultures (Nedergaard 1994).

Although neuronal coupling is a transient event in the cortex, occurring rarely after the third week of life (Connors et al 1983, Peinado et al 1993b), gap junctions that mediate coupling and their constituent connexins were found to be present throughout development and in adult life. Thus, Cx43 was observed in the cortex throughout the life of the animal, and the down-regulation in the production of Cx26 in the third week of life coincided with the onset of expression of Cx32. Significant levels of expression of this protein were detected at the late stages of development and in adult life (Nadarajah et al 1997). Double-labelling experiments with cell-specific markers indicated that Cx32 was localized in oligodendrocytes and in neuronal cell types, whereas Cx43 was expressed predominantly in astrocytes and in a population of cortical neurons (Nadarajah et al 1996).

Early electron microscopical studies in the adult primate and rat neocortex (Sloper & Powell 1978, Peters 1980, Massa & Mugnaini 1982) identified gap junctions between astrocytes, between oligodendrocytes and, rarely, between neurons. In a more recent quantitative analysis of thin sections and freeze-fracture replicas (Nadarajah et al 1996), we found that the frequency and distribution of gap junctions varied in different cytoarchitectonically and functionally distinct regions of the rat cortex. Using well-established ultrastructural criteria, we observed that gap junctions were between glial cells and, less frequently, between neuronal elements. We also identified heterologous junctions between astrocytes and oligodendrocytes and between neurons and glia; the latter category included abundant junctions between astrocytic processes and neurons. However, dye transfer between neurons and astrocytes had not been previously visualized in cortical slice preparations of developing and adult rats, although in a more recent study in slices of adult cortex (Müller 1996), injected astrocytes showed dye coupling with glial fibrillary acidic protein (GFAP)-negative cells.

This observation suggested that astrocytes may also be coupled to neurons, oligodendrocytes or to astrocytes devoid of GFAP. During postnatal cortical development, when astroglial cells are still immature, it is possible that gap junctional communication may largely be restricted to neurons. In this context, it is interesting that our electron microscopical analysis did not reveal neuronal–astroglial gap junctions in the first few weeks after birth (Nadarajah et al 1997). It is also pertinent to note that if neuronal–astroglial gap junctions show an unidirectional intercellular signalling pathway, as demonstrated between oligodendrocytes and astrocytes of the retina (Robinson et al 1993), intracellular injections of cortical neurons may not result in coupling with astrocytes. This has been shown in rat forebrain cell cultures where unidirectional $Ca^{2+}$ signalling was observed from astrocytes to neurons, thereby implying the existence of asymmetrical signalling pathways between such heterologous cell types (Nedergaard 1994). It is also possible that dye injected into neuronal somata may not be detected in other cells if the injected neurons have only few junctional channels at distal dendrites. As for the nature of the connexin make up of astroglial–neuronal gap junctions, studies have reported that Cx43 and Cx32 form heterologous intercellular channels that display asymmetric voltage sensitivity (Swenson et al 1989, Werner et al 1989). However, more recent results have indicated that these two connexins are functionally incompatible (White et al 1995). Alternatively, a third connexin isoform, such as Cx40, now known to be expressed in neuroblasts (Rozental et al 1998) and in astrocytes (Dermietzel 1996) may account for the astrocytic–neuronal gap junctions. Such multiple connexin expression in cortical neurons may enable selective propagation of different second messenger molecules between cells. Since the half-life of these molecules is short, a small difference in the permeability of different connexins could alter intercellular signal transduction (Paul 1995).

Why are neurons in the cortex only transiently coupled and why don't neurons in the adult, shown to express connexins and gap junctions, transfer dye? We speculate that the expression of connexins and the presence of extensive functional coupling during the first two weeks of postnatal life is of physiological significance since it coincides with important events in cortical development. The two-week period after birth is critical in the development of the rat cerebral cortex, when the rates of morphological, neurochemical and functional differentiation are highest (Armstrong-James & Fox 1988, Parnavelas et al 1988, Uylings et al 1990), as is the rate of synapse formation (Blue & Parnavelas 1983). Peinado et al (1993a) have proposed possible roles for gap junctions in the establishment of connectivity between developing cortical neurons. One possibility is that gap junctional coupling may serve to guide chemical synapse formation with groups of neurons rather than with individual cells. Another possibility is that the extensive interneuronal coupling present in the early stages of cortical development culminates in limited coupling between adult cortical neurons through a process involving selection, strengthening and weakening of gap junctions, similar to the selection that occurs during the development of chemical synapses. Since most young neurons show weak electrical coupling (Connors et al 1983), it is unlikely that the primary function of coupling in the developing cortex would be the coordination of electrical activity. However, gap junctions may synchronize neuronal activity by coordinating biochemical activity. A recent study, using calcium-imaging techniques, has provided evidence in support of this notion and elucidated the underlying cellular mechanism that mediates the activation of neuronal domains and the propagation of intercellular calcium waves in neonatal cortex (Kandler & Katz 1998). These authors have proposed that the observed neuronal domains are initiated by glutamate released from developing synapses and acting through metabotropic glutamate receptors. Activation of these receptors may lead to the production of inositol-1,4,5-trisphosphate ($InsP_3$), through a cascade involving G proteins, and to the subsequent release of calcium from intracellular stores. In addition to its role as intracellular messenger, $InsP_3$ appears to function as the intercellular signalling molecule during formation of neuronal domains. Since this and other second messenger molecules (e.g. $Ca^{2+}$, cAMP) are known to have profound influence in many biochemical pathways, neuronal coupling may also regulate the electrical excitability, strength of synaptic transmission, neurite growth and neuronal gene expression (Kater & Mills 1991, Bashir & Collingridge 1992, Ghosh et al 1994, Kandler & Katz 1995). Gap junctional coupling in the adult neocortex has been thought to be involved in the mediation of subthreshold oscillations responsible for synchronizing neuronal firing (Konig et al 1995). Weak coupling between adult cortical neurons may be necessary to orchestrate collective behaviour in neuronal assemblies to ensure that the influence of no single neuron predominates (Peinado et al 1993a).

## Regulation of gap junctional communication and connexin expression

It is emerging that gap junctions are not indiscriminate corridors between cells, but rather specialized communication channels in which the connexin make-up influences the transfer of molecules between cells (Bennett et al 1991). Since most tissues express more than one connexin, it is conceivable that the expression of these proteins would be controlled by multiple and diverse regulatory mechanisms, such that connexins with different functional properties may be expressed as and when needed by the tissue. In most cells, gap junctional communication is modulated by factors that alter the rate of transcription of connexin genes, the stability and translation of connexin transcripts, post-translational processing of these proteins, their insertion into the membrane and assembly into channels, and their removal from the plasma membrane. These factors include applied voltage, phosphorylation, hormones, second messengers and neurotransmitters (Dermietzel & Spray 1993). In addition to these factors, *in vitro* studies involving various cell systems have shown that growth factors also have a modulatory effect on gap junctional communication. For example, it has been shown that epidermal growth factor (Lau et al 1992, Madhukar et al 1989) and transforming growth factor $\beta$ (Maldonado et al 1988) diminish gap junctional coupling, whereas exposure of microvascular endothelial cells and cardiac fibroblasts to fibroblast growth factor (FGF) 2 has been found to enhance cell coupling and expression of Cx43 (Pepper & Meda 1992, Doble & Kardami 1995). Similarly, we have recently shown that short term application of FGF2 to primary cortical cell cultures results in enhanced expression of Cx43 mRNA, protein Cx43 and intercellular dye coupling (Nadarajah et al 1998). From the elevated levels of Cx43 mRNA, it appears that the rate of transcription of this gene is likely to be upregulated through the activation of receptor tyrosine kinase pathway. Taking these observations together with data showing an association between coupling and the cell cycle of cortical neuroepithelial cells (Bittman et al 1997), it is possible that gap junction channels provide a direct conduit for mitogens released on FGF2 induction to effectively regulate proliferation during corticogenesis.

*Acknowledgements*

We wish to thank Howard Evans for providing the connexin antibodies and for useful advice and discussions at every stage of the work. The work was supported by the Wellcome Trust.

## References

Armstrong-James M, Fox K 1988 The physiology of developing cortical neurons. In: Peters A, Jones EG (eds) Cerebral cortex, vol 7: development and maturation of cerebral cortex. Plenum Press, New York, p 237–272

Bashir ZI, Collingridge GL 1992 Synaptic plasticity: long-term potentiation in the hippocampus. Curr Opin Neurobiol 2:328–335

Bennett MVL, Barrio LC, Bargiello TA, Spray DC, Hertzberg E, Saez JC 1991 Gap junctions: new tools, new answers, new questions. Neuron 6:305–320

Bittman K, Owens DF, Kriegstein AR, LoTurco JJ 1997 Cell coupling and uncoupling in the ventricular zone of developing neocortex. J Neurosci 17:7037–7044

Blue ME, Parnavelas JG 1983 The formation and maturation of synapses in the visual cortex of the rat. I. Qualitative analysis. J Neurocytol 12:599–616

Brightman MW, Reese TS 1969 Junctions between intimately apposed cell membranes in the vertebrate brain. J Cell Biol 40:648–677

Connors BW, Benardo LS, Prince DA 1983 Coupling between neurons of the developing rat neocortex. J Neurosci 3:773–782

Dermietzel R 1996 Molecular diversity of gap junction expression in brain tissues. In: Spray DC, Dermietzel R (eds) Gap junctions in the nervous system. Landes Bioscience, Austin, TX, p 13–38

Dermietzel R, Spray DC 1993 Gap junctions in the brain: where, what type, how many and why? Trends Neurosci 16:186–192

Dermietzel R, Traub O, Hwang TK et al 1989 Differential expression of three gap junction proteins in the developing and mature brain tissues. Proc Natl Acad Sci USA 86:10148–10152

Doble BW, Kardami E 1995 Basic fibroblast growth factor stimulates connexin-43 expression and intercellular communication of cardiac fibroblasts. Mol Cell Biochem 143:81–87

Ghosh A, Ginty DD, Bading H, Greenberg ME 1994 Calcium regulation of gene expression in neuronal cells. J Neurobiol 25:294–303

Gilbert CD 1983 Microcircuitry of the visual cortex. Annu Rev Neurosci 6:217–247

Guthrie SC, Gilula NB 1989 Gap junctional communication and development. Trends Neurosci 12:12–15

Gutnick MJ, Prince DA 1981 Dye coupling and possible electrotonic coupling in the guinea pig neocortical slice. Science 211:67–70

Hoh JH, Lal R, John SA, Revel JP, Arnsdorf MF 1991 Atomic force microscopy and dissection of gap junctions. Science 253:1045–1048

Hubel DH, Wiesel TN 1977 Functional architecture of macaque monkey visual cortex. Proc R Soc Lond B 198:1–59

Kandler K, Katz LC 1995 Neuronal coupling and uncoupling in the developing nervous system. Curr Opin Neurobiol 5:98–105

Kandler K, Katz LC 1998 Coordination of neuronal activity in developing visual cortex by gap junction-mediated biochemical communication. J Neurosci 18:1419–1427

Kater SB, Mills LR 1991 Regulation of growth cone behavior by calcium. J Neurosci 11:891–899

Konig P, Engel AK, Singer W 1995 Relation between oscillatory activity and long-range synchronization in cat visual cortex. Proc Natl Acad Sci USA 92:290–294

Kuraoka A, Iida H, Hatae T, Shibata Y, Itoh M, Kurita T 1993 Localisation of gap junction proteins connexins 32 and 26 in rat and guinea pig liver as revealed by quick freeze-deep-etch immunoelectron microscopy. J Histochem Cytochem 4:971–980

Lau AF, Kanemitsu MY, Kurata WE, Danesh S, Boynton AL 1992 Epidermal growth factor disrupts gap-junctional communication and induces phosphorylation of connexin43 on serine. Mol Biol Cell 3:865–874

LoTurco JJ, Kriegstein AR 1991 Clusters of coupled neuroblasts in embryonic neocortex. Science 252:563–566

Lowenstein WR 1981 Junctional intercellular communication: the cell-to-cell membrane channel. Physiol Rev 61:829–913

Luskin MB, Parnavelas JG, Barfield JA 1993 Neurons, astrocytes, and oligodendrocytes of the rat cerebral cortex originate from separate progenitor cells: an ultrastructural analysis of clonally related cells. J Neurosci 13:1730–1750

Madhukar BV, Oh SY, Chang CC, Wade M, Trosko JE 1989 Altered regulation of intercellular communication by epidermal growth factor, transforming growth factor-beta and peptide hormones in normal human keratinocytes. Carcinogenesis 10:13–20

Maldonado PE, Rose B, Loewenstein WR 1988 Growth factors modulate junctional cell-to-cell communication. J Membr Biol 106:203–210

Massa PT, Mugnaini E 1982 Cell junctions and intramembrane particles of astrocytes and oligodendrocytes: a freeze fracture study. Neuroscience 7:523–538

Micevych PE, Abelson L 1991 Distribution of mRNAs coding for liver and heart gap junction proteins in the rat central nervous system. J Comp Neurol 305:96–118

Mione MC, Cavanagh JFR, Harris B, Parnavelas JG 1997 Cell fate specification and symmetrical/asymmetrical divisions in the developing cerebral cortex. J Neurosci 17:2018–2029

Müller CM 1996 Development, topography, and modulation of astroglial coupling in mammalian cortex and hippocampus. International symposium on gap junctions in the nervous system, Seeon, Germany, p 34

Nadarajah B, Thomaidou D, Evans WH, Parnavelas JG 1996 Gap junctions in the adult cerebral cortex: regional differences in their distribution and cellular expression of connexins. J Comp Neurol 376:326–342

Nadarajah B, Jones AM, Evans WH, Parnavelas JG 1997 Differential expression of connexins during neocortical development and neuronal circuit formation. J Neurosci 17:3096–3111

Nadarajah B, Makarenkova H, Becker DL, Evans WH, Parnavelas JG 1998 Basic FGF increases communication between cells in the developing neocortex. J Neurosci 18:7881–7890

Nedergaard M 1994 Direct signaling from astrocytes to neurons in culture of mammalian brain cells. Science 263:1768–1771

Parnavelas JG, Papadopoulos GC, Cavanagh ME 1988 Changes in neurotransmitters during development. In: Peters A, Jones EG (eds) Cerebral cortex, vol 7: development and maturation of cerebral cortex. Plenum Press, New York, p 177–209

Paul DL 1995 New functions for gap junctions. Curr Opin Neurobiol 7:665–672

Peinado A, Yuste R, Katz LC 1993a Gap junctional communication and the development of local circuits in neocortex. Cerebr Cortex 3:488–498

Peinado A, Yuste R, Katz LC 1993b Extensive dye coupling between rat neocortical neurons during the period of circuit formation. Neuron 10:103–114

Pepper MS, Meda P 1992 Basic fibroblast growth factor increases junctional communication and connexin 43 expression in microvascular endothelial cells. J Cell Physiol 153:196–205

Peters A 1980 Morphological correlates of epilepsy: cells in the cerebral cortex. In: Glaser GH, Penry JK, Woodbury DM (eds) Antiepileptic drugs: mechanisms of action. Raven Press, New York, p 21–48

Rakic P 1981 Developmental events leading to laminar and areal organization of the neocortex. In: Schmit FO, Worden FG, Adelman G, Dennis SG (eds) The organization of the cerebral cortex. MIT Press, Cambridge, MA, p 7–28

Rakic P 1988 Specification of cerebral cortical areas. Science 241:170–176

Revel J-P, Karnovsky MJ 1967 Hexagonal array of subunits in intercellular junctions of the mouse heart and liver. J Cell Biol 33:c7–c12

Robinson SR, Hampson ECGM, Munro MN, Vaney DI 1993 Unidirectional coupling of gap junctions between neuroglia. Science 262:1072–1074

Rockel AJ, Hiorns RW, Powell TPS 1980 The basic uniformity in structure of the neocortex. Brain 103:221–244

Rozental R, Morales M, Mehler MF et al 1998 Changes in the properties of gap junctions during neuronal differentiation of hippocampal progenitor cells. J Neurosci 18:1753–1762

Ruangvoravat CP, Lo CW 1992 Connexin 43 expression in the mouse embryo: localization of transcripts within developmentally significant domains. Dev Dynamics 194:261–281

Skoff RP, Knapp PE 1991 Division of astroblasts and oligodendroblasts in postnatal rodent brain: evidence for separate astrocyte and oligodendrocyte lineages. Glia 4:165–174

Sloper JJ, Powell TPS 1978 Gap junctions between dendrites and somata of neurons in the primate sensori-motor cortex. Proc R Soc Lond B 203:39–47

Swenson KI, Jordan JR, Beyer EC, Paul DL 1989 Formation of gap junctions by expression of connexins in Xenopus oocyte pairs. Cell 57:145–155

Unger VM, Kumar NM, Gilula NB, Yeager M 1999 Electron cryo-crystallography of a recombinant cardiac gap junction channel. In: Gap junction-mediated intercellular signalling in health and disease. Wiley, Chichester (Novartis Found Symp 219) p 22–37

Uylings HBM, Van Eden CG, Parnavelas JG, Kalsbeek A 1990 The prenatal and postnatal development of rat cerebral cortex. In: Kolb B, Tees RC (eds) The cerebral cortex of the rat. MIT Press, Cambridge, MA, p 35–76
Werner R, Levine E, Rabadan-Diehl C, Dahl G 1989 Formation of hybrid cellcell channels. Proc Natl Acad Sci USA 86:5380–5384
White TW, Paul DL, Goodenough DA, Bruzzone R 1995 Functional analysis of selective interactions among rodent connexins. Mol Biol Cell 6:459–470
Yancey SB, Biswal S, Revel J-P 1992 Spatial and temporal patterns of distribution of the gap junction protein connexin43 during mouse gastrulation and organogenesis. Development 114:203–212
Yuste R, Nelson DA, Rubin WW, Katz LC 1995 Neuronal domains in developing neocortex: mechanisms of coactivation. Neuron 14:7–17

## DISCUSSION

*Dermietzel:* I'm surprised that you observed such a high level of connexin26 (Cx26; $\beta_2$) staining in postnatal neurons, because we observed Cx26 only in leptomeningeal cells of adult rats. In which type of neurons did you see the staining?

*Parnavelas:* The expression of Cx26 reaches a peak at embryonic day (E) 16 and then diminishes considerably until birth. A second peak in expression of this connexin occurs at postnatal day (P) 14 after which it declines until it is hardly detected at the end of the fourth week of life. Prenatally, Cx26 is expressed by differentiated neurons, identified by the presence of MAP2, as well as by some MAP2-negative cells. Postnatally, Cx26 was found to be expressed in both cortical pyramidal and non-pyramidal neurons.

*Dermietzel:* It is possible that we did not see postnatal expression of Cx26 because we didn't do such a close follow up. Peinado et al (1993) showed that there is an increase in postnatal neuronal coupling between P7 and P14, and he suggests that this is important for the modular organization of the cerebral cortex.

*Parnavelas:* We were guided by two sets of dye-coupling studies. The first, performed by LoTurco & Kriegstein (1991), showed extensive dye coupling at the peak of cortical neurogenesis in rats (E15), with the number of cells coupled in clusters reduced significantly by the end of gestation. In a separate series of studies, Larry Katz and colleagues focused on the early postnatal development and demonstrated extensive dye coupling between cortical neurons at P5–P12 (Peinado et al 1993). They found only few small clusters of labelled cells in the third week. The levels of expression of Cx26 I described here matched closely this pattern of postnatal development of neuronal coupling, suggesting that this connexin protein takes part in the establishment of functional coupling at this stage of development. Katz and colleagues have shown that this coupling is important in the formation of neuronal domains in the cortex (Yuste et al 1995).

*Kistler:* Are hearing loss and peripheral neuropathies the only defects in patients with Cx32 ($\beta_1$) or Cx26 mutations? Because these are essentially knockouts, so I am surprised that their brains work at all.

*Warner:* We don't have any evidence that all those mutations are functional knockouts.

*Gilula:* Joerg Kistler has asked a question that doesn't require an immediate response from anyone because a systematic evaluation of all the neurological consequences has not been made.

*Scherer:* There is one report of slowing in the central auditory pathway of X-linked Charcot-Marie-Tooth (CMTX) patients (Nicholson & Corbett 1996), but cognitively these patients are described as normal. Whether CMTX patients have subtle cognitive defects that a skilled neuropsychologist could identify remains an open question.

*Gilula:* The levels of antigen you detect with your anti-Cx26 antibodies are high. Have you detected similar levels of antigen with independent reagents or is the same reagent used in all your images?

*Parnavelas:* We have used antibodies obtained from Zymed (Zymed Laboratories Inc, South San Francisco, CA) and other commercial sources. Not all these antibodies worked successfully in our hands. The ones that did showed staining comparable to that I have shown here.

*Gilula:* All the antigen you detect in the neuronal cell bodies has to be housed somehow. It is going to be interesting to see how it is housed, and in what state. This may be related to the issues that Howard Evans is studying, although you won't have much of an opportunity to carry out those types of subcellular fractionation experiments in your system.

*Parnavelas:* Why are you so surprised by the high levels of this antigen?

*Gilula:* I don't know of any other cell types where people have found Cx26 present in the cytoplasm at such high levels that then disappear.

*Parnavelas:* We have observed pyramidal neurons loaded with acetylcholine or other molecules at the early stages of cortical development that become devoid of these substances in a matter of a few days.

*Vaney:* I would like to take a different perspective and note that I was rather disappointed that such a low proportion of the neurons in the mature brain expressed immunocytochemically detectable levels of connexins. You said that historically people have not seen much dye coupling between mature neurons, but most of these studies used Lucifer yellow, whereas the use of Neurobiotin and other cationic tracers is quite recent. Also, most studies were performed on brain slices, and we know that much less tracer coupling is seen in retinal slices than in retinal whole mounts. The number of studies in which Neurobiotin has been injected into neurons in the intact brain is limited. These two points should be taken on board when considering the incidence of neuronal coupling in the mature brain.

*Parnavelas:* Peinado et al (1993) have compared Neurobiotin and Lucifer yellow injections in cortical neurons at different stages of postnatal development. They demonstrated that injections of Neurobiotin showed more extensive dye coupling than injections of Lucifer yellow. I should say that earlier studies in embryonic cortex (LoTurco & Kriegstein 1991) showed that injections of Lucifer yellow into single cells resulted in clusters that contained up to 90 cells.

*Vaney:* I would like to make the additional point that blind intracellular injections in the intact brain would strongly favour neurons with large somata. There may well be classes of neurons that have been missed entirely using such injection paradigms.

*Parnavelas:* I also worry about the use of brain slices and how these preparations can affect the results. However, I do not know of another way around this, because these experiments will be extremely difficult to perform *in vivo*.

*Gilula:* Another perspective I can offer is that both you and David Vaney have identified gap junctions in the nervous system, which is thought provoking. Rolf Dermietzel was one of the first people here to appreciate the opportunities that connexin antibodies were going to give us to look at connexin utilization, especially because previously many neurophysiologists had difficulties in finding gap junctions. David Vaney has given us the first insights as to how connexins may be utilized in the retina, and John Parnavelas has provided us with a basis for considering that the connexin gene products are differentially utilized developmentally in different regions of the cortex. Also, the observations of Connors et al (1983) of decreased dye coupling between neurons after birth should be considered in the context of connexin utilization.

*Dermietzel:* I can envisage some caveats. Not only is there a lack of dye coupling in the adult brain, but there is also a lack of electrical coupling. If there are high levels of connexins, then how many channels are required to couple two neurons? I estimate that only a few channels are required for ionic transfer from one neuron to the other because of the high input resistances of the neuronal membrane. Rash et al (1997) showed that there are more than 300 gap junctions per cell in motoneurons of the spinal cord by counting the plaques on freeze-fracture micrographs. They are only associated with chemical synapses, so they're called mixed synapses.

*Gilula:* There's always been a real appreciation that there are too many channels to explain the amount of observed function. Why is there more antigen present than we think is required to explain what the antigen is doing? This leads us to question whether we really understand what all these connexins are being used for.

*Giaume:* As well as the number and the size of gap junctions, the location of junctional channels is a critical and important element to take into consideration when junctional currents are monitored. This is specially the case for cells with a complex morphology like neurons. For electrophysiologists who record from soma, the possibility to monitor junctional currents depends upon the distance between the coupling and the recording sites. Indeed, due to space clamp problems the ability to control the potential is expected to be rather different for events occurring at somatic or dendritic levels.

*Gilula:* This could explain some of the failure to detect channels.

*Nicholson:* About seven years ago we took a different approach to this by looking at primary cultures of cortical neurons and astrocytes from 17-day-old fetal rat cerebrum. We observed a low level of Cx43 ($\alpha_1$) in the neurons, and we could demonstrate neuronal–glial coupling. The interesting thing was that all of the Cx43 was present at the point where two processes crossed. We did not observe Cx26 in neurons at that time, but we could have inadvertently been selecting for subpopulations of neurons by our culture technique. I also gather that Rolf Dermietzel has some data that show Cx26 and Cx43 are not present in the same neurons.

*Musil:* Have you had an opportunity to look at possible changes in the levels of staining with the three antibodies in animal knockouts?

*Parnavelas:* No, but this is one of the things we would like to do in the next few months.

*Green:* We looked at the globus pallidus and caudate nucleus of the human brain from normal and Huntington's disease patients. We found that Cx26 and Cx32 co-localized in similar amounts, with more in the globus pallidus; there were only small amounts of these connexins in the caudate nucleus. Cx43 is the predominant connexin, and in the globus pallidus it was present in similar amounts in Huntington's disease patients and controls. There was, however, a marked increase in Cx43 in the caudate nucleus of Huntington's disease patients; which corresponds with a marked increase in GFAP (glial fibrillary acidic protein) labelling. This may not have anything to do with causing increased numbers of astrocytes, of course, but may simply reflect increased astrocytic activity in this region as indicated by the increased levels of GFAP.

## References

Connors BW, Benardo LS, Prince DA 1983 Coupling between neurons of the developing rat neocortex. J Neurosci 3:773–782

LoTurco JJ, Kriegstein AR 1991 Clusters of coupled neuroblasts in embryonic neocortex. Science 252:563–566

Nicholson G, Corbett A 1996 Slowing of central conduction in X-linked Charcot-Marie-Tooth neuropathy shown by brain auditory evoked responses. J Neurol Neurosurg Psych 61:43–46

Peinado A, Yuste R, Katz LC 1993 Extensive dye coupling between rat neocortical neurons during the period of circuit formation. Neuron 10:103–114

Rash JE, Duffy HS, Dudek FE, Bilhartz BL, Whalen LR, Yasumura T 1997 Grid-mapped freeze-fracture analysis of gap junctions in gray and white matter of adult rat central nervous system, with evidence for a 'panglial syncytium' that is not coupled to neurons. J Comp Neurol 388:265–292

Yuste R, Nelson DA, Rubin WW, Katz LC 1995 Neuronal domains in developing neocortex: mechanisms of coactivation. Neuron 14:7–17

# The role of the gap junction protein connexin32 in the pathogenesis of X-linked Charcot-Marie-Tooth disease

Steven S. Scherer, Linda Jo Bone†, Suzanne M. Deschênes†, Annette Abel, Rita J. Balice-Gordon* and Kenneth H. Fischbeck

*Departments of Neurology, *Neuroscience and †The Cell and Molecular Biology Graduate Group, University of Pennsylvania School of Medicine, Philadelphia, PA 19104, USA*

*Abstract.* Mutations in the gene encoding the gap junction protein connexin32 (Cx32; $\beta_1$) cause the X-linked form of Charcot-Marie-Tooth disease (CMTX), a common form of inherited demyelinating neuropathy. Cx32 is localized to the paranodes and incisures of myelinating Schwann cells, and probably participates in the formation of gap junctions at these locations, thereby allowing the diffusion of ions and small molecules directly across the myelin sheath. In transfected cells different CMTX mutations have different effects on the ability of the mutant protein to form functional gap junctions; some mutant proteins cannot be detected within the cell, other mutant proteins accumulate within the cell but do not reach the cell membrane, while other mutants reach the cell membrane and some of these form functional gap junctions. In transgenic mice two mutants, R142W and 175 frameshift, have similar effects on protein trafficking as in transfected cells: the R142W mutant protein remains in the perinuclear region and does not reach the paranodes or incisures, and the 175 frameshift protein cannot be detected. Thus, different CMTX mutations have different effects on Cx32 protein, and these differences may help to explain the phenotypic differences seen in CMTX kindreds.

*1999 Gap junction-mediated intercellular signalling in health and disease. Wiley, Chichester (Novartis Foundation Symposium 219) p 175–187*

## Mutations in connexin32 cause X-linked Charcot-Marie-Tooth disease

Charcot-Marie-Tooth disease (CMT) is the eponym for a genetically heterogenous group of dominantly inherited peripheral neuropathies. The discovery that many CMT kindreds had electrophysiological and pathological evidence of a primary demyelinating neuropathy led to the subdivisions of type 1/demyelinating and type 2/axonal CMT (Dyck et al 1993). The combination of genetic linkage analysis of CMT1 kindreds and the mapping of the genes encoding peripheral myelin protein 22 kDa (PMP22) and protein zero ($P_0$) revealed that some kindreds have mutations in *PMP22* (CMT1A) whereas other kindreds have mutations in $P_0$ (CMT1B; Suter & Snipes 1995). Whether the X-linked form of CMT (CMTX) is a demyelinating or an axonal form has been debated, as the electrophysiological (slowed nerve conduction velocities) and pathological (demyelination, remyelination and onion bulb formation) features that are characteristic of CMT1 are not as pronounced in CMTX patients (Hahn et al 1990, Nicholson & Nash 1993, Rozear et al 1987, Sander et al 1998).

Following the mapping of CMTX, we screened candidate genes that had been previously mapped to the same region of the X chromosome by northern blot analysis, reasoning that if CMTX was a demyelinating neuropathy then the gene should be expressed by myelinating Schwann cells. Of three candidate genes, the mRNA of the gene encoding connexin32 (Cx32; $\beta_1$) was the only one detected in peripheral nerve. Sequencing the Cx32 gene in eight families led to the discovery of seven different mutations (Bergoffen et al 1993). Many groups have confirmed and extended this finding, and more than 130 different mutations that affect the open reading frame of the Cx32 gene have been described (Bone et al 1997). As shown in Fig. 1, these mutations include missense (amino acid substitutions) and nonsense (premature stop codons) mutations, as well as deletions and frameshifts, affecting all regions of Cx32. Many of the mutations have been reported more than once; some of these probably represent founder effects, whereas others may represent mutational 'hot spots' in the Cx32 gene. Beginning with our initial report, there are CMTX kindreds that do not have a mutation in the open reading frame. Ionasescu et al (1996) reported non-coding region mutations in two CMTX families that had moderate neuropathy and no mutations in the open reading frame of the Cx32 gene. One mutation was just proximal to the start site of transcription, and hence could affect the transcription of the Cx32 gene. The other was in the 5′ untranslated region, and could affect the transcription, the stability or the translation of Cx32 mRNA.

## Analysis of connexin32 mutations

Cx32 belongs to a gene family of 13 members (in mammals), all of which encode gap junction proteins (Bruzzone et al 1996). All gap junctions proteins are highly homologous, with the overall structure shown in Fig. 1. Six connexin proteins oligomerize to form a hemichannel (or connexon), which can form a channel when properly apposed to another hemichannel on an adjacent membrane, allowing the diffusion of ions and small molecules, typically with a molecular mass less than 1000 Da. There can be complex interactions between different connexins, as individual hemichannels can be composed of more than one connexin (heterotypic connexons), and hemichannels composed of different

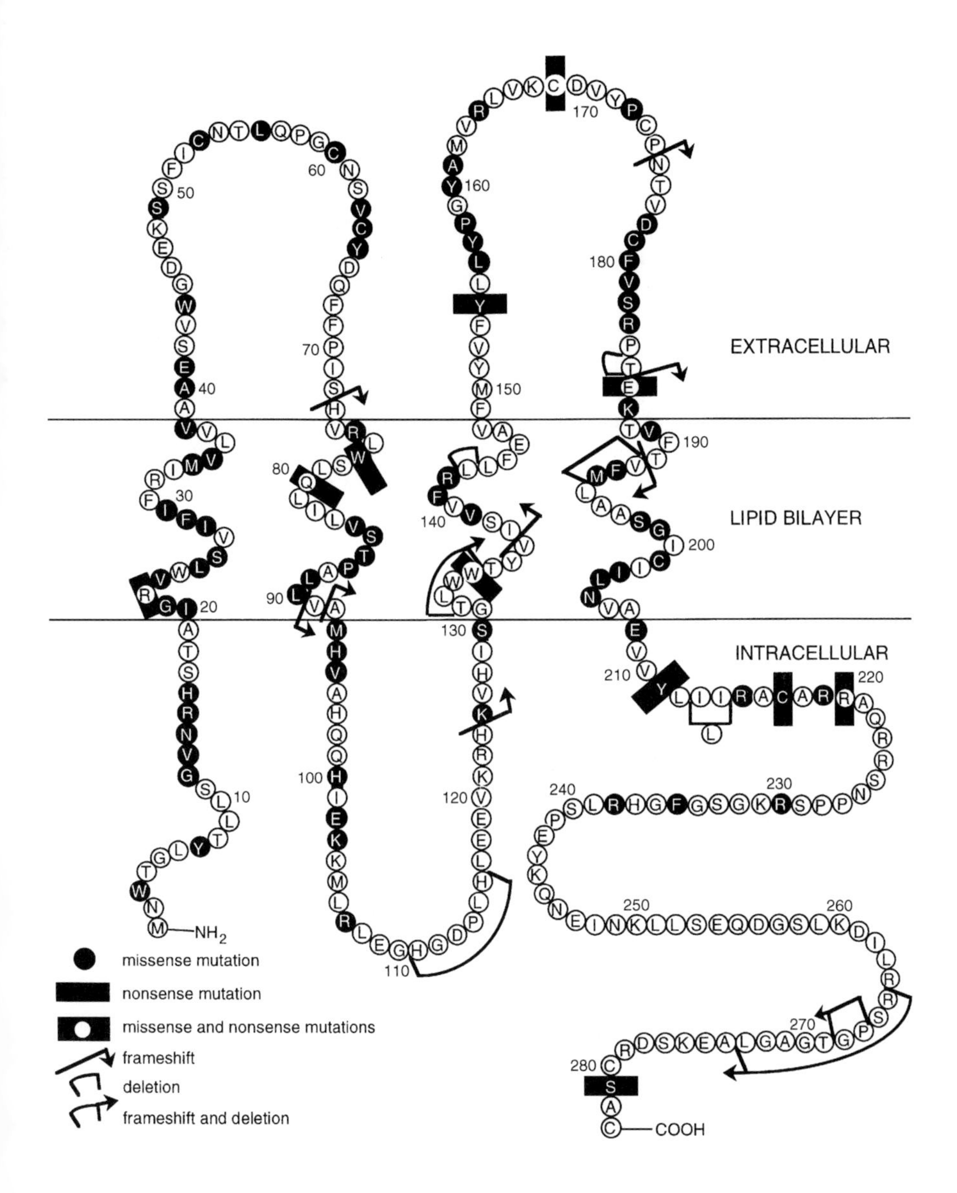


FIG. 1. Mutations in connexin32 (Cx32; $\beta_1$) associated with X-linked Charcot-Marie-Tooth disease. Cx32 is an intrinsic membrane protein with four transmembrane domains, one intracellular and two extracellular loops, and an amino- and carboxy-terminal cytoplasmic tail. The third transmembrane domain contains hydrophilic residues and lines the pore of the hemichannel. The positions of the amino acids affected by known mutations in the Cx32 gene are indicated.

connexins can also form channels (heterotypic gap junctions). If myelinating Schwann cells express other connexins in addition to Cx32 (Chandross et al 1996, Scherer et al 1995), then these interactions may be involved in the pathogenesis of CMTX, as some Cx32 mutants could have dominant negative effects on these other connexins.

The effects of different mutations in the Cx32 gene have been studied in model systems. In pairs of *Xenopus* oocytes the R220stop mutation formed functional gap junctions; this activity was lost with further truncation of Cx32 (Rabadan-Diehl et al 1994). The R142W, E186K and 175 frameshift mutations did not form functional channels with wild-type Cx32 or Cx26 ($\beta_2$), even though the mutant proteins were synthesized and reached the cell membrane (Bruzzone et al 1994). Further, the R142W, E186K and 175 frameshift mutants had dominant negative effects on wild-type Cx32 and Cx26 in this system, demonstrating the potential for such interactions in CMTX. The electrophysiological properties of G12S, S26L, I30N, M34T, V35M, V38M, P87A, E102G and the Δ111–116 deletion have been examined in oocytes (Oh et al 1997). The macroscopic properties of P87A, E102G and the Δ111–116 deletion were normal, the S26L and I30N mutants had modestly altered conductance–voltage relations, and G12S failed to form junctional currents.

In HeLa cells Omori et al (1996) observed dye coupling between cells expressing wild-type Cx32 and R220stop, but not between cells expressing C60F, V139M or R215W. Cells that expressed missense mutations had less Cx32 immunoreactivity on their cell surface than cells expressing wild-type Cx32 or R220stop. They also looked for dominant negative interactions with Cx32 by expressing these CMTX mutations in HeLa cells that already expressed wild-type Cx32. All three missense mutations had significant dominant negative effects, whereas R220stop did not decrease coupling.

We examined 10 Cx32 mutations in rat adrenal pheochromocytoma PC12 cells (Deschênes et al 1997), and found three patterns of protein trafficking (Fig. 2). No Cx32 was detected in cells transfected with the 175 frameshift mutation, even though Cx32 mRNA was detected. In contrast, Cx32 was expressed in a wild-type pattern, with punctate cell surface staining, in clones

FIG. 2. Immunocytochemical localization of connexin32 (Cx32; $\beta_1$) mutants in rat adrenal pheochromocytoma PC12 cells. Cells were transfected with plasmid constructs to express wild-type Cx32 or the indicated X-linked Charcot-Marie-Tooth disease (CMTX) mutations, and immunostained with monoclonal antibodies against Cx32. Cells transfected with wild-type Cx32, as well as R15Q, V63I, V139M and R220stop, have punctate Cx32 immunoreactivity on their surface. Cells transfected with G12S, R142W and E208K have cytoplasmic staining only. Parental PC12J cells have no staining. Bars = 10 $\mu$m. From Deschênes et al (1997), with permission from The Society for Neuroscience.

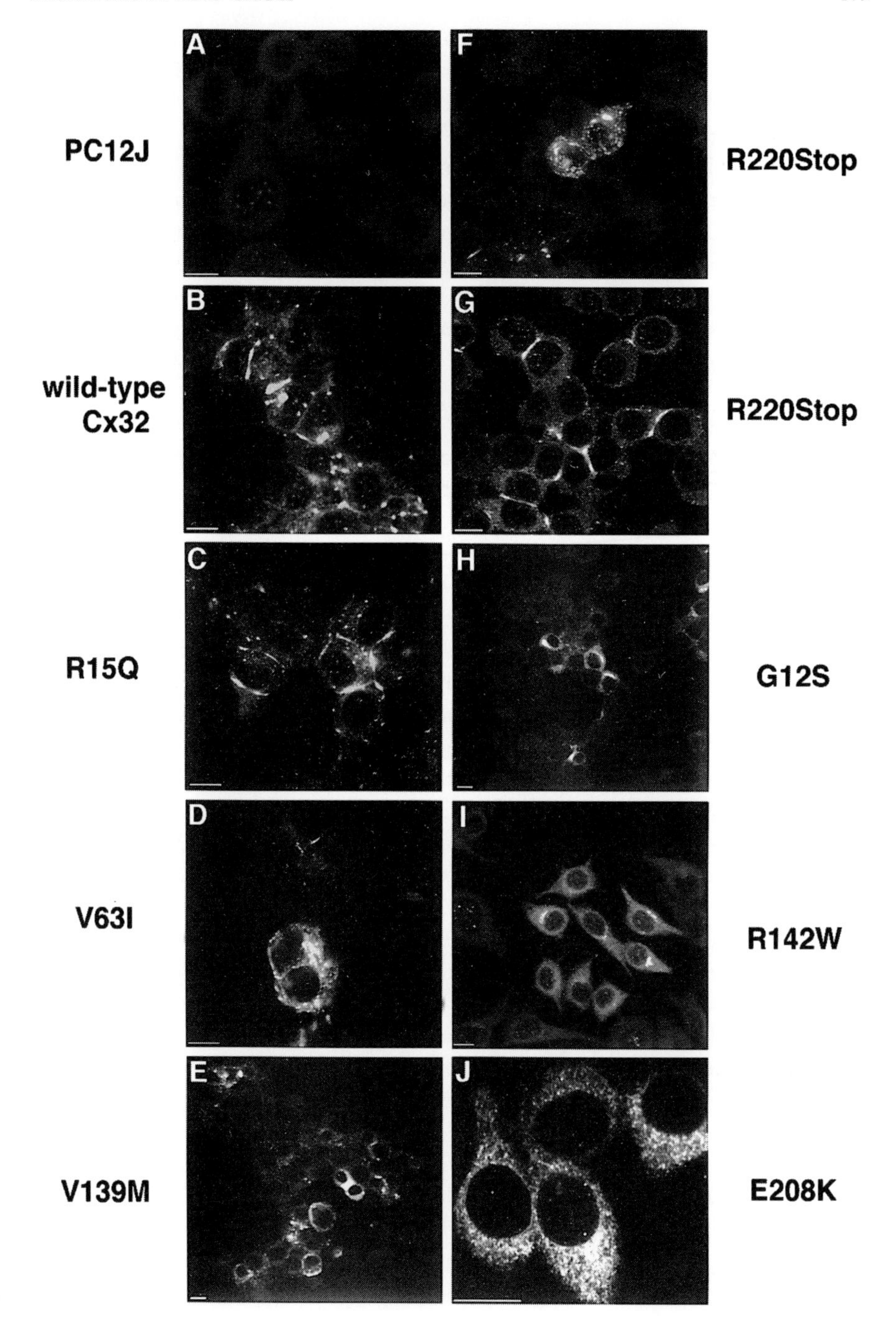

A
PC12J
B
wild-type
Cx32
C
R15Q
D
V63I
E
V139M
F
R220Stop
G
R220Stop
H
G12S
I
R142W
J
E208K

expressing R15Q, V63I, V139M, R215W and R220stop. Cx32 immunoreactivity was entirely cytoplasmic in clones expressing the G12S, R142W, E186K and E208K mutants; G12S, R14S, E186K and, less frequently, E208K reached the Golgi apparatus. Similarly, C53S and P172R mutants do not reach the cell surface of C6 glioma cells, but appear confined to the cytoplasm (Yoshimura et al 1998). These results indicate that mammalian cells may have more stringent requirements than oocytes for the normal trafficking of mutant proteins. In addition, there may be a relationship between clinical severity and the trafficking of Cx32 mutant protein, as mutant proteins that do not reach the cell surface seemed to be associated with a more severe phenotype.

## Animal models of X-linked Charcot-Marie-Tooth disease

All of the inherited demyelinating neuropathies identified to date are caused by mutations in genes expressed by myelinating Schwann cells. Similarly, in the CNS, mutations in genes expressed by oligodendrocytes, such as proteolipid protein (PLP), cause inherited demyelinating diseases in humans and other animals (Nave & Boespflug-Tanguy 1996). There is emerging evidence, largely based on the analysis of knockout mice, that the phenotypes of most naturally occurring mutations in *PMP22*/*Pmp22*, $P_0$ and *PLP*/*Plp* result from toxic effects and not a simple loss of function (Scherer 1997).

Loss-of-function Cx32 gene mutations have been created by the targeted deletion of Cx32 in mice. These mice develop a progressive, demyelinating peripheral neuropathy beginning at about three months of age (Anzini et al 1997, Nelles et al 1996, Scherer et al 1998). For unknown reasons, motor fibres are much more affected than sensory fibres (Fig. 3), a feature not noted in CMTX patients. Like other genes of the X chromosome, the Cx32 gene appears to be randomly inactivated, since in heterozygous $Cx32^{+/-}$ females only some myelinating Schwann cells express Cx32. Heterozygous $Cx32^{+/-}$ females have fewer demyelinated and remyelinated axons than age-matched homozygous $Cx32^{-/-}$ females and $Cx32^{-/Y}$ males, which may explain why women who are obligate carriers of CMTX are often affected themselves.

To determine whether some Cx32 mutations have more than a simple loss of function, we generated transgenic mice expressing the 175 frameshift or the R142W mutation (Bone et al 1998). No Cx32 protein could be detected and no peripheral neuropathy was noted in 26 lines of mice expressing the 175 transgene, even though transgene mRNA was highly expressed in some lines (A. Abel, L. J. Bone, A. Messing, S. Scherer & K. Fischbeck, unpublished observations 1998). Mice expressing the R142W mutation, in contrast, expressed mutant Cx32 and developed a demyelinating peripheral neuropathy (Bone et al 1998). The mutant protein remained in the perinuclear region and did not reach incisures or paranodes

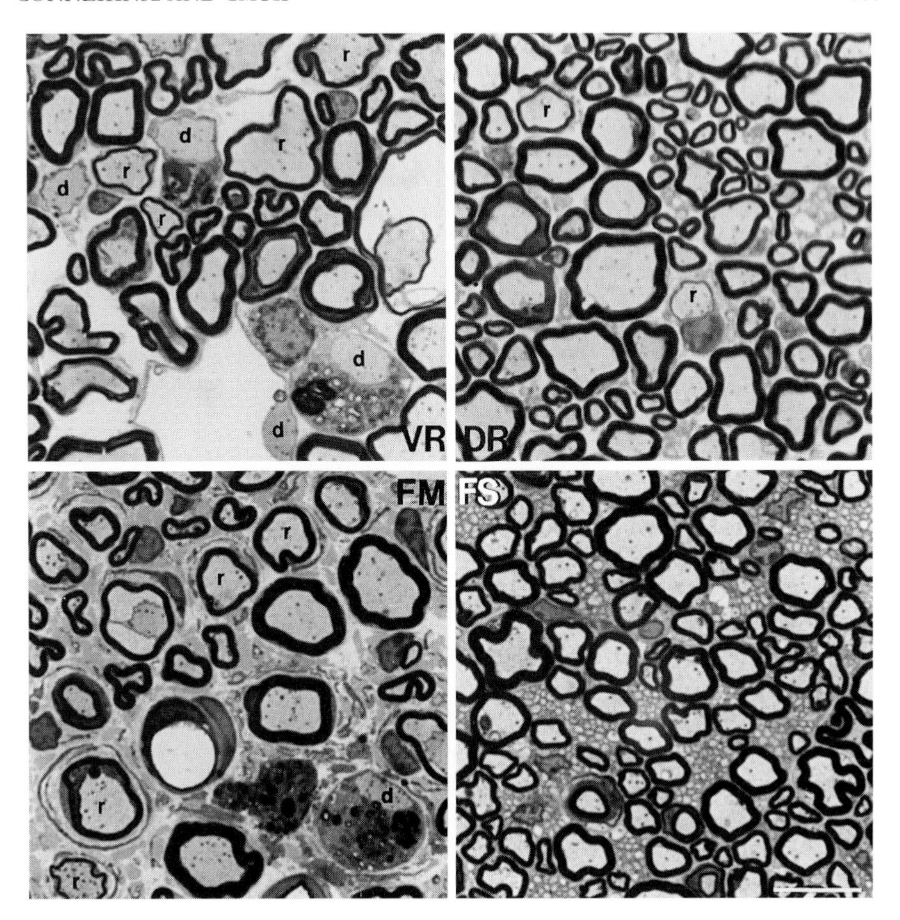


FIG. 3. Cx32-null mice develop a demyelinating neuropathy. These photomicrographs of semi-thin sections stained with toluidine blue are taken from Cx32-null mice that are one year old. The ventral roots (VR) and dorsal roots (DR), as well as the femoral motor (FM) and femoral sensory (FS) nerves, are indicated. Note that the ventral roots and femoral motor nerves contain more demyelinated (d) and thinly remyelinated (r) fibres their sensory counterparts. Bar = 10 $\mu$m.

(Fig. 4), the normal sites of Cx32 immunoreactivity in the myelin sheath (Chandross et al 1996, Scherer et al 1995). Further, the expression of the mutant Cx32 reduced the level of the endogenous/wild-type Cx32, indicating that this mutant may have dominant negative interactions with endogenous Cx32. Thus, the expressing and trafficking of these two Cx32 mutants is remarkably similar to that observed in transfected PC-12 cells (Deschênes et al 1997), demonstrating that

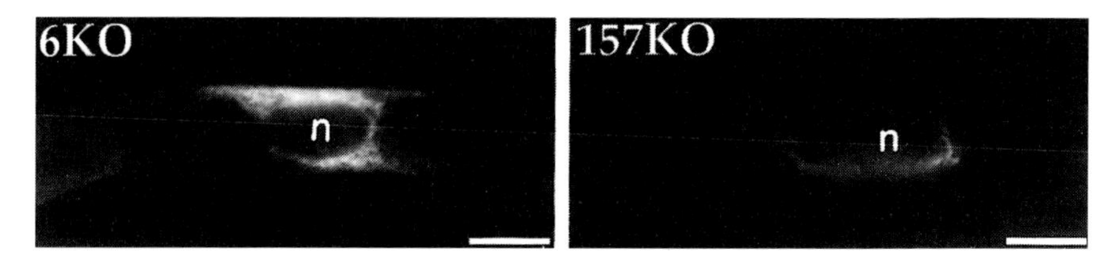


FIG. 4. Perinuclear localization of R142W mutant Cx32 in teased myelinated fibres. R142W transgene-positive, Cx32 knockout (6KO and 157KO) males were generated by crossing transgenic males from transgenic lines #6 and #157, respectively, with Cx32-null females. One-half of the resulting male offspring were transgene positive, and in these mice all of the Cx32 immunoreactivity comes from the R142W mutant protein. n, nucleus. Bars = 10 $\mu$m. From Bone et al (1998).

the altered trafficking of mutant connexins may be one of fundamental perturbations caused by mutations in the Cx32 gene.

## Functional gap junctions in the myelin sheath

The localization of Cx32 to the incisures and paranodes suggested that Cx32 forms gap junctions between the layers of the Schwann cell myelin sheath (Bergoffen et al 1993). In addition, small collections of gap junction-like particles have been observed by freeze-fracture electron microscopy (Sandri et al 1982, Tetzlaff 1982). The potential radial pathway formed by these gap junctions would be 1000-fold shorter than the circumferential pathway within the Schwann cell cytoplasm (Scherer et al 1995).

To determine whether there is a direct radial pathway, we injected living myelinating Schwann cells with 5,6-carboxyfluorescein (Balice-Gordon et al 1998), a fluorescent dye of low molecular mass (376 Da) that passes through gap junctions (Bruzzone et al 1996). By injecting the perinuclear region and monitoring the injections by fluorescence microscopy, we could follow the

---

FIG. 5. Functional gap junctions in the myelin sheath. A living, myelinating Schwann cell viewed in polarized light (A), then in fluorescein optics immediately following an injection of 5,6-carboxyfluorescein (C, D, E), then in rhodamine optics after fixing and immunostaining for myelin-associated glycoprotein (MAG; B). Incisures are seen in polarized light (A) and after MAG immunostaining (C). At a focal plane midway through the depth of the cell, 5,6-carboxyfluorescein fills the outer and inner collars of Schwann cell cytoplasm, creating a double train track pattern, demonstrating that the dye has diffused across the myelin sheath (C). The bracketed region is enlarged in E. This double train track pattern is distinct from the pattern of created by the filling of cytoplasmic channels on the surface of the myelin sheath (D), seen in a plane 5–60 $\mu$m above plane shown in C. Bars = 10 $\mu$m. From Balice-Gordon et al (1998).

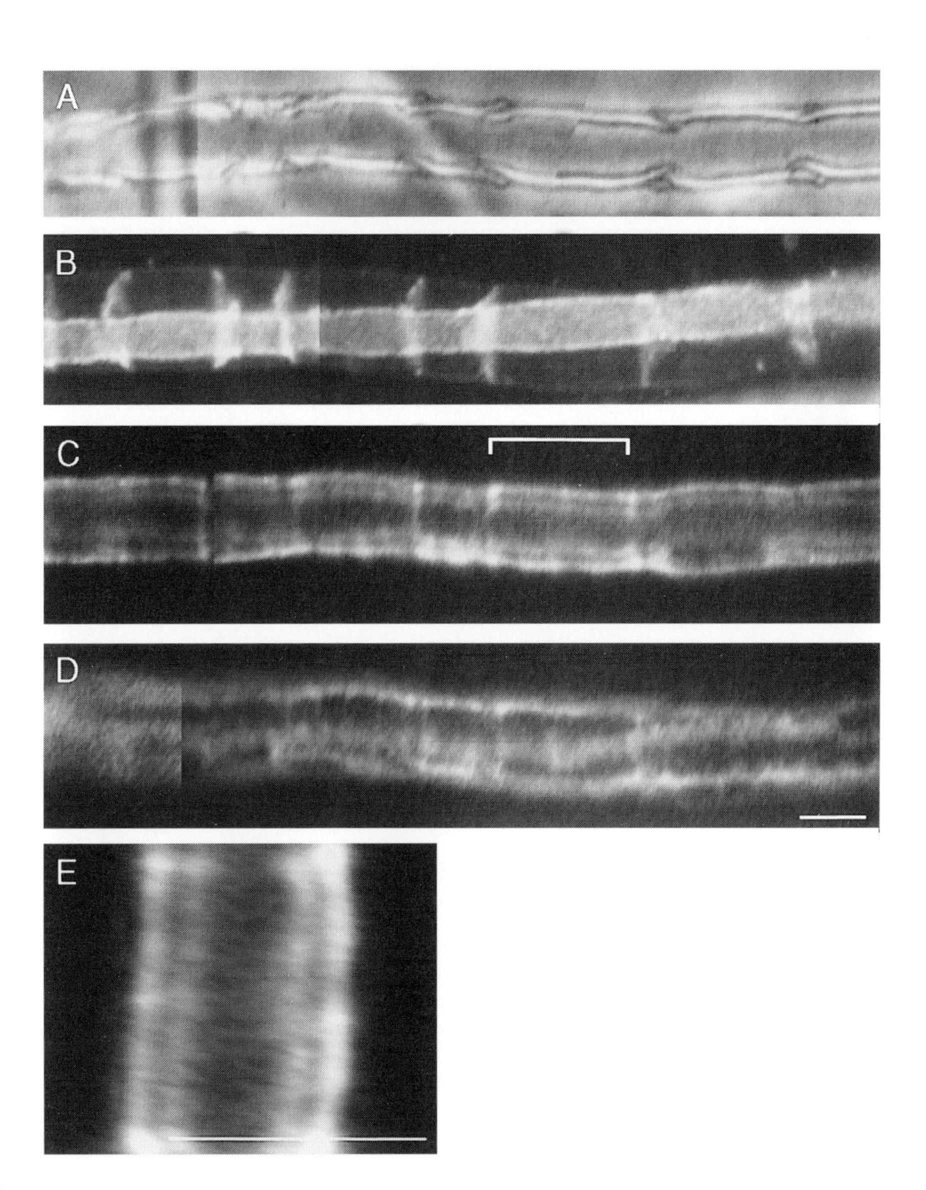
A
B
C
D
E

pathway of dye diffusion within the myelin sheath. An example is shown in Fig. 5. The outer collar of cytoplasm has been filled with 5,6-carboxyfluorescein, and some of the dye has crossed an incisure, reaching the inner collar of cytoplasm. In contrast, fluorescent dyes of large molecular mass, such as rhodamine-conjugated dextran (10 kDa) did not reach the inner collar of cytoplasm. Further, preincubating the fibres in 75 $\mu$M 18-$\alpha$-glycerrhetinic acid (AGA), a pharmacological blocker of gap junctions, prevented 5,6-carboxyfluorescein from reaching the inner collar of cytoplasm. These results demonstrate that there are functional gap junctions within incisures that mediate the diffusion of small molecules across the myelin sheath. If Cx32 mediates diffusion across this radial pathway, then the loss of this radial pathway may damage myelinating Schwann cells and their axons, thereby causing neuropathy. We find, however, that 5,6-carboxyfluorescein diffuses across the myelin sheath in Cx32-null mice, which we interpret as evidence for the existence of another connexin in the Schwann cell myelin sheath.

*Acknowledgements*

We thank our collaborators, especially Roberto Bruzzone, Albee Messing, David Paul and Klaus Willecke. Our work was supported by grants from the National Institutes of Health (NS08075 to S.S.S., K.F. and R.B.-G.) and Muscular Dystrophy Association (K.F. and S.S.S.).

## References

Anzini P, Neuberg DH-H, Schachner M et al 1997 Structural abnormalities and deficient maintenance of peripheral nerve myelin in mice lacking the gap junction protein connexin32. J Neurosci 17:4545–4561

Balice-Gordon RJ, Bone LJ, Scherer SS 1998 Functional gap junctions in the Schwann cell myelin sheath. J Cell Biol 142:1095–1104

Bergoffen J, Scherer SS, Wang S et al 1993 Connexin mutations in X-linked Charcot-Marie-Tooth disease. Science 262:2039–2042

Bone LJ, Deschênes SM, Balice-Gordon RJ, Fischbeck KH, Scherer SS 1997 Connexin32 and X-linked Charcot-Marie-Tooth disease. Neurobiol Dis 4:221–230

Bone LJ, Messing A, Balice-Gordon R, Fischbeck KH, Scherer SS 1998 Dominant effects of a *connexin32* mutation in myelinating Schwann cells of transgenic mice, submitted

Bruzzone R, White TW, Scherer SS, Fischbeck KH, Paul DL 1994 Null mutations of connexin32 in patients with X-linked Charcot-Marie-Tooth disease. Neuron 13:1253–1260

Bruzzone R, White TW, Paul DL 1996 Connections with connexins: the molecular basis of direct intercellular signaling. Eur J Biochem 238:1–27

Chandross KJ, Kessler JA, Cohen RI et al 1996 Altered connexin expression after peripheral nerve injury. Mol Cell Neurosci 7:501–518

Deschênes SM, Walcott JL, Wexler TL, Scherer SS, Fischbeck KH 1997 Altered trafficking of mutant connexin32. J Neurosci 17:9077–9084

Dyck PJ, Chance P, Lebo R, Carney JA 1993 Hereditary motor and sensory neuropathies. In: Dyck PJ, Thomas PK, Griffin JW, Low PA, Poduslo JF (eds) Peripheral neuropathy, 3rd edn. Saunders, Philadelphia, p 1094–1136

Hahn AF, Brown WF, Koopman WJ, Feasby TE 1990 X-linked dominant hereditary motor and sensory neuropathy. Brain 113:1511–1525
Ionasescu VV, Searby C, Ionasescu R, Neuhaus IM, Werner R 1996 Mutations of noncoding region of the connexin32 gene in X-linked dominant Charcot-Marie-Tooth neuropathy. Neurology 47:541–544
Nave K-A, Boespflug-Tanguy O 1996 Developmental defects of myelin formation: from X-linked mutations to human dysmyelinating diseases. Neuroscientist 2:33–43
Nelles E, Bützler C, Jung D et al 1996 Defective propagation of signals generated by sympathetic nerve stimulation in the liver of connexin32-deficient mice. Proc Natl Acad Sci USA 93:9565–9570
Nicholson G, Nash J 1993 Intermediate nerve conduction velocities define X-linked Charcot-Marie-Tooth neuropathy families. Neurology 43:2558–2564
Oh S, Ri Y, Bennett MVL, Trexler EB, Verselis VK, Bargiello TA 1997 Changes in permeability caused by connexin 32 mutations underlie X-linked Charcot-Marie-Tooth disease. Neuron 19:927–938
Omori Y, Mesnil M, Yamasaki H 1996 Connexin 32 mutations from X-linked Charcot-Marie-Tooth disease patients: functional defects and dominant negative effects. Mol Biol Cell 7:907–916
Rabadan-Diehl C, Dahl G, Werner R 1994 A connexin-32 mutation associated with Charcot-Marie-Tooth disease does not affect channel formation in oocytes. FEBS Lett 351:90–94
Rozear MP, Pericak-Vance MA, Fischbeck K et al 1987 Hereditary motor and sensory neuropathy, X-linked: a half century follow-up. Neurology 37:1460–1465
Sander S, Nicholson GA, Ouvrier RA, McLeod JG, Pollard JD 1998 Charcot-Marie-Tooth disease: histopathological features of the peripheral myelin protein (PMP22) duplication (CMT1A) and connexin32 mutations (CMTX1). Muscle Nerve 21:217–225
Sandri C, Van Buren JM, Akert K 1982 Membrane morphology of the vertebrate nervous system. Prog Brain Res 46:201–265
Scherer SS 1997 Molecular genetics of demyelination: new wrinkles on an old membrane. Neuron 18:13–16
Scherer SS, Deschênes SM, Xu Y-T et al 1995 Connexin32 is a myelin-related protein in the PNS and CNS. J Neurosci 15:8281–8294
Scherer SS, Xu Y-T, Nelles E, Fischbeck K, Willecke K, Bone LJ 1998 Connexin32-null mice develop a demyelinating peripheral neuropathy. Glia 24:8–20
Suter U, Snipes GJ 1995 Biology and genetics of hereditary motor and sensory neuropathies. Annu Rev Neurosci 18:45–75
Tetzlaff W 1982 Tight junction contact events and temporary gap junctions in the sciatic nerve fibres of the chicken during Wallerian degeneration and subsequent regeneration. J Neurocytol 11:839–858
Yoshimura T, Satake M, Ohnishi A, Tsutsumi Y, Fujikura Y 1998 Mutations of connexin32 in Charcot-Marie-Tooth disease type X interfere with cell-to-cell communication but not cell proliferation and myelin-specific gene expression. J Neurosci Res 51:154–161

## DISCUSSION

*Gilula:* This is a well-known subject in the gap junction field, and it is probably the best example of where some clinical information drove the development of specific experiments to address connexin gene utilization. Steve Scherer has personally taken this to a point well beyond the initial descriptive observations to appreciating the importance of understanding how, following the mutation of a sequence, a cell in the relevant context handles that mutated sequence. Hiroshi Yamasaki has made some similar observations independently, have you any comments on this?

*Yamasaki:* We have looked at the function and localization of various X-linked Charcot-Marie-Tooth disease (CMTX) connexin32 (Cx32; $\beta_1$) mutants, i.e. a truncation of codon 220 as well as amino acid substitutions at codons 139 and 143. We found that the codon 220 truncation mutants were localized normally in HeLa cells, and communication was restored. In contrast, the two amino acid substitution mutants were localized abnormally and communication was not restored. Moreover, these two mutants abrogated the function of wild-type Cx32 in a dominant negative fashion.

*Dermietzel:* Steve Scherer mentioned one of the first papers on gap junctions in peripheral myelin (Tetzlaff 1982), but he didn't point out that the freeze-fracture studies were done on regenerating Schwann cells and not on non-myelinated Schwann cells. Regenerating Schwann cells up-regulate the expression of gap junctions, but gap junction plaques have rarely been observed in Schwann cells (Sandri et al 1982). We found that in regenerating Schwann cells there is a down-regulation of Cx32 followed by the appearance of Cx46 ($\alpha_3$), so it appears that under regenerating conditions Schwann cells can switch expression from one connexin to another (Chandross et al 1996).

*Scherer:* It has been a problem to find freeze-fracture images that correspond to what I believe must be there. Gap junctions in non-compact myelin (incisures and paranodes) have been difficult to find.

*Kistler:* There are so many mutants of Cx32 that have defects in trafficking and therefore fail to reach the plasma membrane. Is there any indication that these mutations are clustered within the connexin sequence?

*Scherer:* My hunch is that that they are not clustered within one particular spot, but that's based on a handful of mutations; there are more than 130 different mutations to be considered. We will have to look at many more mutants to get a clear idea of whether there is a correlation between the location of the mutation and protein trafficking.

*Werner:* In collaboration with Victor Ionasescu, we have characterized two promoter mutations that cause CMTX (Ionasescu et al 1996). One of these mutations is in the 5′ untranslated region, and was also found by the neurologist Michele d'Urso (International Institute of Genetics and Biophysics, Naples, Italy). He did a sural nerve biopsy on a female patient, who was a physician and therefore agreed to have it done. The sural nerve is in the foot, and if it is removed you only lose sensory function. The patient was heterozygous, and expressed only wild-type mRNA but not the mutant mRNA. We speculated that this mutation produced a new splice donor site, such that the entire 5′ untranslated region of the mRNA was missing, which resulted in it becoming unstable. One of my students recreated these mutations in the mouse, and we are just starting to analyse the F1 generation. Within a few months we should find out what happens to the mutant Cx32 mRNA.

*Scherer:* There are a number of CMTX families that don't have mutations in the coding region, suggesting that they have non-coding region mutations of Cx32. I know of one CMTX family in which the entire gene encoding Cx32 is deleted (Ainsworth et al 1998).

*Evans:* I have a comment about the ability of these mutant connexins to traffic. If these mutant connexins can't oligomerize then they won't form a channel and they may not even go any further than the Golgi apparatus. For example, we studied a tryptophan 3 to tyrosine mutation in Cx32 that doesn't oligomerize, and it just stays in the endoplasmic reticulum. This also indicates that is part of the targeting sequence responsible for directing this protein to the correct location in the cell.

*Lench:* Are there cases where patients with the same mutation show different degrees of disease severity?

*Scherer:* Yes. The severity of disease in CMTX patients varies considerably, even within the same kindred. This variability also obscures our ability to determine whether different mutations cause different degrees of neuropathy.

*Yamasaki:* The codon 220 truncation mutation is functional in cultured cells, so it would be worthwhile to make transgenic mice with that mutation.

*Scherer:* That's a good point. Transgenic mice expressing the 220 truncation would be interesting, because your group showed that this mutation forms functional gap junctions in mammalian cells. We chose the 175 frameshift and Arg142Trp mutations for our transgenic experiments, however, before this was known. I expect that the 220 truncation will be localized to incisures; this raises the interesting question of how this mutation causes demyelination.

## References

Ainsworth PJ, Bolton CF, Murphy BC, Stuart JA, Hahn AF 1998 Genotype/phenotype correlation in affected individuals of a family with a deletion of the entire coding sequence of the connexin 32 gene. Hum Genet 103:242–244

Chandross KJ, Kessler JA, Cohen RI et al 1996 Altered connexin expression after peripheral nerve injury. Mol Cell Neurosci 7:501–518

Ionasescu VV, Searby Ch, Ionasescu R, Neuhaus IM, Werner R 1996 Mutations of the nerve-specific promoter of the connexin32 gene in two families with X-linked dominant Charcot-Marie-Tooth neuropathy. Neurology 47:541–544

Sandri C, Van Buren JM, Akert K 1982 Membrane morphology of the vertebrate nervous system. Prog Brain Res 46:201–265

Tetzlaff W 1982 Tight junction contact events and temporary gap junctions in the sciatic nerve fibres of the chicken during Wallerian degeneration and subsequent regeneration. J Neurocytol 11:839–858

# Cardiovascular disease

Nicholas J. Severs

*National Heart and Lung Institute, Imperial College of Science, Technology and Medicine, Royal Brompton Hospital, Sydney Street, London SW3 6NP, UK*

*Abstract.* Gap junctions play essential roles in the normal function of the heart and arteries, mediating the spread of the electrical impulse that stimulates synchronized contraction of the cardiac chambers, and contributing to co-ordination of function between cells of the arterial wall. Altered gap junctional coupling is implicated in the genesis of arrhythmia, a major cause of death in heart disease. Two abnormalities in myocardial gap junctions identified in human ischaemic heart disease — localized disordering of gap junction distribution at the border zone of infarcts and reduced levels of connexin43 (Cx43; $\alpha_1$) — may lead to heterogeneous wavefront propagation and lowered conduction velocity, key factors that precipitate arrhythmia. In the major arteries, endothelial cells express Cx40 ($\alpha_5$) and Cx37 ($\alpha_4$) and, in some instances, also Cx43, whereas underlying medial smooth muscle cells express only Cx43. Increased Cx43 expression between medial smooth muscle cells is intimately linked to phenotypic transformation to the synthetic state in both early human coronary atherosclerosis, and in the response of the arterial wall to injury. The accumulating evidence suggests that gap junctions in both their guises — as pathways for cell-to-cell signalling in the vessel wall and as pathways for impulse conduction in the heart — may have key roles in the initial pathogenesis and eventual clinical manifestation of human cardiovascular disease.

*1999 Gap junction-mediated intercellular signalling in health and disease. Wiley, Chichester (Novartis Foundation Symposium 219) p 188–211*

Cardiovascular disease is the leading cause of death and disability in most industrialized countries of the developed and developing worlds. Arrhythmias (i.e. disturbances of the normal heart rhythm) are a common, serious and often fatal complication of many forms of heart disease. As gap junctions mediate the orderly cell-to-cell transmission of the action potentials that governs regular synchronous contraction in the healthy heart, derangements of these junctions constitute one possible cause of arrhythmia in heart disease (Smith et al 1991, Saffitz et al 1992, Severs et al 1996). The most prevalent form of cardiovascular disease — ischaemic heart disease — starts long before cardiac symptoms become apparent, with the growth of atherosclerotic lesions in the coronary arteries. Atherosclerosis is the process by which the inner layer (intima) of the arterial wall becomes thickened as a result of smooth muscle cell proliferation, overproduction of extracellular matrix, lipid deposition and calcification. Narrowing of the vessel lumen by advanced atherosclerosis with resultant restriction of blood flow leads to cardiac ischaemia, often accompanied by angina (heart pain) on exercise. Apart from causing progressive narrowing of the arterial lumen, atherosclerotic lesions are prone to sudden rupture, resulting in coronary thrombosis, blockage of the vessel lumen and myocardial infarction (death and scarring of the portion of heart tissue served by the blocked vessel). The pathogenesis of atherosclerosis involves a complex set of interactions among cells of the arterial wall (Ross 1995), and recent findings raise the possibility that gap junctions participate in these interactions (Navab et al 1991, Blackburn et al 1995). This chapter aims to survey selected recent findings on the possible roles of gap junctions in acquired adult cardiovascular disease, set in the context of current knowledge of these junctions and their component connexins in the normal heart and arterial wall.

## Antipeptide antibodies to detect cardiovascular connexins

Four main connexins—connexin43 (Cx43; $\alpha_1$), Cx40 ($\alpha_5$), Cx45 ($\alpha_6$) and Cx37 ($\alpha_4$)—are expressed in cardiovascular cells. To investigate these connexins, we have raised a series of connexin-specific polyclonal antibodies using peptide antigens that match unique sequences of each connexin type (Table 1). Use of different host species for producing these antibodies makes it possible to undertake multiple immunolabelling with highly specific secondary detection systems (based on secondary antibodies against the different host species), thereby permitting simultaneous visualization of the expression patterns of different combinations of connexins. Although antibody characterization is routine, there are potential pitfalls; a rigorous approach is therefore essential. All the new polyclonal antibodies from which results are presented here were affinity purified, characterized by western blotting and immunolabelling of HeLa cell transfectants expressing the different connexin types (kindly provided by Klaus Willecke), and by immunogold labelling at the electron microscopic level to confirm binding of the antibodies to morphologically visualized gap junctions. This last approach necessitates low denaturation techniques to preserve antigenicity, in particular post-embedding immunogold labelling of ultra-thin sections of specimens embedded at low temperature in acrylic resins (Fig. 1a), and freeze-fracture cytochemistry (Fig. 1b). In the latter, physical fixation by freezing replaces chemical fixation, and immunogold-labelled connexins are viewed superimposed upon *en face* freeze-fracture views of the gap junction plaque (Severs 1995, Fujimoto 1997).

**TABLE 1 Examples of peptide antigens used for generating antibodies against connexin43 (Cx43; $\alpha_1$), Cx40 ($\alpha_5$), Cx45 ($\alpha_6$) and Cx37 ($\alpha_4$)**

| *Connexin* | | *Peptide sequence* | | | | | | | | | | | | | | | | | *Peptide abbreviation* | *Host* |
|---|---|---|---|---|---|---|---|---|---|---|---|---|---|---|---|---|---|---|---|---|
| Cx43 | Rat | 131 | E | I | K | K | F | K | Y | G | I | E | E | H | 142 | | | | E12H | Rabbit |
| | Human | | – | – | – | – | – | – | – | – | – | – | – | – | | | | | | HJ* |
| | Mouse | | – | – | – | – | – | – | – | – | – | – | – | – | | | | | | |
| Cx40 | Rat | 254 | S | L | V | Q | G | L | T | P | P | P | D | F | N | Q | C | 268 | S15C | Rabbit |
| | Mouse | | – | – | – | – | S | – | – | – | – | – | – | – | – | – | – | | | R83*, R84, R85 |
| | Human | | A | I | – | – | S | C | – | – | – | – | – | – | – | – | – | | | |
| | Rat | 256 | V | Q | G | L | T | P | P | P | D | F | N | Q | C | L | K | 270 | V15K | Guinea pig |
| | Mouse | | – | – | S | – | – | – | – | – | – | – | – | – | – | – | – | | | GP318*, GP319* |
| | Human | | – | – | S | C | – | – | – | – | – | – | – | – | – | – | E | | | Rabbit R404 |
| Cx45 | Human | 354 | Q | A | Y | S | H | Q | N | N | P | H | G | P | R | E | 367 | | Q14E | Guinea pig |
| | Mouse | | – | – | – | H | – | – | – | – | – | – | – | – | – | – | | | | GP42*, GP43 |
| | Dog | | – | – | – | – | – | – | – | – | – | – | – | – | – | – | | | | |
| | Human | 300 | Q | Y | T | E | L | S | N | A | K | I | A | Y | K | Q | N | 314 | Q15N | Rabbit |
| | Mouse | | – | – | – | – | – | – | – | – | – | – | – | – | – | – | – | | | R401, R402* |
| | Dog | | – | – | – | – | – | – | – | – | – | – | – | – | – | – | – | | | |
| | Human | 141 | Y | P | E | M | E | L | E | S | E | K | E | N | K | E | Q | 155 | Y15Q | Sheep |
| | Mouse | | – | – | – | – | – | – | – | – | – | – | – | – | – | – | – | | | SH45 |
| | Dog | | – | – | – | – | – | – | – | – | – | – | – | – | – | – | – | | | |

| Connexin | Species | Start | | | | | | | | | | | | | | | | End | Peptide | Animal / antibodies |
|---|---|---|---|---|---|---|---|---|---|---|---|---|---|---|---|---|---|---|---|---|
| Cx37 | Human | 253 | G | T | S | S | D | P | Y | T | D | Q | G | | | | | 263 | G11G | Guinea pig |
| | Rat | | – | S | A | – | – | – | – | P | E | – | V | | | | | | | GP40*, GP41 |
| | Mouse | | – | S | A | – | – | – | – | P | E | – | V | | | | | | | |
| | Rat | 244 | R | D | H | D | T | R | P | A | Q | G | S | A | | | | | 255 | R12A | Rabbit |
| | Mouse | | – | – | – | – | A | – | – | – | – | – | – | – | | | | | | | R53, R54 |
| | Human | | Q | G | Q | – | A | P | – | T | – | – | T | S | | | | | | | |
| | Rat | 131 | E | H | Q | M | A | K | I | S | V | A | E | D | G | R | L | R | 146 | E16R | Sheep |
| | Mouse | | – | – | – | – | – | – | – | – | – | – | – | – | – | – | – | – | | | SH37 |
| | Human | | – | L | – | – | – | – | – | – | – | – | – | – | – | – | – | – | | | |
| | Rat | 266 | Y | L | P | M | G | E | G | P | S | S | P | P | C | P | T | Y | 281 | Y16Y | Rabbit |
| | Mouse | | – | – | – | – | – | – | – | – | – | – | – | – | – | – | – | – | | | R3*, R4 |
| | Human | | – | P | – | R | – | Q | – | – | – | – | – | – | – | – | – | – | | | |

The start and end residue numbers for the species from which the sequence is taken are given in each case, together with sequence comparisons for the same connexin type in other species. The abbreviated designations used for each peptide consist of letters denoting the first and last amino acids, with an intervening number indicating the number of amino acids. In describing individual antibodies, current practice is to use a two-part nomenclature, comprising the abbreviated peptide sequence and the code of the specific animal in which it was raised (e.g. antibody S15C[R83] is the antibody raised against sequence S15C in rabbit number 83). Asterisks indicate antibodies used for the findings discussed in this review. The anti-connexin antibodies are routinely affinity purified and characterized by western blot analysis and immunofluorescence labelling of transfected cells, and by immunogold labelling at the electron microscopic level. (For the peptide sequences, – indicates same amino acid as in sequence above.)

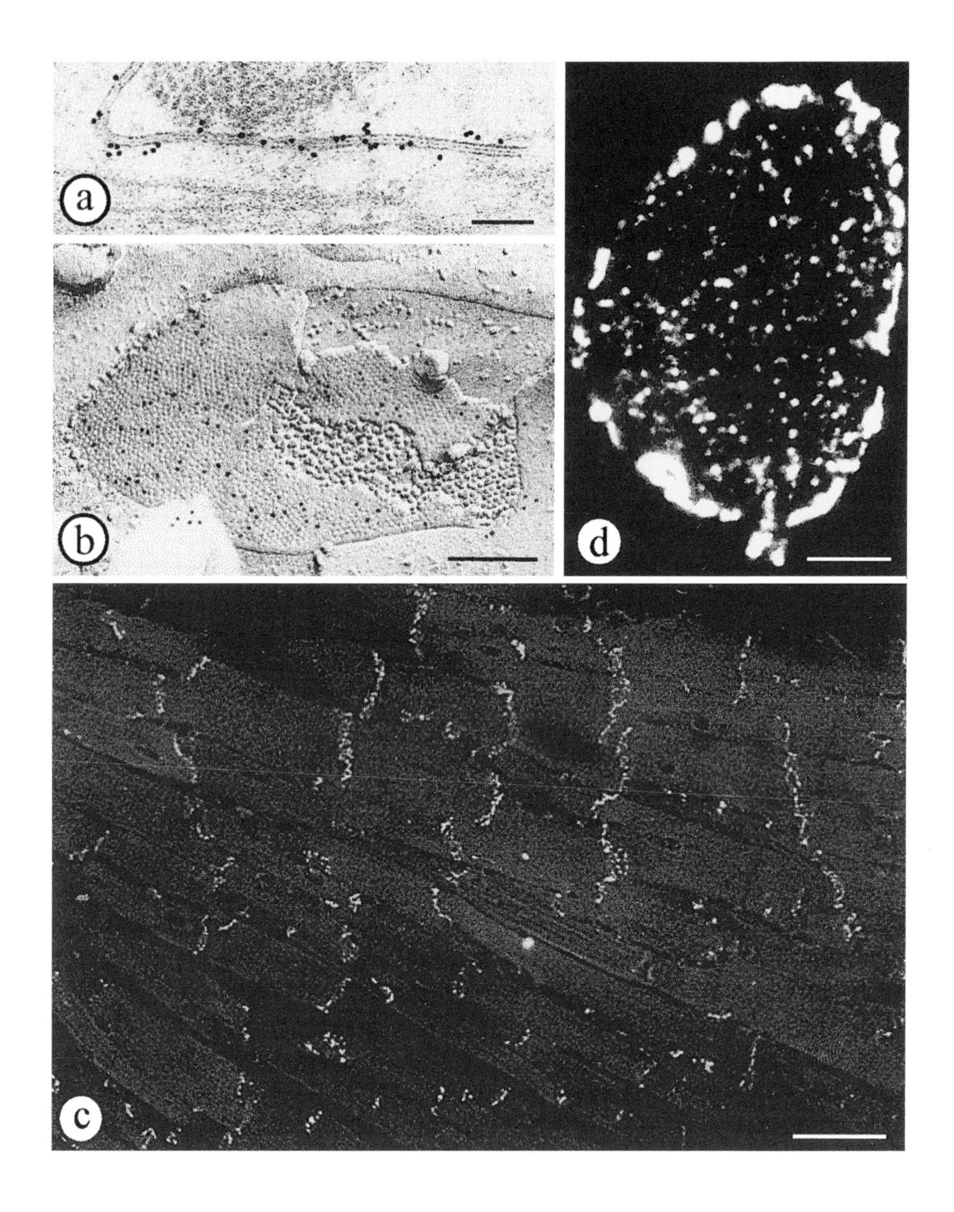
a
b
c
d

## Gap junction organization in normal heart

Cardiac myocytes of the working ventricular and atrial myocardium are elongated, branching, contractile cells, extensively interconnected by clusters of gap junctions organized in intercalated disks (Fig. 1c). The intercalated disks also contain two types of anchoring junction, the fascia adherens and desmosome, which mediate cell-to-cell linkage of the contractile filaments and intermediate filament cytoskeleton, respectively, thereby acting in concert to integrate cardiac electromechanical function. In the intercalated disks of working ventricular myocardium, a population of particularly large gap junctions typically circumscribes the disk periphery (Fig. 1d), in a plane parallel with the lateral surfaces of adjoining myocytes (Gourdie et al 1991). The polar organization of gap junctions into rings at the intercalated disks is thought to favour propagation of the impulse in the longitudinal axis, contributing, together with other features of cellular and junction organization, to the normal pattern of anisotropic spread of the impulse of healthy contractile myocardium. Variations in size, abundance, distribution and connexin make-up of gap junctions in different subsets of myocytes are hypothesized to act as major determinants of the distinctive electrophysiological properties of different regions of the heart (Saffitz et al 1995, Gros & Jongsma 1996, Severs et al 1996). In the atrioventricular node, for example, where slowing of conduction ensures sequential contraction of atria and ventricles, gap junction organization is quite unlike that of working myocardium, the junctions being sparse, small and dispersed in distribution.

---

FIG. 1. Features of connexin43 (Cx43; $\alpha_1$) gap junctions in normal myocardium. (a) and (b) Specificity of anti-connexin antibodies is routinely demonstrated by immunogold labelling at the electron microscopic level, prior to application in immunoconfocal microscopy. (a) Shows immunogold labelling of an ultra-thin section of myocardium embedded at low temperature in Lowicryl. Bar = 100 nm. (b) Illustrates immunogold labelling of myocardium by freeze-fracture cytochemistry. In the example in (b), a thin layer of proteins has been retained on the replica by judicious treatment with sodium dodecyl sulphate, making it possible to carry out immunogold labelling of the replica. Bar = 100 nm. Note that in both (a) and (b) the gold labels (black dots) are specifically associated with morphologically visualized gap junctions. Both examples show labelling for Cx43 in left ventricular myocardial gap junctions. (c) Immunoconfocal microscopy of longitudinally sectioned left ventricular myocardium illustrates the characteristic organization of gap junctions in clusters at the intercalated disks between myocytes. The cells have a branching structure such that each myocyte is on average linked to about 10 others by means of intercalated disks. Atrial myocytes are slender compared with ventricular myocytes, and their gap junctions occur in smaller disks and in groups at lateral contacts between cells. Bar = 50 $\mu$m. (d) In *en face*-viewed intercalated disks reconstructed from stacks of serial optical sections through transversely sectioned myocardium, a conspicuous ring of large gap junctions is seen circumscribing the disk periphery ([c] and [d] are from human left ventricular myocardium; antibody E12H[HJ]). From Smith et al (1991), Peters et al (1993). Bar = 10 $\mu$m.

In ventricular myocardium, the organization of gap junctions into intercalated disks does not take place until after birth, reaching completion at around three months of age in the rat (Angst et al 1997) and six years in the human (Peters et al 1994a). In the neonatal heart, gap junctions are distributed over the entire surfaces of the cells (Peters et al 1994a, Angst et al 1997). Examination of the sequence of events in early postnatal growth in the rat suggests that intercalated disks are first established by the anchoring junctions (desmosomes and fasciae adherentes); only then are gap junctions assembled into the preformed disks (Angst et al 1997). The changing pattern of gap junction organization during postnatal development marks a period of adaptive growth accompanied by alterations in myocardial mechanical and electrical properties. In the human heart, corrective surgery for congenital malformations carried out during the period of adaptive growth, prior to the fully formed disk stage, is associated with fewer late postoperative arrhythmias and with improved cardiac function compared with late surgery (Peters et al 1994a).

## Spatial distribution of connexins in normal heart

The predominant connexin of cardiac muscle is Cx43, found in abundance in the working ventricular and atrial myocardium of all mammalian species (Beyer et al 1989; reviews in Saffitz et al 1995, Gros & Jongsma 1996, Severs et al 1996). In addition to Cx43, Cx40 is typically expressed in atrial muscle cells and, as illustrated in Fig. 2a, in specialized cells of the ventricular conduction system (i.e. those of the conducting bundle branches and Purkinje fibres [Bastide et al 1993, Gourdie et al 1993, Gros et al 1994; for species-related variations see Gros et al 1994, van Kempen et al 1995]). As gap junction channels composed of Cx40 have been shown *in vitro* to have high conductance values (typically 160–200 pS) compared with Cx43 channels ($\sim$60 pS; Bukauskas et al 1995), the presence of Cx40 in Purkinje fibre cells is thought to contribute to the fast conduction properties characteristic of these cells which ensure rapid distribution of the impulse throughout the working ventricular myocardium.

FIG. 2. Confocal micrographs showing double labelling for connexin40 (Cx40; $\alpha_5$) and Cx45 ($\alpha_6$) in mouse ventricular myocardium (presented as separate images). (a) Cx40 is expressed by myocytes of the ventricular conduction system (VCS) which runs close to the endocardial surface (Endo). Cx40 is not detectable in underlying working myocardium (WM). (b) Cx45, like Cx40, is expressed towards the endocardial surface, predominantly co-localizing at the same sites as Cx40. Although some Cx45 may occur in the WM close to the conduction cells (arrow), the working ventricular myocardium is otherwise largely immunonegative for Cx45. Antibodies S15C(R83) and Q14E(GP42). For further details see Coppen et al (1998). Bar = 25 $\mu$m.

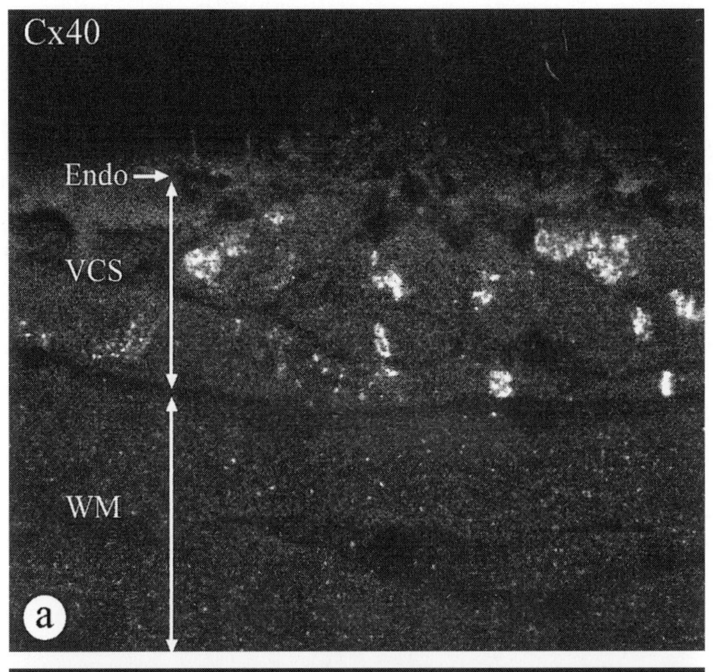
Cx40
Endo
VCS
WM
a

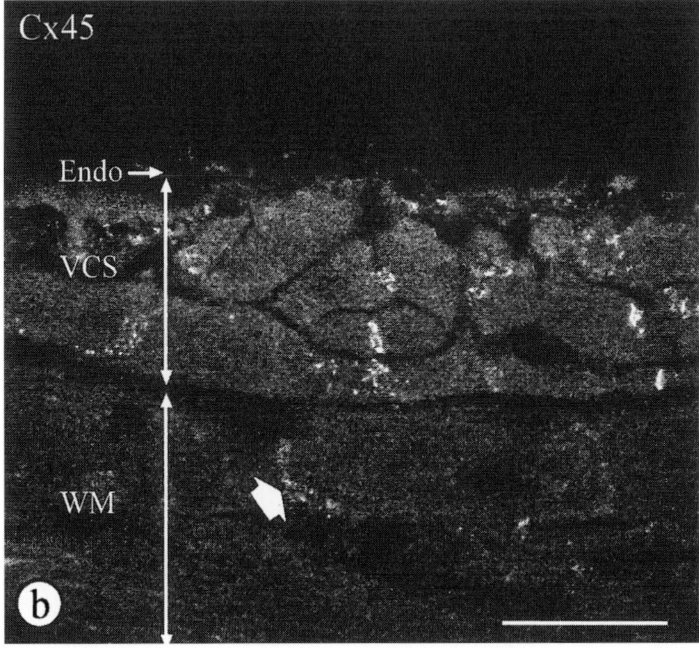
Cx45
Endo
VCS
WM
b

A third connexin, Cx45, is reported to be widely distributed throughout the ventricular and atrial myocardium (Kanter et al 1992, Davis et al 1995; review in Gros & Jongsma 1996). However, using a newly raised and carefully characterized anti-Cx45 antibody (Table 1), we have recently found that Cx45 has a highly restricted distribution in the rodent heart, co-localizing with Cx40-expressing myocytes of the ventricular conduction system (Fig. 2). Although some Cx45 is expressed in immediately adjacent Cx40-negative cells, much of the working rodent ventricle appears immunonegative for Cx45 (Coppen et al 1998). By contrast, a commercially available Cx45 antiserum (raised against the same sequence as used in other published studies on the distribution of cardiac Cx45) gives prominent widespread labelling throughout rodent ventricular myocardium (Coppen et al 1998). The peptide antigen to which this commercially available antiserum was raised has a sequence of four amino acids (PPGY) which are also present in the Cx43 molecule. A six-amino acid peptide corresponding to the part of the Cx43 molecule containing this sequence inhibits the widespread ventricular labelling, suggesting cross-reaction of the commercial Cx45 antiserum with Cx43 in the tissue (Coppen et al 1998). The presence of Cx45 in the conduction system provides an explanation for the observation of first degree atrioventricular block (i.e. impulses continue to reach the ventricle but are delayed) rather than complete heart block in the Cx40 knockout mice (this volume: Willecke et al 1999, Goodenough et al 1999).

## Heart disease, gap junctions and arrhythmia

Most ventricular arrhythmias in human heart disease are due to re-entrant electrical circuits. Re-entry arrhythmias may be classified into two groups: 'macro-reentry' and 'micro-reentry'. Macro-reentry circuits related to an accessory atrioventricular conduction pathway (Wolff-Parkinson White syndrome) appear to arise from an aberrant additional gap junction-coupled link between the atrium and ventricle bypassing the atrioventricular node, as demonstrated in some surgically resected pathways (Peters et al 1994b). The micro-reentry arrhythmias common in patients with ischaemic and hypertrophic heart diseases are precipitated by heterogeneous wavefront propagation, reduced conduction velocity and localized unidirectional block. Two forms of gap junction defect that could plausibly contribute to these conditions have been identified in the diseased human heart: altered gap junction distribution and reduced levels of Cx43 (Smith et al 1991, Peters et al 1993, Severs et al 1996, Kaprielian et al 1998).

Where non-fatal myocardial infarction occurs in ischaemic heart disease, loss of the normal ordered distribution of Cx43 gap junctions is conspicuous in the myocardial zone bordering infarct scar tissue (Smith et al 1991). Gap junctions in these zones are scattered extensively over the lateral borders of the cells, whereas those at more distant sites remain in clearly ordered intercalated disk arrays. Such abnormal gap junction distributions are not solely related to late stages in myocardial degeneration or remodelling associated with fibrosis, but are detectable within a few days after myocardial infarction in rat and dog models (Matsushita et al 1996, Peters et al 1997). By using high resolution electrode arrays to map re-entrant circuits in the epicardial border zone overlying four-day-old infarcts in a canine model of infarction, and then mapping gap junction distribution in precisely the same zones by immunoconfocal microscopy, regions of full thickness gap junction disarray are found to correspond in position to the common central pathway of figure-of-eight re-entrant circuits (Peters et al 1997). Abnormal patterns of gap junction distribution have also been identified in other cardiac abnormalities associated with an increased arrhythmic tendency, notably hypertrophic cardiomyopathy (Sepp et al 1996).

In addition to disrupted patterns of gap junction organization in the immediate vicinity of the infarct, quantitative immunoconfocal microscopy reveals a generalized decrease in immunolabelled Cx43 gap junction content per myocyte throughout the left ventricular myocardium of ischaemic heart disease patients undergoing coronary by-pass operations (Peters et al 1993). Decreased Cx43 protein is similarly detectable in western blots of left ventricular tissue from transplant patients with end-stage ischaemic heart disease, and quantitative northern blotting of these samples demonstrates down-regulation of Cx43 at the level of transcription (Dupont et al 1997). Reduced levels of immunodetectable Cx43 are found in the ventricle both in patients with exercise-induced reversible ischaemia (Peters et al 1993) and in patients with hibernating myocardium (Kaprielian et al 1998). 'Hibernation' defines a state in which a region of the myocardium shows persistent impairment of contractile function at rest owing to reduced coronary blood flow through severely atherosclerotic coronary arteries, but in which contractile function recovers after coronary by-pass surgery (Rahimtoola 1989). However, the reduction in immunolabelled gap junctions is greater in hibernating myocardium than in reversibly ischaemic myocardium, with loss of the large gap junctions at the disk periphery featuring prominently in the hibernating group (Fig. 3). Prognosis in medically treated patients with ischaemic heart disease and impaired ventricular function is poorer when there is evidence for the presence of hibernating myocardium in addition to reversible ischaemia, and the majority of these cardiac deaths are sudden, consistent with a link between reduced Cx43 levels and arrhythmia. Apart from ischaemic heart disease, reduced expression of Cx43 in the left ventricular myocardium may be a factor in arrhythmogenesis in other cardiac disease settings (Campos De Carvalho et al 1994). Whether, in addition to Cx43, altered expression of other connexin types may contribute to a pro-arrhythmic tendency in ischaemic and other human heart disease is still under investigation (Severs et al 1996).

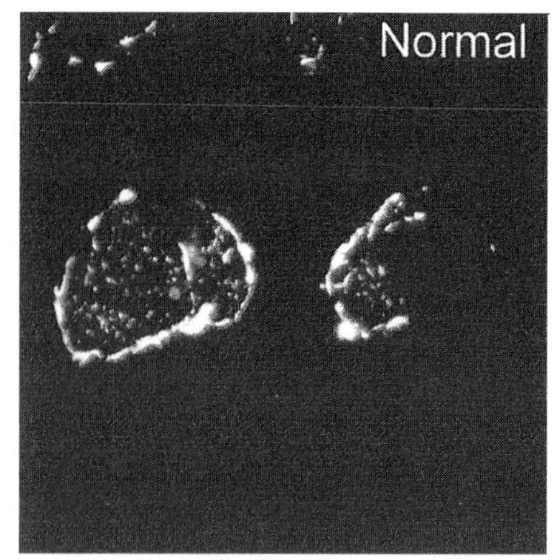
Normal

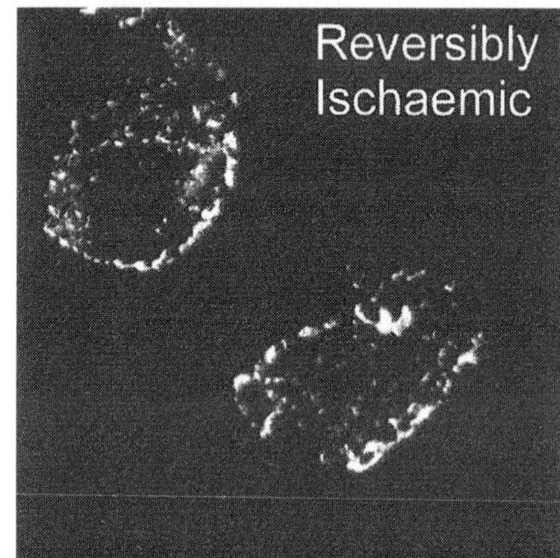
Reversibly
Ischaemic

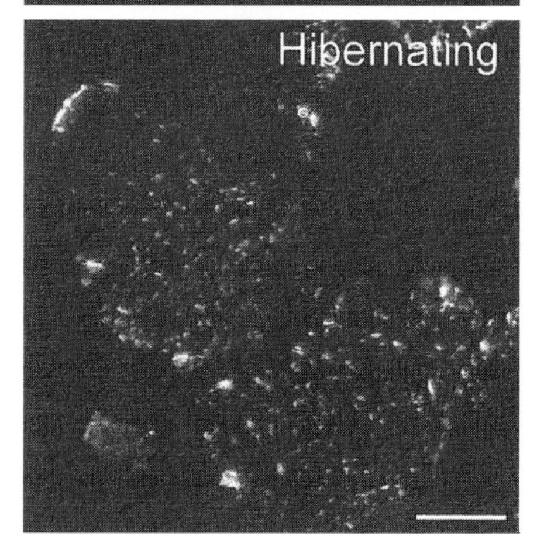
Hibernating

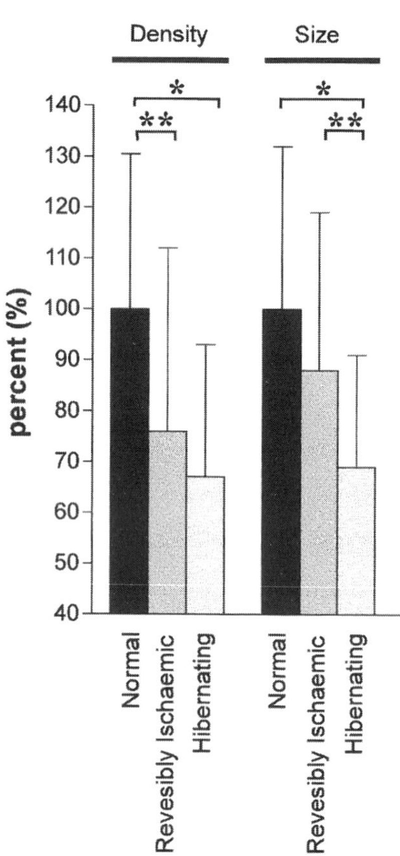
Density
Size
140
130
120
110
100
90
80
70
60
50
40
percent (%)
Normal
Revesibly Ischaemic
Hibernating
Normal
Revesibly Ischaemic
Hibernating

## Gap junctions and connexin expression in the normal arterial wall

In the healthy artery, gap junctional communication between vascular wall cells is thought to contribute to general circulatory homeostasis and the local modulation of vasomotor tone (Christ et al 1996). The endothelium is typically more extensively linked by gap junctions than is the underlying medial smooth muscle. Contacts between endothelial cells and superficial smooth muscle cells occur through discontinuities in the internal elastic lamina of smaller coronary arteries and arterioles, and heterocellular gap junctions are sometimes present at these sites. Dye tracer studies in isolated arterioles indicate that gap junctional coupling between endothelial cells is more extensive than that between the underlying smooth muscle cells, and heterocellular communication may be predominantly unidirectional, from endothelium to smooth muscle (Little et al 1995a).

The connexin composition of vascular cell gap junctions varies according to cell type and location. Smooth muscle cells of larger arteries *in situ* express Cx43. Although Cx40 has also been reported in some arterioles and in the A7r5 aortic smooth muscle cell line (Moore & Burt 1994, Little et al 1995b), this connexin is not detectable in smooth muscle cells of larger arteries *in situ*. However, Cx40 is abundant in endothelial cells of large arteries where it is co-expressed with Cx37 and, in some instances, also Cx43 (Reed et al 1993, Little et al 1995b, Yeh et al 1997a). For example, aortic and pulmonary artery endothelia express Cx40, Cx37 and Cx43, whereas coronary artery endothelium expresses Cx40 and Cx37 but lacks Cx43 (Yeh et al 1997a). Multiple immunogold label electron microscopy demonstrates that where Cx40, Cx37 and Cx43 are co-expressed, as in aortic and pulmonary artery endothelium, individual gap junctional plaques commonly contain all three connexin types (Fig. 4; Ko et al 1997). The diverse connexin make-up of arterial endothelial gap junctions suggests complex regulation and functional differentiation of endothelial intercellular communication properties.

---

FIG. 3. Reduced expression of connexin43 ($\alpha_1$) in left ventricular myocardium of patients with ischaemic heart disease. The panels on the left illustrate gap junctions in *en face*-viewed intercalated disks from normally perfused, reversibly ischaemic and hibernating regions of myocardium identified by thallium scanning and magnetic resonance imaging. Bar = 10 $\mu$m. The histogram on the right shows quantitative analysis of numerical density and size of immunolabelled gap junctions from reconstructed disks in 15 patients. Reversibly ischaemic and hibernating tissues show significant reductions in immunolabelled gap junctional density (ANOVA $P<0.001$). Hibernating tissue contains significantly smaller gap junctions than other groups (ANOVA $P<0.001$). Bars = S.D. $*P<0.001$; $**P=0.012$. Loss of gap junctions in the hibernating group predominantly involves loss of large gap junctions at the periphery of the disk. From Kaprielian et al (1998).

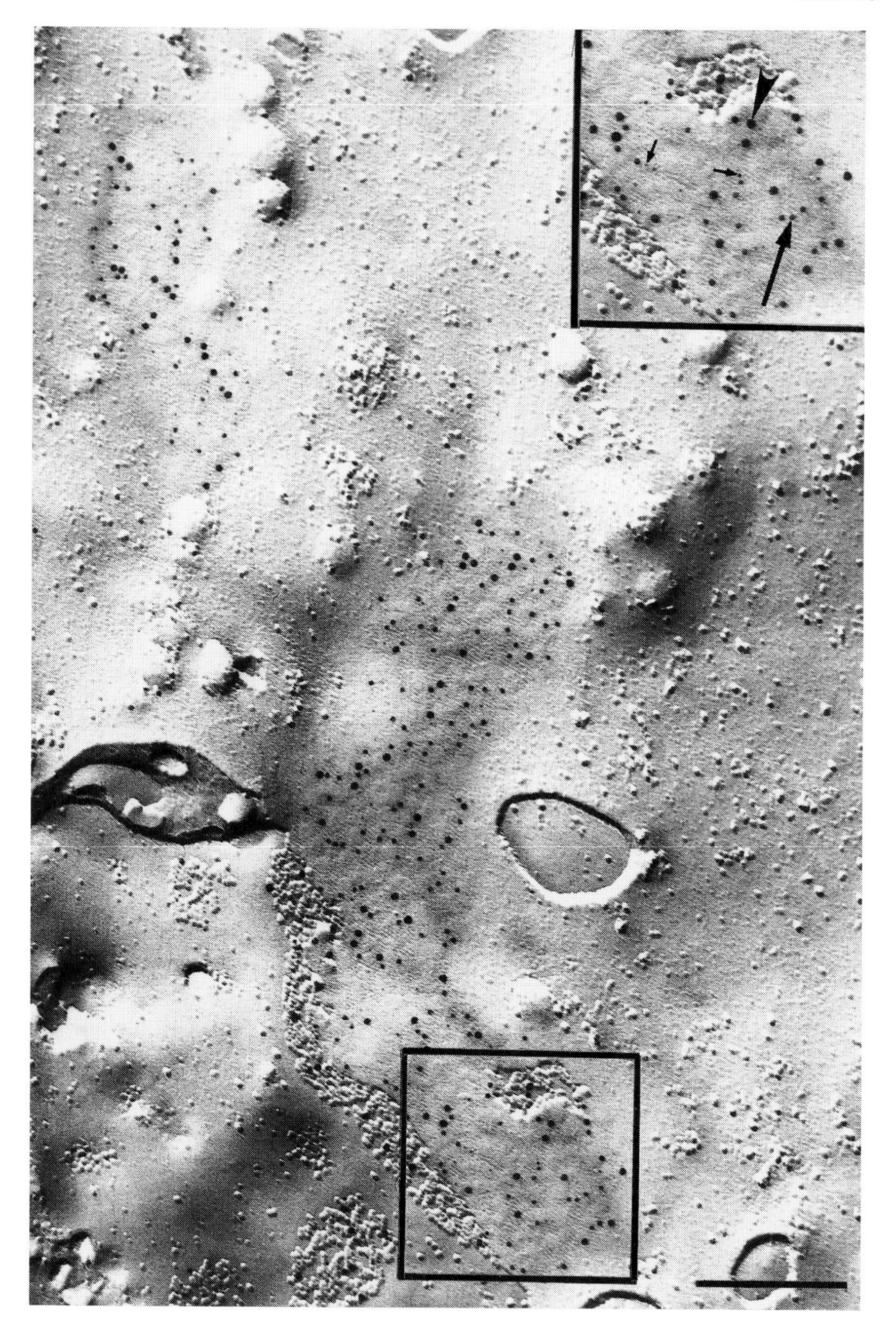

## Gap junctions in diseased and injured arterial walls

Endothelial gap junction coupling has been implicated in the co-ordination of endothelial cell migration and replication after injury and in angiogenesis (Pepper et al 1992, Xie & Hu 1994, Christ et al 1996), but little information is available on whether endothelial connexin expression patterns of the intact artery alter during disease or injury processes. Recent studies on arterial smooth muscle, however, demonstrate that a conspicuous up-regulation of Cx43 gap junctions goes hand-in-hand with the phenotypic transformation of these cells which leads to intimal growth in arterial injury and atherosclerosis. In cultured smooth muscle cells, immunodetectable Cx43 expression is low in contractile-type cells but high in synthetic-type cells (Rennick et al 1993), and this relationship is paralleled in human coronary atherosclerosis (Blackburn et al 1995), where a marked increase in the abundance of Cx43 gap junctions between intimal smooth muscle cells is apparent in the early stages of disease (Fig. 5). With further lesion growth and accumulation of extracellular matrix material, gap junction quantity subsequently declines (Blackburn et al 1995) in line with reduced intercellular communication reported in cultured smooth muscle cells isolated from atherosclerotic lesions (Andreeva et al 1995). A prominent up-regulation of Cx43 gap junctions, similar to that observed in the early stages of atherosclerosis, is also found in smooth muscle cells of the neointima formed after injuring the rat carotid artery with a balloon catheter (Fig. 5). In this procedure, inflation of a balloon within the artery removes the endothelium and stimulates rapid proliferation of smooth muscle cells (Kocher et al 1991). Similar intraluminal balloons are used in balloon angioplasty, an intervention designed to restore coronary artery patency in patients with arteries blocked with atherosclerotic plaque, and a common complication of this procedure is restenosis due to injury-provoked smooth muscle cell proliferation.

In the rat carotid model, a transient increase in smooth muscle cell gap junction expression occurs prior to neointimal formation, in the innermost (subluminal) medial zone, the major site from which the cells subsequently found in the neointima are recruited (Yeh et al 1997b). Enhanced Cx43 gap junction expression by arterial smooth muscle cells *in vivo* thus features as an early event both in the slowly evolving intimal growth of atherosclerotic disease, and in the

FIG. 4. Co-localization of connexin40 (Cx40; $\alpha_5$), Cx37 ($\alpha_4$) and Cx43 ($\alpha_1$) to the same gap junctional plaque in aortic endothelium, as demonstrated by freeze-fracture cytochemistry. A portion of the lower junction (boxed) is shown at high magnification in the inset (top right). In this example, 5 nm gold (small arrows) was used to label Cx37, 10 nm gold (large arrow) to label Cx40 and 15 nm gold (arrowhead) to label Cx43. Bar = 250 $\mu$m. Micrograph by Stephen Rothery.

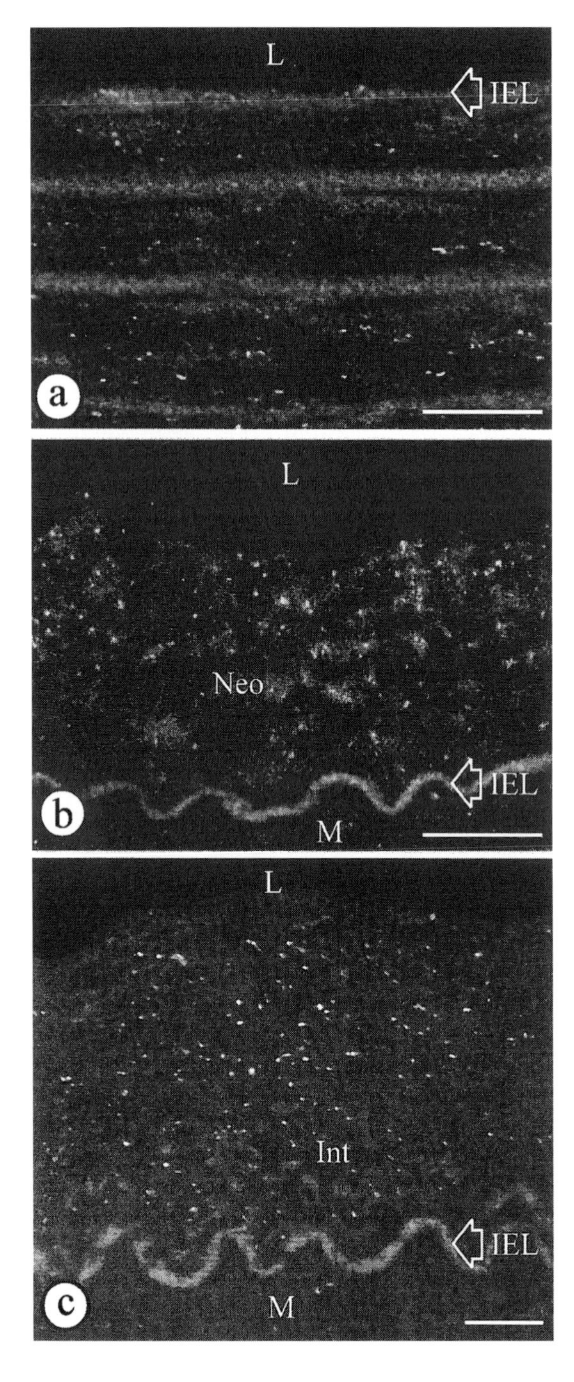

L
IEL
a
L
Neo
IEL
b
M
L
Int
IEL
c
M

much more rapid intimal growth induced by balloon injury. Such an increased potential for gap junctional communication between smooth muscle cells appears to be closely linked both to the phenotypic transition process itself and subsequent maintenance of the synthetic state. Why smooth muscle cells might need to communicate under these circumstances, the nature of the putative signalling molecules involved, and the possible significance of gap junctions as a growth control mechanism in this setting, are key questions for further investigation. Recent evidence that connexin expression in cultured smooth muscle cells is influenced by growth factors (Mensink et al 1996, Yeh et al 1996) raises the possibility of co-operative interactions between extracellular signalling and direct cell-to-cell communication in the modulation of smooth muscle cell behaviour in arterial injury and disease. Thrombin, in particular, has been found to be a potent stimulator of Cx43 expression in cultured arterial smooth muscle cells (Yeh et al 1996), and could plausibly mediate similar effects in prothrombogenic sites denuded of endothelium after balloon injury *in vivo*.

## Concluding comment

The findings presented raise the possibility that gap junctions in both their guises — as pathways for electrical coupling in the myocardium and as pathways for intercellular signalling in arterial cells — could play roles both in arrhythmic dysfunction of the diseased human heart and its most common prelude, pathogenesis of coronary atherosclerosis.

*Acknowledgements*

Work reported and illustrated here from the author's laboratory was supported by the British Heart Foundation (Grant No. PG 97175) and the Wellcome Trust (Grant No. 046218/Z/95). I thank all those who have contributed to the development of this work, in particular C. R. Green, R. G. Gourdie, E. Harfst, N. S. Peters, J. Smith, J. Blackburn, E. Dupont, H.-I. Yeh, R. R. Kaprielian, S. R. Coppen, Y.-S. Ko, C. Vozzi and S. Rothery.

---

FIG. 5. Increased expression of connexin43 (Cx43; $\alpha_1$) gap junctions between intimal smooth muscle cells after balloon catheter injury in the rat carotid artery (a, b) and in the early stages of human coronary atherosclerosis (c). In the uninjured rat carotid artery, only low levels of immunodetectable Cx43 gap junctions are present in the medial smooth muscle (a). After injury (b), abundant gap junctions are detected between smooth muscle cells of the neointima (Neo), formed on the luminal side of the internal elastic lamina (IEL). Gap junctions are similarly abundant in the intima (Int) during the early stages of human coronary atherosclerosis (c). L, lumen; M, media. Bars = 25 $\mu$m.

## References

Andreeva ER, Serebryakov VN, Orekhov AN 1995 Gap junctional communication in primary culture of cells derived from human aortic intima. Tissue Cell 27:591–597

Angst BD, Khan LUR, Severs NJ et al 1997 Dissociated spatial patterning of gap junctions and cell adhesion junctions during postnatal differentiation of ventricular myocardium. Circ Res 80:88–94

Bastide B, Neyses L, Ganten D, Paul M, Willecke K, Traub O 1993 Gap junction protein connexin40 is preferentially expressed in vascular endothelium and conductive bundles of rat myocardium and is increased under hypertensive conditions. Circ Res 73:1138–1149

Beyer EC, Kistler J, Paul DL, Goodenough DA 1989 Antisera directed against connexin43 peptides react with a 43-kd protein localized to gap junctions in myocardium and other tissues. J Cell Biol 108:595–605

Blackburn JP, Peters NS, Yeh H-I, Rothery S, Green CR, Severs NJ 1995 Upregulation of connexin43 gap junctions during early stages of human coronary atherosclerosis. Arterioscler Thromb Vasc Biol 15:1219–1228

Bukauskas FF, Elfgang C, Willecke K, Weingart R 1995 Biophysical properties of gap junction channels formed by mouse connexin40 in induced pairs of transfected human HeLa cells. Biophys J 68:2289–2298

Campos De Carvalho AC, Masuda MO, Tanowitz HB, Wittner M, Goldenberg RCS, Spray DC 1994 Conduction defects and arrhythmias in Chagas' disease: possible role of gap junctions and humoral mechanisms. J Cardiovasc Electrophysiol 5:686–698

Christ GJ, Spray DC, el-Sabban M, Moore LK, Brink PR 1996 Gap junctions in vascular tissues. Evaluating the role of intercellular communication in the modulation of vasomotor tone. Circ Res 79:631–646

Coppen SR, Dupont E, Rothery S, Severs NJ 1998 Connexin45 expression is preferentially associated with the ventricular conduction system in mouse and rat heart. Circ Res 82:232–243

Davis LM, Rodefeld ME, Green K, Beyer EC, Saffitz JE 1995 Gap junction protein phenotypes of the human heart and conduction system. J Cardiovasc Electrophysiol 6:813–822

Dupont E, Vozzi C, Coppen SR, Kaprielian RR, Severs NJ 1997 Connexin mRNA and protein expression is altered in human heart disease. J Mol Cell Cardiol 29:A108 (abstr)

Fujimoto K 1997 SDS-digested freeze-fracture replica labelling electron microscopy to study the two-dimensional distribution of integral membrane proteins and phospholipids in biomembrane: practical procedure, interpretation and application. Histochem Cell Biol 107:87–96

Goodenough DA, Simon AM, Paul DL 1999 Gap junctional intercellular communication in the mouse ovarian follicle. In: Gap junction-mediated intercellular signalling in health and disease. Wiley, Chichester (Novartis Found Symp 219) p 226–240

Gourdie RG, Green CR, Severs NJ 1991 Gap junction distribution in adult mammalian myocardium revealed by an antipeptide antibody and laser scanning confocal microscopy. J Cell Sci 99:41–55

Gourdie RG, Severs NJ, Green CR, Rothery S, Germroth P, Thompson RP 1993 The spatial distribution and relative abundance of gap-junctional connexin40 and connexin43 correlate to functional properties of the cardiac atrioventricular conduction system. J Cell Sci 105:985–991

Gros DB, Jongsma HJ 1996 Connexins in mammalian heart function. BioEssays 18:719–730

Gros DB, Jarry-Guichard T, ten Velde I et al 1994 Restricted distribution of connexin40, a gap junctional protein, in mammalian heart. Circ Res 74:839–851

Kanter HL, Saffitz JE, Beyer EC 1992 Cardiac myocytes express multiple gap junction proteins. Circ Res 70:438–444

Kaprielian RR, Gunning M, Dupont E et al 1998 Down-regulation of immunodetectable connexin43 and decreased gap junction size in the pathogenesis of chronic hibernation in the human left ventricle. Circulation 97:651–660

Ko Y-S, Yeh H-I, Rothery S et al 1997 Simultaneous localization of connexins 40, 37 and 43 in pulmonary artery endothelial gap junctions. In: Werner R (ed) Gap junctions. IOS Press, Amsterdam, p 183–187

Kocher O, Gabbiani F, Gabbiani G et al 1991 Phenotypic features of smooth muscle cells during the evolution of experimental carotid artery intimal thickening. Biochemical and morphologic studies. Lab Invest 65:459–470

Little TL, Xia J, Duling BR 1995a Dye tracers define differential endothelial and smooth muscle coupling patterns within the arteriolar wall. Circ Res 76:498–504

Little TL, Beyer EC, Duling BR 1995b Connexin 43 and connexin 40 gap junctional proteins are present in arteriolar smooth muscle and endothelium *in vivo*. Am J Physiol 268:H729–H739

Matsushita T, Takahashi K, Yokoyama K, Takamatsu T 1996 Three dimensional and temporal expression of connexin43 in ischemic rat heart. Mol Biol Cell 7 (suppl):464a (abstr)

Mensink A, Brouwer A, Van den Burg EH et al 1996 Modulation of intercellular communication between smooth muscle cells by growth factors and cytokines. Eur J Pharmacol 310:73–81

Moore LK, Burt JM 1994 Selective block of gap junction channel expression with connexin-specific antisense oligodeoxynucleotides. Am J Physiol 267:C1371–C1380

Navab M, Ross LA, Hama S et al 1991 Interactions of human aortic wall cells in co-culture. Atheroscler Rev 23:153–160

Pepper MS, Montesano R, El Aoumari A, Gros D, Orci L, Meda P 1992 Coupling and connexin 43 expression in microvascular and large vessel endothelial cells. Am J Physiol 262:C1246–C1257

Peters NS, Green CR, Poole-Wilson PA, Severs NJ 1993 Reduced content of connexin43 gap junctions in ventricular myocardium from hypertrophied and ischaemic human hearts. Circulation 88:864–875

Peters NS, Severs NJ, Rothery SM, Lincoln C, Yacoub MH, Green CR 1994a Spatiotemporal relation between gap junctions and fascia adherens junctions during postnatal development of human ventricular myocardium. Circulation 90:713–725

Peters NS, Rowland E, Bennett JG, Green CR, Anderson RH, Severs NJ 1994b The Wolff-Parkinson-White syndrome: the cellular substrate for conduction in the accessory atrioventricular pathway. Eur Heart J 15:981–987

Peters NS, Severs NJ, Coromilas J, Wit AL 1997 Disturbed connexin43 gap junction distribution correlates with the location of reentrant circuits in the epicardial border zone of healing canine infarcts that cause ventricular tachycardia. Circulation 95:988–996

Rahimtoola SH 1989 The hibernating myocardium. Am Heart J 117:211–221

Reed KE, Westphale EM, Larson DM, Wang H-Z, Veenstra RD, Beyer EC 1993 Molecular cloning and functional expression of human connexin37, an endothelial cell gap junction protein. J Clin Invest 91:997–1004

Rennick RE, Connat J-L, Burnstock G, Rothery S, Severs NJ, Green CR 1993 Expression of connexin43 gap junctions between cultured vascular smooth muscle cells is dependent upon phenotype. Cell Tissue Res 271:323–332

Ross R 1995 Cell biology of atherosclerosis. Annu Rev Physiol 57:791–804

Saffitz JE, Hoyt RH, Luke RA, Kanter HL, Beyer EC 1992 Cardiac myocyte interconnections at gap junctions—role in normal and abnormal electrical conduction. Trends Cardiovasc Med 2:56–60

Saffitz JE, Davis LM, Darrow BJ, Kanter HL, Laing JG, Beyer EC 1995 The molecular basis of anisotropy; role of gap junctions. J Cardiovasc Electrophysiol 6:498–510

Sepp R, Severs NJ, Gourdie RG 1996 Altered patterns of cardiac intercellular junction distribution in hypertrophic cardiomyopathy. Heart 76:412–417
Severs NJ 1995 Freeze-fracture cytochemistry: an explanatory survey of methods. In: Severs NJ, Shotton DM (eds) Rapid freezing, freeze fracture and deep etching. Wiley-Liss Inc, New York, p 173–208
Severs NJ, Dupont E, Kaprielian RR et al 1996 Gap junctions and connexins in the cardiovascular system. In: Yacoub MH, Carpentier A, Pepper J et al (eds) Annual of cardiac surgery, 9th edn. Current Science, London, p 31–44
Smith JH, Green CR, Peters NS, Rothery S, Severs NJ 1991 Altered patterns of gap junction distribution in ischemic heart disease. An immunohistochemical study of human myocardium using laser scanning confocal microscopy. Am J Pathol 139:801–821
van Kempen MJ, ten Velde I, Wessels A et al 1995 Differential connexin distribution accommodates cardiac function in different species. Microsc Res Tech 31:420–436
Willecke K, Kirchhoff S, Plum A, Temme A, Thönnissen E, Ott T 1999 Biological functions of connexin genes revealed by human genetic defects, dominant negative approaches and targeted deletions in the mouse. In: Gap junction-mediated intercellular signalling in health and disease. Wiley, Chichester (Novartis Found Symp 219) p 76–96
Xie H, Hu VW 1994 Modulation of gap junctions in senescent endothelial cells. Exp Cell Res 214:172–176
Yeh H-I, Kanthou C, Dupont E, Lupu F, Severs NJ 1996 Differential effects of growth factors on gap junction expression in cultured human aortic smooth muscle cells. Eur Heart J 17 (suppl):397 (abstr)
Yeh H-I, Dupont E, Coppen S, Rothery S, Severs NJ 1997a Gap junction localization and connexin expression in cytochemically identified endothelial cells from arterial tissue. J Histochem Cytochem 45:539–550
Yeh H-I, Lupu F, Dupont E, Severs NJ 1997b Upregulation of connexin43 gap junctions between smooth muscle cells after balloon catheter injury in the rat carotid artery. Arterioscler Thromb Vasc Biol 17:3174–3184

## DISCUSSION

*Gilula:* I would like to ask Mark Yeager to comment on these changes in gap junction utilization in the heart.

*Yeager:* I am impressed with how complicated the process of arrhythmogenesis is and how challenging it is to identify specific pathways for targeting therapies. The levels of complication range from variations in the expression of different connexins to changes in the size number and distribution of gap junctions, as well as variations in the proportion of open and closed channels. The conduction properties of the tissue will therefore depend not only on the anatomical distribution of gap junctions but also on their regulation. An interplay of these factors is likely to be involved in arrhythmogenesis that occurs in different pathophysiological states such as unstable angina pectoris, acute myocardial infarction and post-myocardial infarction, as well as cardiomyopathies resulting from various aetiologies.

*Beyer:* This also raises a question that we have struggled with. If we start with the assumption that the arrangement of gap junctions in areas such as the infarct border zones contribute to the re-entry of arrhythmias, should we try to isolate those regions so that they don't conduct and contribute to re-entry or should we try to improve coupling so that there is improved passage of current through those regions?

*Green:* We put neonatal myocytes into culture and used a potential-sensitive dye to follow the depolarization of the cells in the presence of uncoupling agents such as butyrate. In this way we were able to create a culture of arrhythmic cells and mimic miniature re-entrant arrhythmias, although this is by precisely manipulating the levels of coupling.

*Severs:* A similar difficulty arises if you assume that the increased gap junction expression in the arterial wall is linked with early stages of atherogenesis. If you try to imagine a therapeutic intervention that blocks this increase then you may also block a fundamental process, which is the repair of a damaged tissue, so there could be many unforeseen deleterious consequences. Perhaps one conclusion is that existing therapies, such as cardiac transplantation or by-pass surgery, deal with the problem in an effective way because they solve a multitude of problems all at once.

*Sanderson:* One can make comparisons with asthma, in which there is a hypertrophy of smooth muscle in the airways. Cultured smooth muscles from various sources change their secretory phenotype, and one often observes the development of calcium waves. Therefore, the culture system, in which damage is induced, is useful for looking at changes in cell proliferation, gap junction morphology and calcium signalling in response to injury.

*Gilula:* Cecilia Lo has some relevant information on dominant negatives in the heart, so I would like to ask her to comment on this.

*Lo:* We have recently performed some studies using transgenic models to address the role of connexin43 (Cx43; $\alpha_1$) gap junctions in heart morphogenesis. The Cx43 knockout mouse generated by Jerry Kidder and Janet Rossant dies at birth as a result of conotruncal malformation (Reaume et al 1995). The defect is located in the region of the right ventricle where the pulmonary outflow tract emerges. One typically sees a double outpouching around the base of the outflow tract. This results in pulmonary outflow obstruction and death of the mice at birth. We were puzzled by this phenotype, since during development, there is little expression of Cx43 in this region of the heart. So what is the role of Cx43? To address this question, we generated a number of different transgenic mice. From these studies, we now think that the basis for this heart phenotype arises from a perturbation of neural crest cells. This conclusion was arrived at by examining two transgenic mouse models, one in which Cx43 was overexpressed, the CMV43 transgenic mice (Ewart et al 1997), and a dominant negative transgenic animal in which a Cx43-lacZ fusion protein was expressed. We had previously shown that this fusion protein inhibits gap junctional communication in cultured cells (Sullivan & Lo 1995). It is important to appreciate that in both types of transgenic models, we targeted the expression of the transgene to the dorsal neural tube, the region from which neural crest cells are derived, and also in a subpopulation of neural crest cells. In both transgenic models, we observed heart defects in the conotruncal region of the heart, the same region affected in the Cx43 knockout mice. These results suggest that the level of Cx43 function in cardiac neural crest cells is of critical importance in heart morphogenesis, but the question remains as to the role of gap junctions in the development of cardiac crest cells. We now have both *in vitro* and *in vivo* data which suggest that the migration behaviour of neural crest cells is directly modulated by the level of Cx43 function. We took neural folds from the postotic hindbrain region, which is where cardiac crest cells originate, and explanted them in culture prior to the time when cardiac crest cells emerge. By monitoring the area of the neural crest outgrowth emerging from the explant, we were able to obtain an indication of the rate of neural crest migration. With overexpression of Cx43 in homozygous CMV43 transgenic mice, we observe an apparent increased migration rate. Significantly, this increase was transgene dosage dependent. In contrast, in the dominant negative mice that expressed the Cx43-lacZ fusion protein, and in the Cx43 knockout mice, migration decreased. In the Cx43 knockout mice, migration decreased in step with Cx43 gene dosage. Moreover, recently we have obtained evidence indicating similar migratory behavioural changes *in vivo*. This analysis was made possible using a transgenic mouse containing a lacZ reporter gene driven by the Cx43 promoter. This reporter construct labels all neural crest cells and thus, by breeding this transgene into the CMV43 and Cx43 knockout mice, we were able to examine the migration of crest cells in these animals (Lo et al 1997). Such studies showed increased and decreased abundance of lacZ-expressing cells in the heart outflow tract in the CMV43 and Cx43 knockout mice, respectively. To address whether this change in migration behaviour may be related to functional changes in gap junctional communication, we carried out *in vivo* dye injections to quantitate dye coupling in presumptive neural crest cells situated lateral to the dorsal hindbrain neural fold. In parallel, dye coupling analysis was carried out to quantitate gap junctional communication in the neural crest outgrowths *in vitro*. These studies showed that *in vitro* and *in vivo* dye coupling is elevated in the neural crest cells of CMV43 transgenic embryos, whereas dye coupling was reduced in the dominant negative transgenic and Cx43 knockout mouse embryos. It is interesting to note that dye coupling persisted in the homozygous knockout embryo, albeit at reduced levels. This would suggest that other connexins are present in neural crest cells. Overall, these findings suggest an important role for Cx43 junctions in modulating the migration and development of cardiac neural crest cells.

*Werner:* I am intrigued by the possibility that increased expression of Cx43 might precipitate plaque formation. We found that in transgenic mice that express luciferase under the control of the Cx43 promoter, Cx43 was expressed in the testes and lung. We performed RNase protection assays because these represent a more quantitative way of measuring mRNA levels. As a control we also looked at the expression of endogenous Cx43 in the same animal. We did this on untreated and oestrogen-treated mice, and found that there was about a three- to fivefold reduction in the levels of endogenous Cx43 mRNA in the hearts of oestrogen-treated mice. This may have some relevance to why women suffer less from atherosclerosis than men.

*Gilula:* The effects of oestrogen in females are certainly of interest to cardiologists. It would be interesting to explore the relevance of these observations in terms of cardiac physiology in the animal model.

*Severs:* Arterial smooth muscle is not responsive to oestrogen or progesterone. This is solely a property of uterine smooth muscle.

*Werner:* Yes, in the uterus Cx43 expression is up-regulated by oestrogen, but it is puzzling why it is down-regulated in the myocardium. The next step is for us to do the experiments in non-transgenic mice.

*Goodenough:* Nick Severs, in the *en face* staining patterns of the rat aorta that you showed, did you see a mixed expression of all three connexins no matter where you looked?

*Severs:* Cx40 ($\alpha_5$) is the most abundant connexin in the aortic endothelium, and although Cx43 and Cx37 ($\alpha_4$) are found together, there tend to be patches in which one or the other predominates.

*Goodenough:* The reason I'm asking is that Joe Gabriels and David Paul have done a similar experiment and have obtained similar results to yours, except that they observed Cx43 in the vicinity of branch points and at ostia of vessels coming out of the aorta. In the non-branching areas of the aorta they didn't see any Cx43 (unpublished results 1998).

*Severs:* We haven't noticed that particular association, but it does sound plausible. Eric Beyer has been involved in some work which has demonstrated that Cx43 is highly expressed in subconfluent endothelial cell cultures (Larsen et al 1997), but once the cells become confluent the expression of Cx43 decreases and Cx37 increases. Therefore, in the aorta *in vivo* we may actually be seeing regions that have different endothelial turnover rates in response to turbulence.

*Goodenough:* That Cx43 is expressed in areas of turbulence was just the point I wanted to bring out. Joe Gabriels and David Paul have also placed a partially restricting cuff on the aorta, and within a few days can see an up-regulation of Cx43 expression in endothelial cells downstream of the cuff, where one would predict most of the turbulence to be (unpublished results 1998).

*Musil:* In several cell types it has been shown that cell–cell adhesion molecules of the cadherin class might be important in establishing areas where gap junctions are assembled. Have you looked for N-cadherin at the infarct border zone and in the hypertrophied heart? What happens to the intercalated discs and the cadherins when gap junctions are redistributed?

*Severs:* At the time of the original study that showed the disordered distribution of gap junctions at the infarct border zone, we did standard thin-section electron microscopy (Smith et al 1991). It is clear that although some cells maintain intercalated discs of sorts, the arrangement alters and the interactions between the cells change. This is also associated with degeneration or dedifferentiation of the cells in these regions. Fasciae adherentes junctions can become disorganized but we haven't looked specifically at N-cadherin by immunolabelling. However, changes in N-cadherin distribution are important during early postnatal development, as adhesive junctions are established at the discs before gap junctions.

*Lo:* The homozygous Cx43 knockout mouse survives for at least a couple of weeks postnately. You have shown that Cx45 ($\alpha_6$) is present in the myocardium in small amounts. Are the Cx45 junctions localized at the intercalated discs? And secondly, in the absence of Cx43 is enough Cx45 present to perform a partial rescue function?

*Severs:* Where Cx45 occurs in working myocardium, it does co-localize with Cx43. The two connexins are within the same discs, and probably within the same gap junctions. The amount of Cx45 in working myocardium, as judged by immunolabelling, is extremely small, but I wouldn't like to guess if this is sufficient for a partial rescue function in the homozygous Cx43 knockout mouse.

*Beyer:* None of the data are quantitative. We don't know the absolute amounts of any of the connexins in any of the regions or even in a specific cell. We have done some immunostaining of the homozygous null mice, and have found some Cx45 in the ventricles of those animals.

*Goodenough:* There also seems to be a normal amount of Cx43 in the Cx40 knockout.

*Lo:* Nick Severs also showed decreased levels of Cx43 in diseased ischaemic hearts. How do you interpret this?

*Severs:* The reduced levels may result in a slowing of conduction velocity. When we wrote this up five years ago this was just a speculation (Peters et al 1993). Jeffrey Saffitz with Eric Beyer recently published their work showing that in heterozygous Cx43 knockout mice, which reportedly had 50% of the normal level of Cx43, there was a reduction in conduction velocity in the order of 30–40% (Guerrero et al 1997). This seemed to be a direct experimental demonstration that reduced levels of Cx43 do indeed result in decreased conduction velocity. However, we still have to be cautious in view of recent work that Cecilia Lo has been involved with.

*Lo:* We did not observe any changes in conduction velocity in heterozygous Cx43 knockout mice. We have exchanged animals with Jeff Saffitz's group because we thought there may be differences in the strain background of the mice or the techniques used for measuring conduction velocity. However, studies in

Mario Delmar's lab still failed to detect any differences in conduction velocity between wild-type and heterozygous Cx43 knockout mice, and we don't understand why.

*Gilula:* Nick Severs, could you clarify which antibodies you used in your triple gold labelling experiments.

*Severs:* The strategy of our antibody production gave us a panel of antibodies raised in different species. We used antibodies raised in mouse, guinea-pig and rabbit for the triple labelling, so we were able to apply specific secondary detection systems with second antibodies to these different species.

*Gilula:* Using this system you have shown that three connexins are localized in the same plaque. When will we know if these connexins form heteromers with each other?

*Severs:* This is a question that immunogold labelling at the electron microscopic level cannot answer and will have to await further study.

## References

Ewart JL, Cohen MF, Meyer RA et al 1997 Heart and neural tube defects in transgenic mice overexpressing the Cx43 gap junction gene. Development 124:1281–1292

Guerrero PA, Schuessler RB, Davis LM et al 1997 Slow ventricular conduction in mice heterozygous for connexin43 null mutation. J Clin Invest 99:1991–1998

Larson DM, Wrobleski MJ, Sagar GDV, Westphale EM, Beyer EC 1997 Differential regulation of connexin43 and connexin37 in endothelial cells by cell density, growth and TGF-$\beta$1. Am J Physiol Cell Physiol 272:C405–C415

Lo CW, Cohen MF, Huang GY et al 1997 Cx43 gap junction gene expression and gap junctional communication in mouse neural crest cells. Dev Genet 20:119–132

Peters NS, Green CR, Poole-Wilson PA, Severs NJ 1993 Reduced content of connexin43 gap junctions in ventricular myocardium from hypertrophied and ischaemic human hearts. Circulation 88:864–875

Reaume AG, de Sousa PA, Kulkarni S et al 1995 Cardiac malformation in neonatal mice lacking connexin43. Science 267:1831–1834

Smith JH, Green CR, Peters NS, Rothery S, Severs NJ 1991 Altered patterns of gap junction distribution in ischemic heart disease. An immunohistochemical study of human myocardium using laser scanning confocal microscopy. Am J Pathol 139:801–821

Sullivan R, Lo CW 1995 Expression of a connexin 43/beta-galactosidase fusion protein inhibits gap junctional communication in NIH3T3 cells. J Cell Biol 130:419–429

# Misregulation of connexin43 gap junction channels and congenital heart defects

Chiranjib Dasgupta, Bertha Escobar-Poni*, Maithili Shah†, John Duncan* and William H. Fletcher*‡[1]

*Departments of Biochemistry, *Anatomy and †Physiology, Loma Linda University School of Medicine, 11201 Benton Street, and the ‡Jerry L. Pettis Memorial VA Medical Center, Loma Linda CA 92357, USA*

*Abstract.* Although there is general agreement that gap junction channels formed by the connexin43 (Cx43; $\alpha_1$) protein most likely have important roles during heart development, evidence to support this view has been equivocal. Lacking this information, it is difficult to understand the basis of heart malformations found in the Cx43 knockout mice and in children with a severe form of visceroatrial heterotaxia that coincides with missense mutations of the Cx43 gene. To address this issue we used a combination of western blots to follow the emergence of Cx43 in heart, and *in vitro* and *in vivo* phosphorylation to assess the effect of mutations on Cx43 phosphorylation. We evaluated the activity ratios of cAMP-dependent protein kinase and protein kinase C in hearts of 8.5-day-old mouse embryos through to birth. The results demonstrate that Cx43 is present in the native phosphorylated species in day 8.5 hearts and thereafter. Further, the activities of cAMP-dependent protein kinase and protein kinase C are mirror images of each other during the 8.5–10.5 days of early heart development. From these results we conclude that Cx43 gap junction channels are present and capable of being regulated by day 8.5 of embryonic heart development.

*1999 Gap junction-mediated intercellular signalling in health and disease. Wiley, Chichester (Novartis Foundation Symposium 219) p 212–225*

The gap junction proteins (the connexins) are well conserved in their four transmembrane domains and two extracellular loops, but not in their cytoplasmic loop or carboxyl tail. Because of this, it has been frequently suggested that the latter two regions may confer differences in the physiological roles or in the regulation of channels formed by each of the gap junction proteins. Currently, there is no compelling evidence that each of the connexins form channels which mediate

[1]This chapter was presented at the symposium by William H. Fletcher, to whom correspondence should be addressed.

distinctive functions, although this may well occur in specific cells or tissues possibly during development. In contrast, there is a lengthy history of results which demonstrate that gap junction-mediated cell–cell communication can be regulated by cytoplasmic pH, transjunctional voltage, or phosphorylation and dephosphorylation events. Although several of the connexins are known to be phosphoproteins, the best studied of these is connexin43 (Cx43; $\alpha_1$), the dominant gap junction protein of the mammalian heart. Cx43 transcript and protein have been found in 8–16 cell mouse embryos (Nishi et al 1991); however, several studies have failed to detect it in the heart until 9.5–10.5 days post coitum (dpc; Ruangvoravat & Lo 1992). As the heart undergoes looping and commences to beat prior to these times, the role of Cx43 in early heart development and function becomes problematic.

This background discordance between the time of appearance of Cx43 and early heart formation is an important issue in view of the finding that point mutations in the Cx43 gene occur in children diagnosed with visceroatrial heterotaxia (VAH), in which polysplenia or asplenia syndromes are combined with severe heart malformations—including pulmonic stenosis or atresia, double-outlet right ventricle and transposition of the great arteries—along with conduction deficiencies. The type and severity of organ malformations give clues about the timing of the onset of these abnormalities.

Table 1 summarizes the anatomic pathologies and mutations found in these children. As these results illustrate, there is a generalized failure to establish normal left/right asymmetry, which is especially detrimental for the heart, where most defects are focused on the right side (i.e. the conotruncus). The frequency of pulmonic stenosis or atresia (five of the six children) is notable because this is a consistent anomaly in the Cx43 knockout mice (Reaume et al 1995). Most of the mutations replace potential sites for phosphorylation, mainly serines, with amino acids that cannot be phosphorylated. Five of the children have a similar mutation which substitutes Ser364 with proline (S364P), although there is no kinship amongst the families of these children. The S364P mutation has no detectable effect on the production or post-translational processing of Cx43, as cells expressing this mutant protein have gap junction plaques in membrane equivalent in size and number to the junctions of cells transfected with normal Cx43 (Britz-Cunningham et al 1995). However, the open status of S364P mutant Cx43 channels is improperly regulated by cAMP-dependent protein kinase and protein kinase C. Communication (Lucifer yellow dye transfer) through normal Cx43 channels is enhanced by the former enzyme and diminished or unaltered by protein kinase C (Godwin et al 1993). In contrast, communication via channels formed by the S364P mutant Cx43 is unaffected by cAMP-dependent protein kinase, but significantly enhanced by protein kinase C (Britz-Cunningham et al 1995). In addition, there are other, more subtle, differences that distinguish the

**TABLE 1 Mutations in patients with visceroatrial heterotaxia syndrome as diagnosed at Loma Linda**

| *Patient* | *Diagnoses* | *Allele* | *Residue(s) mutated* | *Substitution* |
|---|---|---|---|---|
| A16 | Asplenia syndrome: DORV, VSD, right AI, right BPI, abnormal thoraco-abdominal symmetry, absent spleen, pulmonic atresia | 1 | 364 | TCA (Ser) → CCA (Pro) |
| | | 2 | 352 | GAA (Glu) → GGA (Gly) |
| A13 | Polysplenia syndrome: DORV, VSD, left AI, left BPI, abdominal situs inversus, cleft spleen[a] | 1 | 364 | TCA (Ser) → CCA (Pro) |
| | | 2 | 365 | AGC (Ser) → AAC (Asn) |
| A30 | Polysplenia syndrome: DORV, VSD, left AI, left BPI, abdominal situs inversus, multiple spleens, pulmonic stenosis | 1 | 364 | TCA (Ser) → CCA (Pro) |
| | | 2 | 373 | AGC (Ser) → GGC (Gly) |
| A39 | Asplenia syndrome: right BPI, AV canal, LV hypoplasia, pulmonic atresia, TGA, abnormal thoraco-abdominal symmetry, absent spleen | 1 | 364 | TCA (Ser) → CCA (Pro) |
| | | 2 | [b] | [b] |
| A54 | Asplenia syndrome: Left BPI, common atrium, LV hypoplasia, VSD, DORV, pulmonic and mitral atresia, TAPVR, right-sided stomach, midline liver, absent spleen | 1 | 364 | TCA (Ser) → CCA (Pro) |
| | | 2 | [b] | [b] |
| A27 | No defined syndrome: DORV, VSD, right AI, malrotation of the bowel, single spleen, pulmonic stenosis | 1 | 326 | ACC (Thr) → GCC (Ala) |
| | | 2 | 352 | GAA (Glu) → GGA (Gly) |

[a] Abnormal, deeply cleft spleen demonstrated by CT scan.

[b] Normal throughout the coding sequence.

AI, atrial isomerism; AV, atrioventricle; BPI, bronchopulmonary isomerism; DORV, double-outlet right ventricle; LV, left ventricle; TAPVR, total anomalous pulmonary venous return; TGA, transposition of the great arteries; VSD, ventricular septal defect.

channels formed by the S364P mutant Cx43 from channels formed by the wild-type protein. Western blots reveal that the phosphorylated forms of mutant Cx43 migrate differently than those species of wild-type Cx43, with the 44 kDa species (Cx43P1 by the nomenclature of Musil & Goodenough 1991), migrating at 43.5 kDa and the 46 kDa (Cx43P2; Musil & Goodenough 1991) form migrating at ~47 kDa (Britz-Cunningham et al 1995). These mobility shifts suggest that the S364P mutant protein may be conformationally different from normal Cx43, possibly due to differences in the amount or sites of phosphorylation. If so, this could be why the S364P mutant Cx43 channels mediate communication between contacting cells at only half the level (24% incidence of dye transfer versus 50% incidence for normal Cx43) of that found in cells expressing wild-type Cx43 (Britz-Cunningham et al 1995). The S364P Cx43 thus acts as a weak dominant negative mutation which nevertheless mediates cell–cell communication to a significantly greater extent than that of parental cells which lack Cx43 gap junctions (Britz-Cunningham et al 1995).

## Current studies

The concept that the S364P mutation may have a dominant negative effect would be consistent with the recent observation that the S364P mutant Cx43 causes heterotaxia when expressed in the dorsal blastomeres of *Xenopus* embryos, whereas expression of normal Cx43 has no such effect (Levin & Nascone 1997; M. Levin, personal communication 1998). These results support the concept that communication through Cx43 gap junctions does play a role in left/right patterning. Further, the heart defects found in Cx43 knockout mice and the children with VAH, and the results from frog embryos all indicate that it is not the mere presence or absence of Cx43 that leads to dysmorphias. Rather, it is the misregulation of the open status of Cx43 channels that causes developmental abnormalities.

In order to evaluate this possibility, several ambiguities had to be addressed, the most critical of which is the time of appearance of Cx43 in the developing heart.

*In situ* hybridization using antisense riboprobes did not detect Cx43 transcripts in the heart until 9.5–10.5 dpc (Ruangvoravat & Lo 1992). Even so, transcript has been demonstrated by reverse transcriptase-PCR (Nishi et al 1991), but no assessment has been done of Cx43 protein in the early stages of cardiogenesis. This is of great importance with regard to the contribution, if any, of Cx43 gap junctions in the initiation and maintenance of the heartbeat, which begins about day 8.5 in the mouse embryo (c.f. Ruangvoravat & Lo 1992). Also relevant to this issue is the lack of information about the activities of cAMP-dependent protein kinase and protein kinase C in the developing heart. Evidence clearly shows that Cx43 channels are gated by these enzymes when microinjected into living cells

(Somogyi et al 1989; Godwin et al 1993). Whether this occurs in developing systems has not been reported.

Each of these questions was addressed using hearts from embryonic mice removed at 8.5 dpc through to birth. Western blots revealed the presence of parental Cx43 and two lower mobility, presumably phosphorylated, species in 8.5 and 9.5 dpc hearts. In 10.5 dpc hearts, only parental Cx43 was detectable, but all three forms were present in 11.5 dpc hearts and throughout the remainder of development (Duncan 1998). As expected, the absolute amount of Cx43 increased with progressive enlargement of the heart.

These results indicate that low, but detectable, amounts of Cx43 protein are present in the midline heart tube and throughout heart development. Thus, gap junctions formed by this protein could contribute to the initiation and maintenance of the heartbeat. The fact that protein but not transcript for Cx43 can be detected implies that low levels of transcript must exist and suggests the possibility that transcript half-life may be longer in embryonic heart than in the adult organ.

The activity ratios (enzyme activity in the absence versus the presence of co-factors) of cAMP-dependent protein kinase (activity $\pm$ cAMP) and protein kinase C ($\pm Ca^{2+}$ and phospholipid) were also evaluated using hearts from 8.5 dpc embryos through to birth (Duncan 1998).

Cyclic AMP-dependent protein kinase was about 40% activated in hearts of 8.5 dpc embryos, but rose markedly at 9.5 dpc, and by 10.5 dpc the enzyme was almost fully active, where it remained through to 11.5 dpc. After this time, enzyme activity declined until just prior to birth.

Unexpectedly, protein kinase C was maximally (100%) active in 8.5 and 9.5 dpc hearts, but at 10.5 dpc had declined to half that amount and was reduced even further (to 20%) by 11.5 dpc, after which it slowly declined to levels about 20-fold less than that found in 8.5 dpc hearts.

Together, these results suggest that the absence of the higher molecular weight forms of Cx43 as determined by western analysis of 10.5 dpc hearts may be related to the precipitous drop in protein kinase C activity at the same time. This suggests that protein kinase C may be predominantly responsible for phosphorylation of Cx43 in 8.5 and possibly 9.5 dpc hearts. If this is so, then the S364P mutant Cx43 would form channels that are more likely to be open in 8.5 dpc hearts, at which time normal Cx43 channels would tend to be more often closed than open. At the moment, this is purely speculation, but it is consistent with the experimental results and could explain the malformations found in the children with VAH. At present, we are beginning to evaluate transgenic mice that express the S364P mutant Cx43, which preliminary results suggest does give rise to heart malformations that are largely focused on the conotruncus.

Although the results to date indicate that the cytoplasmic carboxyl tail domain spanning residues 360–376 is a target for the phosphorylation of one or several of the six serine residues in this region, there is no direct evidence which specific amino acids and protein kinases are involved.

To assess this, we used three synthetic peptides and Cx43 purified from rat hearts (a generous gift from Mark Yeager, Scripps Research Institute, La Jolla, CA, USA) or produced by recombinant technology (a generous gift from Norton B. Gilula, Scripps Research Institute, La Jolla, CA, USA). The peptides include one corresponding to residues 360–376 of Cx43, with the other two being identical except one had a S364P conversion and the other a Ser365 to Asn conversion—the latter two peptides both corresponding to mutations found in the children with VAH (Table 1). Each of these compounds was used as a substrate for phosphorylation by cAMP-dependent protein kinase or protein kinase C, both of which were purified to homogeneity as previously described (Godwin et al 1993).

The results showed that each of the enzymes could phosphorylate the peptides. However, peptides corresponding to the Cx43 mutations incorporated one-half or less the amount of $\gamma^{32}$P-ATP incorporated by the peptide corresponding to normal Cx43 (360–376). Notably, the rate of $\gamma^{32}$P-ATP incorporation was extremely slow, suggesting the possibility that either the enzymes had a higher affinity for the phosphorylated substrate than for the non-phosphorylated substrate, or they recognized secondary structure, not just primary sequence (Shah 1998).

This was examined using purified or recombinant full-length Cx43 as a substrate. In these studies we observed that cAMP-dependent protein kinase catalysed the incorporation of $\gamma^{32}$P-ATP into Cx43 with kinetics that were much more rapid than those seen using peptides as substrates (Shah 1998). Although these results indicate that secondary structure within the Cx43 polypeptide chain is important for its rate of phosphorylation, we do not know whether phosphate incorporation was confined to the 360–376 domain or occurred in other regions of Cx43 as well. Also, these results were obtained under *in vitro* conditions, which may or may not accurately reflect what occurs *in vivo*.

To more closely mimic the *in vivo* situation, we used metabolic labelling of Cx43 with $^{32}$P-orthophosphate in cells transfected with plasmid only (null controls), wild-type Cx43 or S364P mutant Cx43. Over a three-hour period, null controls incorporated negligible phosphate. Cells transfected with normal Cx43 incorporated detectable phosphate within one hour, and significantly increasing amounts at two-hour and three-hour intervals. Cells expressing S364P mutant Cx43 also incorporated $^{32}$P-orthophosphate but at a slower rate and markedly reduced amount, about one-fifth or less than that of cells with normal Cx43. Although several interpretations of these results were considered, the most reasonable is that Ser364 is a crucial site at which phosphorylation may be required before subsequent phosphorylations can occur (Shah 1998). Additional studies must be done to test this possibility.

*A proposed mechanism for gating of connexin43 channels*

Based on the results from *in vitro* phosphorylation and the effect of the S364P mutation on regulation of Cx43 channels by microinjected protein kinases, it is reasonable to propose that phosphorylation of serine residues in the 360–376 domain of Cx43 is important in controlling the gating of Cx43 channels. The question is, how could this be accomplished?

The 360–376 domain is highly conserved in the Cx43 coding sequence of all species reported to date, from frogs to humans. This is shown in Fig. 1a, which demonstrates that this domain is fully conserved except for the chick in which the amino acid equivalent to human Ser364 is a proline. Thus, the children with VAH and the associated Cx43 S364P mutation have a regulatory domain corresponding to that of the chick. This is intriguing because Cx43 is absent from chick myocardium, where Cx42 ($\alpha_5$) dominates (Minkoff et al 1993).

In addition to the cytoplasmic tail domain, there is another region wherein Cx43 is conserved, including in the chick. This is the cytoplasmic loop domain spanning residues 101–109 in which the motif basic-basic-X-X-basic-X-X-basic-basic is a consistent feature. This regular spacing of positively charged residues suggests a possible model for tail–loop interaction once serine residues in the 360–376 domain are phosphorylated. Three different secondary structure prediction algorithms were used to test this model (studies done by Kevin Balli, Loma Linda University with the collaboration of Mark Yeager, Scripps Research Institute). All three programs predicted that the 101–109 loop region was a right-handed alpha helix, but no structure could be predicted for the 360–376 domain. However, when serines 364, 368 and 372 were phosphorylated then arranged as an anti-parallel coiled-coil with residues 101–109 and energy minimization carried out, an energetically favourable interaction was predicted. In this case, the

(a)

| | | | |
|---|---|---|---|
| Human | 360 | DQRPSSRASSRASSRPR | 376 |
| Bovine | 361 | DQRPSSRASSRASSRPR | 377 |
| Murine | 360 | DQRPSSRASSRASSRPR | 376 |
| Rat | 360 | DQRPSSRASSRASSRPR | 376 |
| Chick | 359 | DQRPPSRASSRASSRPR | 375 |
| Xenopus | 357 | DQRPSSRASSHASSRPR | 373 |

(b)

| | | | |
|---|---|---|---|
| Human | 101 | RKEEKLNKKEEE | 112 |
| Bovine | 101 | RKEEKLNKKEEE | 113 |
| Murine | 101 | RKEEKLNKKEEE | 112 |
| Rat | 101 | RKEEKLNKKEEE | 112 |
| Chick | 101 | RKEEKLNKREEE | 112 |
| Xenopus | 101 | RKEEKLNRKEEE | 112 |

FIG. 1. Comparative sequences of connexin43 (Cx43; $\alpha_1$) (a) and cytoplasmic loop (b). This demonstrates the extreme conservation of putative regulatory domains of Cx43, one in the cytoplasmic tail (a) and the other in the cytoplasmic loop (b). In panel a, the serine residues (**S**) that are potential sites of phosphorylation are conserved in all species, except the chick in which Ser363 (analogous to Ser364 in humans) is a proline (P). Panel b shows the complete conservation of evenly spaced positively charged residues, arginine (R) or lysine (**K**) that could interact with phosphorylated (negatively charged) serines of the cytoplasmic tail.

phosphorylated serines would be electrostatically bound with lysines 102, 105 and 108 of the cytoplasmic loop. Whether this configuration would open or close Cx43 channels is unknown. Further, it is uncertain whether the tail–loop interaction would be intramolecular or intermolecular, or both. Finally, although it seems likely that the strength of bonding between the tail and loop domains would vary depending on whether one, two or all three proposed serines are phosphorylated, there is no a priori basis for predicting how these variations would affect channel gating or conductance.

To begin testing this model, we constructed expression plasmids in which lysines 102, 105 and 108 were replaced with uncharged amino acids, or the entire 101–109 domain was deleted. Also, a plasmid has been made that codes for Cx43 in which serines 364, 368 and 372 have been replaced with negatively charged residues that should mimic phosphorylated serines. Each of these has been used to transfect SK-Hep1 cells, a human hepatoma line that is communication deficient but produces low levels of Cx45 ($\alpha_6$) transcripts (Laing et al 1994).

Preliminary results suggest that both constructs with cytoplasmic loop mutations have severely reduced cell–cell communication relative to control cells transfected with wild-type Cx43. Interestingly, cells expressing Cx43 missing the three lysine residues (102, 105, 108) have more severely disrupted communication than cells which express Cx43 lacking the entire 101–109 domain. The major difference in these two constructs is the presence of Arg101 and Lys109 in the site-specific mutants, whereas these residues are missing in the 101–109 deletion mutant. It is possible that the remaining basic charges alter this domain's interaction with the cytoplasmic tail such that channels are more prone to closure.

The combined results demonstrate that the carboxyl tail regulatory domain (residues 360–376) of Cx43 is phosphorylated by at least two of the major serine/threonine protein kinases. Further, phosphorylation appears to be hierarchical in that when Ser364 is absent, the phosphorylation of Cx43 is reduced to a greater extent than can be accounted for by a single missing serine residue, unless the missing residue is a site that must be phosphorylated before additional sites for this modification become available. Ongoing studies will test this possibility.

### *Controversy about the connexin43 mutations and visceroatrial heterotaxia*

Since our original observations were published, at least two groups have reported attempts to confirm our results but have been unable to do so. There are several reasons that could account for this discrepancy. The most obvious of these is that the patients evaluated in these other studies do not fit the phenotype of the VAH children with known Cx43 mutations. This is summarized in Table 2, where it can be seen that the reports from Gebbia et al (1996) and Splitt et al (1997) were based on VAH patients that had a far lower proportion of asplenia or polysplenia than did

**TABLE 2 Comparison of patient populations and corresponding diagnoses as examined by four different groups**

| | *Patient pool population* | | | |
|---|---|---|---|---|
| *Diagnoses* | *Gebbia et al (1996)* | *Splitt et al (1997)* | *LLUMC* | *Ivemark (1955) Type B* |
| Asplenia | 4/19 (21%) | 10/48 (21%) | 3/6 (50%) | 41/69 (59%) |
| Polysplenia | 5/19 (26%) | 9/48 (19%) | 3/6 (50%) | ? |
| Outflow tract obstruction | 5/19 (26%) | 28/48 (58%) | 5/6 (83%) | 41/41 (100%) |
| Double outlet right ventricle | ? | ? | 5/6 (83%) | ? |
| Total patients | 10 | 48 | >107 | 69 |
| Total transplanted patients | ? | ? | 57 | 0 |
| Control | ? | ? | <50 | 0 |

LLUMC, Loma Linda University Medical Center.

the Loma Linda cohort, and a greatly reduced incidence of outflow tract obstruction. Also, no mention was made of transplantations, suggesting that the heart malformations were less severe than those of children with Cx43 mutations.

For comparison, Table 2 includes the results of Ivemark's hallmark study of 69 cases of splenic agenesis (Ivemark 1955). The Loma Linda University Medical Center patients are far more similar to those he described than are the patients examined by Gebbia et al (1996) and Penman Splitt et al (1997). Although Ivemark reported on 41 cases he classified as type B ('agenesis of the spleen with pulmonary stenosis or atresia, with or without transposition of the great arteries'; Ivemark 1955), he described no observation of double-outlet right ventricle but did mention that 18 of the 41 patients had a 'right aortic arch'.

*Acknowledgements*

We thank Norton B. Gilula and Mark Yeager for providing the Cx43 proteins used in these studies. Thanks, also, to our clinical colleagues Leonard Bailey, Ranae Larson, Neda Mulla and Joyce Johnston, who provided diagnostic information and access to surgical specimens. Special thanks to Anna-Marie Martinez for preparing the manuscript, including tables and figures. This work is supported by grants from the Veterans Affairs Medical Research Service, the National Institutes of Health and Loma Linda University School of Medicine. W.H.F is a Research Career Scientist of the Veterans Affairs Medical Research Service.

## References

Britz-Cunningham SH, Shah MM, Zuppan CW, Fletcher WH 1995 Mutations of the connexin 43 gap junction gene in patients with heart malformations and defects of laterality. N Engl J Med 332:1323–1329
Duncan J 1998 The role of connexin43 in heart development. PhD thesis, Loma Linda University, Loma Linda, CA, USA
Gebbia M, Towbin JA, Casey B 1996 Failure to detect *Cx43* mutations in 38 cases of sporadic and familial heterotaxy. Circ 94:1909–1912
Godwin AJ, Green LM, Walsh MP, McDonald JR, Walsh DA, Fletcher WH 1993 *In situ* regulation of cell–cell communication by the cAMP-dependent protein kinase and protein kinase C. Mol Cell Biochem 127/128:293–307
Ivemark BI 1955 Implications of agenesis of the spleen on the pathogenesis of conotruncus anomalies in childhood: an analysis of the heart malformations in the splenic agenesis syndrome, with fourteen new cases. Acta Paed, Uppsala (suppl 4)
Laing JG, Westphale EM, Engelmann GL, Beyer EC 1994 Characterization of the gap junction protein, connexin45. J Memb Biol 139:31–40
Levin M, Nascone N 1997 Two molecular models of initial left–right asymmetry generation. Med Hypotheses 49:429–435
Minkoff R, Rundus VR, Parker SB, Beyer EC, Hertzberg EL 1993 Connexin expression in the developing avian cardiovascular system. Circ Res 73:71–78
Musil LS, Goodenough DA 1991 Biochemical analysis of connexin43 intracellular transport, phosphorylation, and assembly into gap junctional plaques. J Cell Biol 115:1357–1374
Nishi M, Kumar NM, Gilula NB 1991 Developmental regulation of gap junction gene expression during mouse embryonic development. Dev Biol 146:117–130
Reaume AG, DeSousa PA, Kulkarni S et al 1995 Cardiac malformation in neonatal mice lacking connexin43. Science 267:1831–1834
Ruangvoravat CP, Lo CW 1992 Connexin43 expression in the mouse embryo: localization of transcripts within developmentally significant domains. Dev Dynam 194:261–281
Shah M 1998 Regulation of connexin43 by multi-site phosphorylation. PhD thesis, Loma Linda University, Loma Linda, CA, USA
Somogyi R, Batzer A, Kolb HA 1989 Inhibition of electrical coupling in pairs of murine pancreatic acinar cells by OAG and isolated protein kinase C. J Memb Biol 108:273–282
Splitt M, Tsai MY, Burn J, Goodship JA 1997 Absence of mutations in the regulatory domain of the gap junction protein connexin 43 in patients with visceroatrial heterotaxy. Heart 77:369–370

## DISCUSSION

*Beyer:* You mentioned a construct termed H7. Is this the dominant negative construct from Bruzzone et al (1994)?

*Fletcher:* Yes. The H7 connexin43 (Cx43; $\alpha_1$) mutant causes a greater disruption of development and a higher degree of heterotaxia than the serine 364 to proline mutant. When Levin and Mercola inject ventral blastomeres with H7, the serine 364 to proline mutant or normal Cx43, the normal Cx43 causes a low incidence of heterotaxia, the H7 construct causes higher incidence of heterotaxia, and the serine 364 to proline (S364P) mutant Cx43 causes about the same high incidence of heterotaxia. They interpret this to be a weak dominant negative mutant. However, the results I showed you were generated from injecting dorsal blastomeres, which normally have gap junctions.

*Warner:* What are they actually injecting?

*Fletcher:* They inject either all dorsal or all ventral cells of two- to four-cell stage embryos with cRNA.

*Warner:* Early events in the *Xenopus* embryo are complicated and to understand the results of those experiments we need to look closely at exactly what they were doing, what they were injecting, when they were injecting it and for how long the RNA persists.

*Gilula:* Is this work published?

*Goodenough:* It is discussed in Levin & Mercola (1998).

*Gilula:* Because if the work isn't available for people here to be able to discuss it confidently, it makes it more complicated for people who might know something about the system to address the issues. We need to have seen it and thought about it before we can discuss it.

*Fletcher:* I would like to add a point of clarification here. True heterotaxia is not situs inversus; it is situs ambiguous. It is a failure to establish left/right asymmetry. It will go right or it will go left.

*Lo:* Kristi Hunter in Peggy Kirby's laboratory has been looking at Cx43 in the chick embryo and she finds that it is expressed asymmetrically in the node (K. Hunter & P. Kirby, personal communication 1998), like activin receptor IIa and other genes that are known to play a role in left/right patterning (Levin et al 1995). This is interesting because this distribution suggests that it is relevant in terms of left/right axis determination. These correlations don't prove that this is the case, but they are suggestive, especially as they are consistent with some of the *Xenopus* studies.

I would also like to make a comment about the Loma Linda subtype of visceroatrial heterotaxia (VAH) patients. It is significant that in this population, double-outlet right ventricle and pulmonary stenosis were found in conjunction with mutations in Cx43. This suggests that Cx43 may have several roles to play, modulating cardiac crest cells and heart morphogenesis and also left/right patterning. Pulmonary stenosis would not be unexpected for neural crest cell perturbations, whereas double-outlet right ventricle is one of the phenotypes often observed with cardiac crest ablations. Therefore, this subpopulation of VAH patients has a preponderance of perturbations that are consistent with the involvement of Cx43 (Kirby & Waldo 1995).

*Gilula:* I would like to ask Nick Severs for his perspective on Bill Fletcher's work in this area. Are you aware of controversy?

*Severs:* I am aware of controversy. What Bill says is plausible, but I have to say I'm not an expert in these congenital abnormalities so I can't comment in detail.

*Lau:* You demonstrated a decreased level of phosphorylation in the S364P mutant. Did you look at the levels of protein in those cells?

*Fletcher:* Yes, we showed that the protein is expressed.

*Lau:* Do you have any evidence that in this mutant the phosphorylated site in the C-terminal tail interacts with a putative 'receptor' possibly located in the cytoplasmic loop?

*Fletcher:* We don't have any hard data on that yet, except that when we mutate those particular residues we observe misregulation of gap junction channels, which is consistent with this hypothesis but does not prove it.

*Yamasaki:* Can you clarify the effect of protein kinase C (PKC) on the communication in those cells transfected with the S364P mutant, because PKC is usually known to be associated with decreased cell–cell communication? Also, have you looked at the effect of TPA (12-*O*-tetradecanoylphorbol 13-acetate)?

*Fletcher:* PKC increases cell–cell communication in the presence of the mutation, but it does not increase communication in the wild-type. We have not looked at the effect of TPA because 12 or 14 years ago, Craig Byus and I published some compelling evidence that TPA in hepatoma cells activates cAMP-dependent protein kinase as well as PKC (Byus et al 1983), so we try to avoid pleiotropic activators if we can. We avoided this problem in this case because we just microinjected purified enzymes into transgenic cells. In most cases PKC activation slows down communication, although this it doesn't happen all the time. For example, Chanson et al (1988) showed that PKC had no negative effects on cell–cell communication in pancreatic acinar cells. This work was done before we knew that pancreatic acinar cells contained Cx32 ($\beta_1$) and Cx26 ($\beta_2$).

*Yamasaki:* What is the total enzyme activity of PKC in the heart?

*Fletcher:* The total enzyme activity increases more or less linearly as the heart grows. For example, 80 8.5 days post coitum (dpc) hearts are needed for each activity ratio assay, which requires three data points, i.e. triplicates. This decreases to 35 hearts per activity ratio assay for 12.5 dpc hearts.

*Musil:* I have a couple of questions on phosphorylation. You talked about the rate of phosphorylation of the recombinant Cx43 by isolated PKA. What was the stoichiometry of this reaction?

*Fletcher:* The $K_m$ is about 140; the $V_{max}$ is terrible—I won't even tell you what it is—and the $K_{cat}$ is about 14. Therefore, phosphorylation of the peptide is about 100-fold slower than the phosphorylation of the Cx43 recombinant.

*Musil:* I was wondering what the mole-to-mole ratio of phosphate to recombinant Cx43 was, compared to the relatively high stoichiometry you showed for the Cx43 peptide.

*Fletcher:* We have done Edman degradation sequencing on the peptide but it's a little too difficult to do with intact Cx43. There are 17 potentially phosphorylatable serines in intact Cx43, so we would be sequencing it forever to try to figure out which ones were relevant.

*Musil:* If you knew the amount of recombinant Cx43 in your phosphorylation assay, you could determine the average number of radiolabelled phosphate molecules incorporated per Cx43 molecule to determine the stoichiometry of the reaction.

*Fletcher:* Recombinant Cx43 has to be denatured at 95 °C for 5 min before it can be used in the assay because it's so highly folded, so all these things contribute to my inability to answer your question.

*Musil:* Have you looked at *in vitro* PKC phosphorylation?

*Fletcher:* Yes, and it's the same except that PKC phosphorylates more rapidly and almost certainly to a higher level. If I had to say which enzyme favours Cx43, I would have to say PKC, although PKA is able to phosphorylate it.

*Musil:* In terms of figuring out the physiological role of Cx43 phosphorylation, have you tested Cx43 constructs that mimic constitutive phosphorylation of Ser364 by substituting an aspartic acid in this position?

*Fletcher:* We have made the expression plasmid, and transfected the cells. The junctions may be permanently open with aspartates replacing serines 364, 368 and 372, but we haven't yet done the microinjection work to look at dye transfer, so give me three months or so and I will be able to tell you what happens.

*Lau:* In collaboration with Paul Lampe, we have been looking at the phosphorylation sites of Cx43 induced by PKC. We have not detected phosphorylation of Ser364, rather the primary phosphorylated site we see is Ser368.

*Fletcher:* Are you using peptide or intact protein?

*Lau:* This is a comparison between phosphorylated, full-length Cx43 *in vitro* and *in vivo*.

*Fletcher:* How did you establish that Ser368 is phosphorylated in the full-length protein? Did you do sequencing?

*Lau:* It was by a combination of tryptic peptide analysis, sequencing and site-directed mutation analysis.

*Gilula:* Now that you have been able to define the interesting relationship between some specific residues and developmental defects, is there any direct value to the clinician of having this information at hand?

*Fletcher:* Yes. Our clinicians are now doing more analyses of these families—for example, of thymus function and interfamily history to see if any attribute correlates with Cx43 mutations. The family inquiry sheet has grown from two to seven pages.

*Beyer:* Do the patients have defects in other tissues that are derived from neural crest cells?

*Lo:* We have looked at the thymuses of the Cx43 knockout mice in collaboration with Lisa Spain, and found that some of the null mutant embryos have hypoplastic thymuses, which is exactly what you would expect from a neural crest cell perturbation.

*Gilula:* So now you can use your transgenic approach with mice to target the expression of specific connexins that may have some therapeutic benefit for developmental abnormalities.

*Fletcher:* You can't do any of these approaches in children.

*Beyer:* But if a patient was just about to have a heart transplant, you could look at the adrenal medulla to see if it was functioning properly.

*Fletcher:* Cardiologists are now doing that, but up to this point in time they didn't do that. These children have messed up hearts and they're going to die. Cardiologists do not care whether their adrenal medulla is fine. Their priority is to keep the children alive.

## References

Bruzzone R, White TW, Paul DL 1994 Expression of chimeric connexins reveals new properties of the formation and gating behavior of gap junction channels. J Cell Sci 107:955–967

Byus CV, Trevillyan JM, Cavit LJ, Fletcher WH 1983 Activation of cyclic adenosine 3′:5′-monophosphate-dependent protein kinase in H35 hepatoma and Chinese hamster ovary cells by a phorbol ester tumor promoter. Cancer Res 43:3321–3326

Chanson M, Bruzzone R, Spray DC, Regazzi R, Meda P 1988 Cell uncoupling and protein kinase C: correlation in a cell line but not in a differentiated tissue. Am J Physiol 255:699–704

Kirby ML, Waldo KL 1995 Neural crest and cardiovascular patterning. Circ Res 77:211–215

Levin M, Mercola M 1998 The compulsion of chirality: toward an understanding of left–right asymmetry. Genes Dev 12:763–769

Levin M, Johnson RL, Stern CD, Kuehn M, Tabin C 1995 A molecular pathway determining left–right asymmetry in chick embryogenesis. Cell 82:803–814

# Gap junctional intercellular communication in the mouse ovarian follicle

Daniel A. Goodenough*, Alex M. Simon† and David L. Paul†

*Departments of Cell Biology* and Neurobiology†, Harvard Medical School, 240 Longwood Avenue, Boston, MA 02115, USA*

*Abstract.* A targeted disruption of the gene encoding the gap junction protein connexin37 (Cx37; $\alpha_4$) results in female infertility. Mutant follicles are not observed to develop beyond early antral stages, and there is a lack of both observable mature Graafian follicles and ovulation. The oocytes are unable to acquire meiotic competence. Following oocyte failure, the residual follicular cells do not undergo atresia but rather transdifferentiate into luteal cells, resulting in a mutant ovary populated with numerous, inappropriate corpora lutea. These results indicate that the Cx37-containing gap junctions formed between oocyte and follicular cells permit bidirectional signalling between the two cell types. These junctions are required for oocyte growth and development during preantral stages of the follicle, and for the inhibition of follicle cell luteinization. An additional role for these junctions may be to permit transfer of cytoplasmic signals required to hold oocytes in meiotic arrest. Since the mutant follicles never acquire meiotic competence, this latter role for gap junctional communication cannot be tested in this model.

*1999 Gap junction-mediated intercellular signalling in health and disease. Wiley, Chichester (Novartis Foundation Symposium 219) p 226–240*

## Early follicular development

Beginning at the early primitive streak stage of mouse embryogenesis (7–7.5 days post coitum, dpc) the primordial germ cells begin their migration from the yolk sac mesoderm at the base of the allantois. By 8 dpc the cells populate the hindgut endoderm and by 10 dpc they are in the hindgut mesentery and coelomic wall, already beginning to arrive at the gonadal ridges (Ginsburg et al 1990, McLaren 1992). Within the female gonadal ridges, these cells are then surrounded by follicular cells and differentiate into oogonia. The cells undergo proliferation by mitosis, some of them differentiating into primary oocytes. Most oogonia and primary oocytes undergo cell death (atresia), whereas a subset that have entered the first meiotic prophase arrest in the diplotene stage of meiotic prophase I. The arrested primary oocytes, together with their surrounding simple layer of follicular cells, are called primordial follicles (Sadler 1997). These follicles rest until the onset of puberty, at which time a subset of the resting follicles are recruited to undergo folliculogenesis, which takes about 20 days in the mouse. Many of the recruited follicles also undergo atresia, whereas a select few complete folliculogenesis and are ovulated (Wassarman & Albertini 1994).

## Communication between oocyte and granulosa cells

There is a complex interplay between oocyte and follicular (granulosa) cells during folliculogenesis (Gougeon 1996, Buccione et al 1990) that is mediated by growth factors and hormones. The process of folliculogenesis may be divided into two phases: preantral, when the oocyte grows in mass and the granulosa cells proliferate by mitosis (Cecconi et al 1991); and antral, when the oocyte ceases growth and acquires competence to complete the first meiotic division. Intraovarian growth factor signalling between granulosa cells and oocytes dominates the preantral phase (McGrath et al 1995), whereas the antral phase is dominated by extraovarian hormonal influences (Albertini 1992). During both phases, oocyte and granulosa cells are in direct communication with each other via gap junctional intercellular pathways.

## Gap junctions in the mouse ovarian follicle

Gap junctions are macular collections of intercellular channels that facilitate the exchange of low molecular weight molecules ($<1000$ Da) and ions between the cytoplasms of adjacent cells. Each intercellular channel is composed of paired hexamers (connexons) formed from a family of integral membrane proteins, the connexins (Goodenough et al 1996, Kumar & Gilula 1996). Gap junctions form between follicular cells, and between follicular cells and the oocyte, during follicle development (Albertini & Anderson 1974, Anderson & Albertini 1976; for reviews see Wassarman & Albertini 1994, Grazul-Bilska et al 1997). The junctions joining the follicular cells are formed primarily of connexin43 (Cx43; $\alpha_1$; Beyer et al 1989) and Cx45 ($\alpha_6$; Okuma et al 1996), whereas those joining the follicular cells with the oocyte were recently shown to be composed at least partly of Cx37 ($\alpha_4$; Simon et al 1997). These connections will be discussed in detail below.

### Gap junctions and oocyte growth

One of the functions of the granulosa cells in the preantral follicle is to support a 34-fold increase in oocyte mass (Schultz & Wassarman 1977, Bachvarova et al 1980). Preantral oocytes denuded of their granulosa cells fail to grow in culture as compared with intact follicles (Eppig 1979); indeed, there is a direct correlation between the rate of oocyte growth and the number of conjoined follicular cells (Herlands & Schultz 1984). Oocytes are known to lack many transport systems for nutrient uptake, and numerous studies have demonstrated that gap junctional transfer of nutrients from granulosa cell to oocyte is required for oocyte growth (reviewed in Wassarman & Albertini 1994, Eppig et al 1996).

### Gap junctions and meiotic competence

During the antral phase, the oocytes become competent to resume the first meiotic division (Sorensen & Wassarman 1976). This developmental change can be triggered earlier in follicle development, since immature oocytes removed from the follicle can be induced to undergo germinal vesicle breakdown (GVB, a standard assay for meiotic competence) by fibroblast-conditioned medium (Chesnel et al 1994). This indicates that gap junctional communication is not required for the acquisition of meiotic competence. Granulosa/oocyte gap junctional communication does help maintain meiotic arrest in antral follicles once competence is acquired (Cho et al 1974). Culture of antral oocytes that have had their granulosa cells removed spontaneously resume meiosis, as evidenced by GVB and condensation of chromosomes.

### Connexin types in ovarian follicles

Since mRNAs encoding Cx37, Cx40 ($\alpha_5$) and Cx43 have been previously detected in the ovary (Beyer et al 1989, Haefliger et al 1992, Reed et al 1993), antibodies specific for these connexins have been used to localize the corresponding proteins. Immunohistochemical study of wild-type mouse ovaries using anti-Cx37 antibodies reveals macular staining in follicles at all stages of development except primordial follicles (Simon et al 1997). The macular staining can be resolved to reside exclusively on the oocyte side of the zona pellucida (see Fig. 1), adjacent to the oolemma, where previous electron microscopic studies have demonstrated the location of oocyte–granulosa cell gap junctions (Anderson & Albertini 1976, Gilula et al 1978). Anti-Cx40 antibodies do not label ovarian follicles. Antibodies to Cx43 label granulosa–granulosa gap junctions, but in contrast to anti-Cx37 antibodies they do not label the surface of the oocyte. This is remarkable in that gap junctions form between granulosa cell processes at the oolemmal surface in addition to forming between granulosa processes and the oocyte (Gilula et al

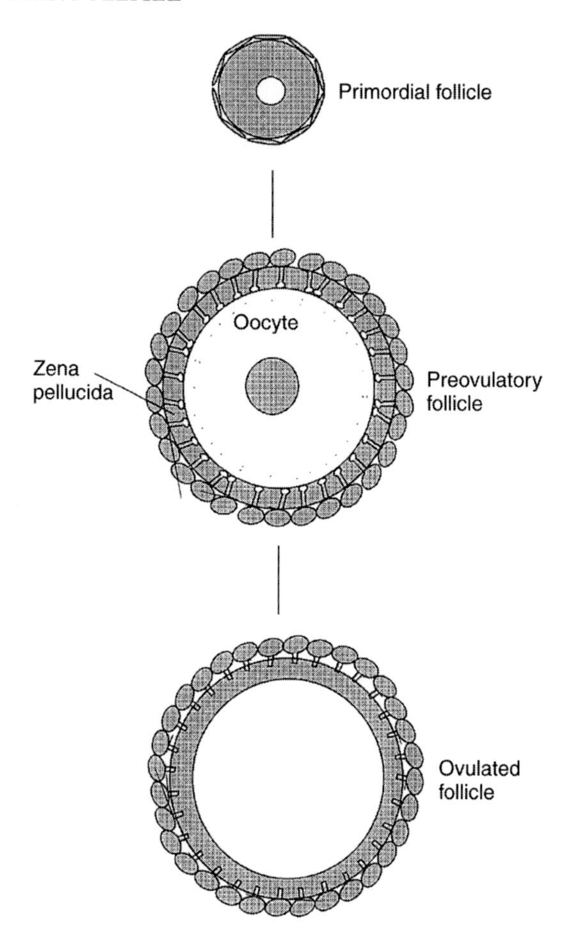


FIG. 1. A diagram of three stages in folliculogenesis. The primordial follicle consists of a primary oocyte arrested in meiotic prophase, surrounded by a squamous layer of follicular (granulosa) cells. In the preovulatory follicle only the innermost layer of granulosa cells are drawn, which have now proliferated by mitosis and extended processes through the zona pellucida to make gap junctions with the oocyte, and with each other. At ovulation, these granulosa cell processes are withdrawn and the gap junctions between oocyte and granulosa cell are broken.

1978). If these juxtaoolemmal granulosa–granulosa gap junctions contain Cx43, they may be too rare to present a consistent immunofluorescent signal (Beyer et al 1989). Alternatively, they may be composed of another member of the connexin family. It is clear, however, that there is not an anti-Cx43 antibody signal at the surface of the oocyte which corresponds to the intensity or distribution of the anti-Cx37 antibody signal, indicating that the majority of the oocyte/granulosa gap junctions are not composed of heterotypic connexin channels composed of Cx43 and Cx37. It is not known whether the granulosa cells express another, as yet undocumented, connexin within their processes at the oocyte surface. Thus, the granulosa cells use Cx43 for homocellular interactions and a different connexin for their heterocellular interactions with the oocyte. This asymmetry of Cx43 expression in the granulosa cells provides a challenging opportunity to study specific connexin targeting by the granulosa cell to different membrane domains.

## Targeted disruption of the mouse connexin37 gene

The mouse gene encoding Cx37 has been disrupted (Simon et al 1997). This connexin, together with Cx40, is found in abundance joining endothelial cells in blood vessels in addition to its location at the oocyte/granulosa cell interface. $Cx37^{-/-}$ mice are viable and phenotypically grossly normal. Although subtle changes in the propagation of excitation along arterioles have been detected (S. Kumer, A. M. Simon, B. R. Duling & D. L. Paul, unpublished results 1998), endothelia in blood vessels also appear grossly normal. The most striking phenotypic change was that $Cx37^{-/-}$ females are sterile, whereas the males are fertile. $Cx37^{-/-}$ females are incapable of ovulation, even in response to pregnant mare serum gonadotropin stimulation. In ovaries from $Cx37^{-/-}$ females follicular development is consistently arrested before full maturation. At sexual maturity ($\sim$5 weeks) $Cx37^{-/-}$ ovaries are smooth, lacking the distensions of the surface due to large follicles easily visible on wild-type ovaries. Histological analysis of two-week-old females reveals that both $Cx37^{-/-}$ and wild-type ovaries contain many developing preantral follicles. At 5 weeks, however, $Cx37^{-/-}$ ovaries never display the numerous Graafian follicles that are typical of wild-type ovaries. Only occasional antral follicles are observed in the $Cx37^{-/-}$ ovaries. $Cx37^{-/-}$ females do not develop normal follicles with a slower time course since Graafian follicles are still not detected by four months.

Female infertility correlates with a loss of oocyte–granulosa gap junctions and gap junctional intercellular communication. In the $Cx37^{-/-}$ follicles granulosa cell processes extend across the zona pellucida, making adherens junctions with the oolemma in a manner similar to that in wild-type animals (Gilula et al 1978). In the mutant follicles, however, gap junctions cannot be found by electron microscopy, nor can any anti-Cx37 antibody staining be seen at the oolemma by immunohistochemistry. Cx43 staining typical of granulosa–granulosa gap junctions is normal in $Cx37^{-/-}$ follicles; however, no Cx43 is seen to replace the missing Cx37 signal at the oocyte surface. Neurobiotin can be injected into oocytes in preantral follicles from $Cx37^{+/-}$ and $Cx37^{-/-}$ ovaries. In the heterozygotes,

Neurobiotin is transferred to surrounding granulosa cells in 10 out of 11 instances, whereas no Neurobiotin transfer is observed in the Cx37$^{-/-}$ follicles (eight injections). In Cx37$^{-/-}$ follicles, direct injection of Neurobiotin into granulosa cells results in transfer to neighbouring granulosa cells, consistent with the normal distribution of Cx43 between these cells.

## Signalling from oocytes to granulosa cells

Inappropriate formation of corpora lutea is observed in Cx37$^{-/-}$ females. In the normal ovarian cycle, ovulation triggers mural granulosa cells to differentiate into progesterone-secreting (luteal) cells. Cx37$^{-/-}$ ovaries contain small corpora lutea that are five- to 10-fold more abundant than those seen in wild-type animals, suggesting premature luteinization. These observations suggest that gap junctional communication also regulates corpus luteum formation. Surgical removal of oocytes from rabbit Graafian follicles causes morphological luteinization (el-Fouly et al 1970), in support of the model that the oocyte transmits a signal through gap junctions to the granulosa cells resulting in the inhibition of luteinization.

## Signalling from granulosa cells to oocytes

Oocyte development is also influenced by gap junctional communication. Competence to resume meiosis is not achieved until the antral phase of follicular development, and it correlates with oocyte size and the appearance of a chromatin rim around the nucleolus. This competence can be experimentally assayed by observing an oocyte's ability to resume meiosis spontaneously when experimentally removed from the follicle (Mattson & Albertini 1990, Sorensen & Wassarman 1976).

The oocyte sizes, chromatin state and ability to undergo GVB are summarized in Table 1. Although the mutant oocytes are smaller than wild-type oocytes, they nonetheless grow larger than those in primordial follicles, indicating that gap

**TABLE 1 Comparison of oocyte parameters between normal and connexin37$^{-/-}$ oocytes**

| *Phenotype* | *Oocyte diameter* | *Rimmed chromatin* | *Germinal vesicle breakdown* |
|---|---|---|---|
| Cx37$^{+/+}$ | 85 ± 3 μm (*n* = 19) | 87% (*n* = 23) | 83% (*n* = 12) |
| Cx37$^{-/-}$ | 57 ± 2 μm (*n* = 21)[a] | 18% (*n* = 33) | 9% (*n* = 34) |

[a]Below the size normally associated with meiotic competence (Sorensen & Wassarman 1976).

junctional communication is only partially required for oocyte growth. Although the majority of $Cx37^{-/-}$ oocytes are blocked in their ability to acquire meiotic competence, the possibility remains that the oocytes slowly achieve competence and are then rapidly eliminated by an unknown mechanism. Gap junctional communication thus appears necessary for the acquisition of meiotic competence by the oocyte. This conclusion is not consistent with the observation, reviewed above, that GVB could be induced in immature, denuded oocytes by fibroblast-conditioned medium (Chesnel et al 1994), indicating that a diffusable factor can trigger GVB. Although these results cannot be currently reconciled, it is possible that gap junctional communication is required during the preantral phase of folliculogenesis to stimulate the expression or surface deployment in the oocyte of a critical receptor for a diffusable factor, and that these two forms of communication exist in series.

As reviewed above, previous studies suggest that gap junctional communication maintains oocytes arrested in the first meiotic prophase by delivering inhibitory signals, in particular cAMP (Cho et al 1974, Gilula et al 1978, Bornslaeger & Schultz 1985) together with other cAMP-dependent inhibitory factors (Eppig et al 1983). If this model were true, then competent oocytes lacking junctional communication with granulosa cells would undergo premature meiotic resumption. Since oocytes in the $Cx37^{-/-}$ females never acquire meiotic competence, these follicles *in vivo* do not permit a test of this hypothesis.

## Future experiments

Methods are available that permit follicles to grow and acquire meiotic competence in cell culture. Developing mouse ovarian follicles can be placed in tissue culture at early stages (Eppig & O'Brien 1996, Eppig et al 1996). In these methods, preantral follicles are dissected from 8-, 9-, 10-, 11- and 12-day-old female mice (Eppig & Schroeder 1989) and maintained in culture for 14, 13, 12, 11 and 10 days, respectively (Eppig et al 1992). Following these culture conditions, almost 100% of wild-type follicles will undergo GVB, indicating that they have acquired meiotic competence (Sorensen & Wassarman 1976). Oocyte–granulosa cell complexes from $Cx37^{-/-}$ mice could be cultured together with wild-type follicles (Eppig et al 1992). By testing and comparing both sets of follicles for GVB, it should be possible to determine if the mutant follicles remain unable to grow at wild-type rates and acquire meiotic competence in culture as they do *in situ*. The culturing of the follicles will permit their direct observation and determination if the mutant follicles acquire meiotic competence but then undergo a rapid elimination not observable in the whole animal. Granulosa cells in the $Cx37^{-/-}$ follicles may also be studied in culture. *In vivo* these cells differentiate morphologically into luteal cells. If the granulosa cells in the cultured $Cx37^{-/-}$ follicles also differentiate into luteal cells according to morphological criteria (Fields & Fields 1996) as they do *invivo*, this would offer an experimental opportunity to attempt to determine the factors originating in the oocyte that inhibit granulosa cell luteinization.

Oocytes in cultured follicles could be injected with Cx37 mRNA to see if the development of meiotic competence and the inhibition of luteinization can be rescued *in vitro*. If the oocyte/cumulus junctions must be composed of solely Cx37, then injection of Cx37 mRNA on the oocyte side of the cell–cell interaction will be insufficient for rescue, since the cumulus cells will continue to lack this protein. If, however, granulosa cells are using a different member of the connexin gene family, which can form heterotypic intercellular channels with Cx37, then providing Cx37 to the oocyte side of the junction may be sufficient for rescue. If rescue does occur, this would provide an opportunity to study the ability of connexins to functionally substitute for Cx37 by injecting mRNA coding for other members of the connexin family.

While it remains to be determined whether defects in Cx37 cause human infertility or other pathologies, Cx37 distribution in human ovarian follicles is similar to the mouse (Simon et al 1997). In addition, abnormalities observed in $Cx37^{-/-}$ animals are similar to those exhibited in karyotypically normal spontaneous premature ovarian failure, a human disorder of unknown aetiology (Nelson et al 1994). In particular, inappropriate luteinization is a highly specific feature of this disease, suggesting that alterations in connexin-mediated intercellular communication could be responsible.

*Acknowledgements*

We gratefully acknowledge grant support from GM 18974 and GM 37751 from the National Institutes of Health and a grant from the American Heart Association.

## References

Albertini DF 1992 Regulation of meiotic maturation in the mammalian oocyte: interplay between exogenous cues and the microtubule cytoskeleton. BioEssays 14:97–103

Albertini DF, Anderson E 1974 The appearance and structure of intercellular connections during the ontogeny of the rabbit ovarian follicle with particular reference to gap junctions. J Cell Biol 63:234–250

Anderson E, Albertini DF 1976 Gap junctions between the oocyte and companion follicle cells in the mammalian ovary. J Cell Biol 71:680–686

Bachvarova R, Baran MM, Tejblum A 1980 Development of naked growing mouse oocytes *in vitro*. J Exp Zool 211:159–169

Beyer EC, Kistler J, Paul DL, Goodenough DA 1989 Antisera directed against connexin43 peptides react with a 43-kD protein localized to gap junctions in myocardium and other tissues. J Cell Biol 108:595–605

Bornslaeger EA, Schultz RM 1985 Regulation of mouse oocyte maturation: effect of elevating cumulus cell cAMP on oocyte cAMP levels. Biol Reprod 33:698–704

Buccione R, Schroeder AC, Eppig JJ 1990 Interactions between somatic cells and germ cells throughout mammalian oogenesis. Biol Reprod 43:543–547

Cecconi S, Tatone C, Buccione R, Mangia F, Colonna R 1991 Granulosa cell–oocyte interactions: the phosphorylation of specific proteins in mouse oocytes at the germinal vesicle stage is dependent upon the differentiative state of companion somatic cells. J Exp Zool 258:249–254

Chesnel F, Wigglesworth K, Eppig JJ 1994 Acquisition of meiotic competence by denuded mouse oocytes: participation of somatic-cell product(s) and cAMP. Dev Biol 161:285–295

Cho WK, Stern S, Biggers JD 1974 Inhibitory effect of dibutyryl cAMP on mouse oocyte maturation *in vitro*. J Exp Zool 187:383–386

el-Fouly MA, Cook B, Nekola M, Nalbandov AV 1970 Role of the ovum in follicular luteinization. Endocrinology 87:286–293

Eppig JJ 1979 A comparison between oocyte growth in coculture with granulosa cells and oocytes with granulosa cell–oocyte junctional contact maintained *in vitro*. J Exp Zool 209:345–353

Eppig JJ, O'Brien MJ 1996 Development *in vitro* of mouse oocytes from primordial follicles. Biol Reprod 54:197–207

Eppig JJ, Schroeder AC 1989 Capacity of mouse oocytes from preantral follicles to undergo embryogenesis and development to live young after growth, maturation and fertilization *in vitro*. Biol Reprod 41:268–276

Eppig JJ, Freter RR, Ward-Bailey PF, Schultz RM 1983 Inhibition of oocyte maturation in the mouse: participation of cAMP, steroid hormones, and a putative maturation inhibitory factor. Dev Biol 100:39–49

Eppig JJ, Wigglesworth K, O'Brien MJ 1992 Comparison of embryonic developmental competence of mouse oocytes grown with and without serum. Mol Reprod Dev 32:33–40

Eppig JJ, O'Brien MJ, Wigglesworth K 1996 Mammalian oocyte growth, and development *in vitro*. Mol Reprod Dev 44:260–273

Fields MJ, Fields PA 1996 Morphological characteristics of the bovine corpus luteum during the estrous cycle and pregnancy. Theriogenology 45:1295–1325

Gilula NB, Epstein ML, Beers WH 1978 Cell-to-cell communication and ovulation. A study of the cumulus–oocyte complex. J Cell Biol 78:58–75

Ginsburg M, Snow MH, McLaren A 1990 Primordial germ cells in the mouse embryo during gastrulation. Development 110:521–528

Goodenough DA, Goliger JA, Paul DL 1996 Connexins, connexons and intercellular communication. Annu Rev Biochem 65:475–502

Gougeon A 1996 Regulation of ovarian follicular development in primates: facts and hypotheses. Endocr Rev 17:121–155

Grazul-Bilska AT, Reynolds LP, Redmer DA 1997 Gap junctions in the ovaries. Biol Reprod 57:947–957

Haefliger J-A, Bruzzone R, Jenkins NA, Gilbert DJ, Copeland NG, Paul DL 1992 Four novel members of the connexin family of gap junction proteins. Molecular cloning, expression, and chromosome mapping. J Biol Chem 267:2057–2064

Herlands RL, Schultz RM 1984 Regulation of mouse oocyte growth: probable nutritional role for intercellular communication between follicle cells and oocytes in oocyte growth. J Exp Zool 229:317–325

Kumar N, Gilula NB 1996 The gap junction communication channel. Cell 84:381–388

Mattson BA, Albertini DF 1990 Oogenesis: chromatin and microtubule dynamics during meiotic prophase. Mol Reprod Dev 25:374–383

McGrath SA, Esquela AF, Lee SJ 1995 Oocyte-specific expression of growth/differentiation factor-9. Mol Endocrinol 9:131–136

McLaren A 1992 Development of primordial germ cells in the mouse. Andrologia 24:243–247

Nelson LM, Anasti JN, Kimzey LM et al 1994 Development of luteinized Graafian follicles in patients with karyotypically normal spontaneous premature ovarian failure. J Clin Endocrinol Metab 79:1470–1475

Okuma A, Kuraoka A, Iida H, Inai T, Wasano K, Shibata Y 1996 Colocalization of connexin 43 and connexin 45 but absence of connexin 40 in granulosa cell gap junctions of rat ovary. J Reprod Fertil 107:255–264

Reed KE, Westphale EM, Larson DM, Wang Z, Veenstra RD, Beyer EC 1993 Molecular cloning and functional expression of human Connexin37, an endothelial cell gap junction protein. J Clin Invest 91:997–1004

Sadler TW 1997 Langman's Medical Embryology, 7th edn. Williams & Wilkins, Baltimore, MD

Schultz RM, Wassarman PM 1977 Biochemical studies of mammalian oogenesis: protein synthesis during oocyte growth and meiotic maturation in the mouse. J Cell Sci 24:167–194

Simon AM, Goodenough DA, Li E, Paul DL 1997 Female infertility in mice lacking connexin 37. Nature 385:525–529

Sorensen RA, Wassarman PM 1976 Relationship between growth and meiotic maturation of the mouse oocyte. Dev Biol 50:531–536

Wassarman PM, Albertini DF 1994 The mammalian ovum. In: Knobil E, Neill JD (eds) The physiology of reproduction. Raven Press Ltd, New York, p 79–122

## DISCUSSION

*Willecke:* Did you look at whether the passage of dye from the granulosa cells to the oocyte is impaired?

*Goodenough:* This is difficult to look at because you have to inject a small cell, and the dye is transferred to a much larger cell, so the signal is diluted.

*Willecke:* Could you inject Neurobiotin close to the edge of the oocyte?

*Goodenough:* Theoretically, this could be done but I haven't systematically tried to do it because the dye spreads rapidly throughout all of the surrounding granulosa cells. It would be worth finding out if there is some asymmetry in the passage.

*Willecke:* Are special markers available for those granulosa cells that are making junctions with oocytes?

*Goodenough:* I don't know. I do know that granulosa cells which make junctions with oocytes are not just those which are immediately adjacent to the oocyte. They go back in tiers of two, three and sometimes four, and they reach past each other to make contact with the oocyte. No one has looked at how many make direct contact and whether this defines the cumulus field. There are low levels of connexin37 (Cx37; $\alpha_4$) only in those granulosa cells that are next to the oocytes, which may turn out to be a marker.

*Gilula:* Have you carried out electrical coupling studies to rule out the possibility that another connexin may be present? If the cumulus cells use Cx37 to associate or establish communication with the oocyte surface, would you propose that Cx37 is conferring some special channel activity on the oocyte that Cx43 ($\alpha_1$) can't provide?

*Goodenough:* I haven't done any electrical coupling experiments, and I don't have any insights into whether Cx37 has a special channel activity. It may be possible to answer this, however, because the oocyte can be injected with mRNA coding for other connexins. With the culture systems that have been set up by a number of laboratories, one can systematically try to rescue the oocytes. We have cultured the follicles and they fail at the same time as they do *in vivo*, so they're not rescued simply by being in serum. It should be possible to rescue the oocytes by injecting mRNA encoding Cx37, then one could see whether other connexins can also rescue them.

*Gilula:* When the granulosa cells are being primed for ovulation they take on certain phenotypes, one of which is the production of plasminogen activator. Have you looked at whether the production of plasminogen activator is altered as a result of communication interference?

*Goodenough:* No, I haven't looked at plasminogen activator.

*Gilula:* The formation of atresic follicles and corpus lutea is interesting because there are proposals that these events in the pre-ovulatory follicle regulate other events in the ovary, and not necessarily events in the same follicle.

*Goodenough:* It's important to point out that although these follicles have the morphology of steroid-secreting cells, they may not be luteal cells. They could be oestrogen-producing cells.

*Warner:* Apart from the oocyte system, do you see defects in other systems? Some years ago there was a report of a point mutation in Cx37 in lung tissue that had an effect on junction formation.

*Goodenough:* I have not looked in any detail at the lung. There are no gross morphological problems, and the animals live to ripe old ages. In collaboration with Brian Duling, David Paul and Alex Simon, we have started to look at the propagation of excitation along arterioles in the Cx37 knockout (unpublished results 1998). There are some interesting differences in that relaxation responses can be propagated but contractile responses cannot.

*Gilula:* Are there any changes in the testis? And is Cx37 used there?

*Goodenough:* There are no changes that I can detect histologically. To my knowledge Cx37 is not used in the testis. It is used in the blood vessels.

*Werner:* Cx37 is expressed in the spermatogonia, whereas Cx43 is expressed in Sertoli cells. It is possible that there are heterologous junctions between Cx43 and Cx37 (Chang et al 1996). Could you test this by determining whether there is rectification? Because Cx37 is similar to Cx38 ($\alpha_2$), and we have shown that Cx38 and Cx43 form rectifying junctions.

*Goodenough:* It is unlikely that heterologous junctions are present between these two connexins because the Cx43 staining on the oocyte side of the zona does not match the Cx37 staining on the oocyte side of the zona. There may be other heterologous junctions, because I don't know what connexin the granulosa cell is using in order to contact oocytes, although I'm relatively certain that it's not Cx43.

*Warner:* Presumably, there's likely to be a heterologous junction somewhere in the system, because cells expressing Cx37 in order to make homologous junctions with the oocyte could also be making heterologous junctions with other cells.

*Gilula:* A point of clarification. In this context heterologous junctions are already established between the oocyte and the granulosa cells. We are talking about whether or not there are heterotypic connexin interactions.

*Beyer:* Are there human forms of infertility that look like this?

*Becker:* Yes. The condition is called polycystic ovary syndrome, and it affects about 20% of women to different degrees.

*Beyer:* Did you breed the different connexin knockouts with each other?

*Goodenough:* The initial goal of this exercise was to generate Cx37/Cx40 ($\alpha_5$) double knockouts. The animals die perinatally, and they're covered with bruises. Histological inspection of the connective tissue reveals a large increase in the number of lymphatic vessels, and sections through the mediastinum shows a thoracic duct that is larger than the aorta and full of blood. The conclusion is that the blood vasculature system has become extremely leaky, and that lymphatic system compensates by up-regulating so that the fluid is returned and the animals don't become oedematous.

I would also like to take this opportunity to mention briefly some interesting results we have on the lens system. We have knocked out Cx50 ($\alpha_8$) in the mouse. The homozygous knockout animals are normal until birth. Following birth, the animals develop microphthalmia and have small ocular lenses. By the end of the first week, a zonular pulverulent cataract becomes visible, which increases with age. Western analysis of the lenses reveals that the levels of Cx43 and Cx46 ($\alpha_3$) are not detectably different from those in the wild-type, indicating that there is not a compensatory up-regulation of those connexins in the absence of Cx50. Cx50 knockout lenses from embryonic day 15.5 show intraepithelial fibre-to-fibre and epithelial cell-to-fibre dye transfer patterns that are similar to dye transfer patterns in the wild-type embryonic day 15.5 lenses. These patterns may change after birth, and most certainly will change in the affected regions of the lens following cataract formation. Western analysis of soluble and insoluble fractions from three-month-old lenses reveals that $\alpha$A crystallin and $\alpha$B crystallin are detected in the insoluble fraction of Cx50 knockout lenses, but not wild-type lenses. Differences in the solubility of $\gamma$ crystallin are also detected, although no lower molecular weight proteolytic fragments of $\gamma$ crystallin are detected, in contrast to the Cx46 knockout mice.

*Gilula:* Joerg Kistler, would you like to comment on these interesting results in the lens?

*Kistler:* I was initially puzzled that the two connexins could not substitute for each other. Jiang & Goodenough (1996) showed that most of the channels in the fibre cells are heteromeric, so one would think that if one were removed the other would take its place. The lens phenotype in the Cx50 knockout mouse could be explained by a rate-limiting effect caused by halving the number of normal channels, so that there are insufficient numbers of channels to sustain the massive flux of fluid which is predicted from the lens circulation model.

*Beyer:* I'm not sure I agree, in the sense that it cannot be just that the number of total channels is reduced. There has to be something specific about the roles of Cx50 and Cx46, because you get a phenotype with each of those but not from just having half as much.

*Fletcher:* The Cx40 knockout is interesting because in polysplenia syndrome there is no SA or AV mode. There are multiple independent pacemakers. However, polysplenic patients, who do have QRS and P wave problems, don't have quite the same problem. They appear to have a better heart rhythm and ejection fraction than the Cx40 knockout. This suggests that polysplenic patients have an alternative compensatory pathway using another connexin. This may not strictly involve myocardial cells, but if it does this means they have increased their speed of propagation of potential.

*Goodenough:* I can't imagine that there is another pathway, unless there is a separate anatomical pathway through the valve structure of the heart, which forms the insulating layer between the atrium and the ventricles. The only way for the signal to pass, to my knowledge, is through the conduction system between the two. There may be other connexins present that are up-regulated or down-regulated, which may change the property of propagation through the conduction systems, or there may be breaks in the skeleton of the heart and short circuits across this break.

*Willecke:* The conduction system may be a specialized to ensure high and efficient conduction, and if it is partially or completely removed, the other cells may be able to take over, but they will be much less efficient. Therefore, I wouldn't be surprised if this does not lead to lethality, but to other complications.

*Fletcher:* That doesn't fit with the data.

*Willecke:* But in other species there does not appear to exist such a clear-cut conduction system in the myocardium.

*Kumar:* I would like to ask Dan Goodenough whether the ratio of the soluble and insoluble crystallins in the lens is the same in the knockout mouse as it is in the wild-type?

*Goodenough:* The amounts of $\alpha$A crystallin and $\alpha$B crystallin in the insoluble fraction differed between the knockout and the wild-type, as did the amount of $\gamma$ crystallin. The mouse lenses form cataracts almost as soon as they are removed from the animal, so they are difficult to handle. An interesting point is that we don't see any proteolytic processing of $\gamma$ crystallin in our animals, as compared to the Cx46 knockout animals.

*Gilula:* I have been closely related to this lens project with Xiaohua Gong. Although they both have effects in the lens, there are some dramatic differences between the Cx46 knockout and the Cx50 knockout. Micropthalmia raises a set of interesting developmental issues that have to be explored carefully. The nature and timing of the appearance of a cataract is significantly different in the two knockouts. I'm not sure that there is a zonular pulverulent cataractous phenotype that would apply to what we have seen. From my experience with the human zonular pulverulent condition I would say they are distinctly different. Would you agree with that Eric Beyer?

*Beyer:* Whenever I ask ophthalmologists this question, they say it's difficult to compare the morphological characteristics of the cataracts in mice and humans.

*Gilula:* The time frame of cataract progression in the Cx46 knockout is more along the lines of age-dependent cataractogenesis. The molecular change that we have observed as a result of this defect is in the nuclear region, which suggests that there is a unique cleavage of the $\gamma$ crystallins that has not been observed before, so now we should investigate whether this $\gamma$ crystallin cleavage takes place in humans. This cleavage is of a sort that has been associated with caspase-like enzymes, as opposed to some of the other enzymes that have been identified by others. In addition, the electrophysiological impedance measurements indicate that there's a complete loss of electrical coupling from the zone between the differentiating fibres and the mature fibres in the cortex. This is at a distance from the cataract, and there is no immunohistochemical evidence that intact Cx50 extends into this region. There is no electrophysiological evidence of a functional pathway for electrical coupling in the mature fibre region in the Cx46 knockout. Our conclusion is that in the Cx46 knockout there is a region of the lens that is completely devoid of electrical coupling, and we propose that this region normally uses Cx46. We have no evidence that Cx50 can provide the channel activity for that region in the lens. I can appreciate that if there are heteromeric channels throughout the lens, one would expect to find that Cx50 is still functional in the nuclear region, but this isn't what we've seen physiologically.

*Beyer:* You said you have no evidence that Cx50 is present. What is that statement based on?

*Gilula:* It's based on immunohistochemical analysis with antibodies that recognize the C-terminal domain as an intact domain. We don't have antibodies that recognize the integral membrane portion of Cx50, but we know it's there because we can isolate it and identify it in gels. Our conclusion is that if it's there in a fragmented form, it's not active and does not contribute directly to the channel.

*Beyer:* Is that region devoid of Cx50 in wild-type animals?

*Gilula:* We presume that there are no changes in distribution. This presumption is based on the same sort of analogy.

*Nicholson:* Is Cx50 in the clipped form functional when expressed in exogenous systems such as oocytes?

*Kistler:* Yes. Truncated recombinant Cx50 makes functional channels in oocytes. The voltage gating is the same but the pH gating is different.

## References

Chang M, Werner R, Dahl G 1996 A role for an inhibitory connexin in testis? Dev Biol 175: 50–56

Jiang JX, Goodenough DA 1996 Heteromeric connexons in lens gap junction channels. Proc Natl Acad Sci USA 93:1287–1291

# Connexins in tumour suppression and cancer therapy

Hiroshi Yamasaki, Yasufumi Omori, Vladimir Krutovskikh, Weibin Zhu, Nikolai Mironov, Kohji Yamakage and Marc Mesnil

*Multistage Carcinogenesis Unit, International Agency for Research on Cancer, 150 Cours Albert Thomas, F-69372, Lyon Cedex 08, France*

*Abstract.* Malignant cells usually show altered gap junctional intercellular communication and are often associated with aberrant expression or localization of connexins. Transfection of connexin genes into tumorigenic cells restores normal cell growth, suggesting that connexins form a family of tumour suppressor genes. Some studies have also shown that specific connexins may be necessary to control growth of specific cell types. Although we have found that genes encoding connexin32 (Cx32; $\beta_1$), Cx37 ($\alpha_4$) and Cx43 ($\alpha_1$) are rarely mutated in tumours, our recent studies suggest that methylation of the connexin gene promoter may be a mechanism by which connexin gene expression is down-regulated in certain tumours. We have produced various dominant negative mutants of the genes encoding Cx26 ($\beta_2$), Cx32 and Cx43, some of which prevent the growth control exerted by the corresponding wild-type genes. A decade ago, we proposed a method to enhance killing of cancer cells by diffusion of therapeutic agents through gap junctions. Recently, we and others have shown that gap junctional intercellular communication is responsible for the bystander effect seen in herpes simplex virus thymidine kinase/ganciclovir gene therapy. Thus, connexin genes can exert dual effects in tumour control: tumour suppression and a bystander effect for cancer therapy.

*1999 Gap junction-mediated intercellular signalling in health and disease. Wiley, Chichester (Novartis Foundation Symposium 219) p 241–260*

## Gap junctional intercellular communication is altered in tumour cells and by carcinogenic agents

During the process of carcinogenesis, a normal cell acquires multiple genetic changes to become malignant (Kinzler & Vogelstein 1997). The critical genes specifically altered in carcinogenesis are directly or indirectly involved in cell growth control. Thus, most oncogenes and tumour suppressor genes so far identified are involved in cell cycle control and signal transduction. These molecules function to control individual cell growth. However, there is another set of molecules that control cell growth at a higher level of hierarchy, i.e. that of tissue homeostasis, mediated by various forms of intercellular communication, and it is reasonable to assume that such genes are also altered during carcinogenesis.

Gap junctional intercellular communication is the only means for multicellular organisms by which cells can exchange signals directly from the inside of one cell to those of neighbouring ones (Loewenstein 1987, Pitts & Finbow 1986). Since the hallmark of tumours is their deviation from tissue homeostasis, it has long been postulated that gap junctional intercellular communication is disturbed in cancers.

The first evidence for the involvement of aberrant gap junctional intercellular communication in carcinogenesis was the observation that a human cancer showed a reduced level of gap junctional intercellular communication (Loewenstein & Kanno 1966); this observation has now been extended to almost all tumour cells (reviewed by Yamasaki 1990). Cell lines established from tumours as well as cells transformed *in vitro* usually show impaired gap junctional intercellular communication ability. These include not only rodent but also human cell lines. Gap junctional intercellular communication between transformed cells and neighbouring normal counterparts is clearly defective in murine embryonic BALB/c3T3 cells. A similar lack of heterologous gap junctional intercellular communication between transformed and normal cells has also been observed with rat liver epithelial cell lines (reviewed in Yamasaki 1990). To extend these *in vitro* results, we developed a simple method by which gap junctional intercellular communication can be examined in slices of liver freshly removed from the rat (Krutovskikh & Yamasaki 1995). Using this method, we have shown a reduced level of gap junctional intercellular communication in rat liver tumours. We also applied such a method for measuring gap junctional intercellular communication in human liver freshly removed surgically in a hospital. Again, we found that the gap junctional intercellular communication level was much reduced in hepatocellular carcinoma cells in comparison with normal cells or surrounding lesions. It therefore appears that reduced gap junctional intercellular communication is a common feature of many tumour cells (reviewed in Krutovskikh & Yamasaki 1997).

Further studies of gap junctional intercellular communication changes using multistage models of rat liver and mouse skin carcinogenesis have revealed that there is, in general, a progressive decrease in the level of gap junctional intercellular communication during tumour progression. For example, in rat liver carcinogenesis, gap junctional intercellular communication is already reduced in many preneoplastic foci and further reduction is evident in tumours (Krutovskikh et al 1991). Kamibayashi et al (1995) observed a progressive decrease in connexin26 (Cx26; $\beta_2$) and Cx43 ($\alpha_1$) staining during skin carcinogenesis.

Another line of evidence that suggests a causal role of blockage of intercellular communication in carcinogenesis has come from studies in which agents or genes involved in carcinogenesis were shown to modulate gap junctional intercellular communication. First, the mouse skin tumour-promoting agent 12-*O*-tetradecanoylphorbol 13-acetate (TPA) has been shown to inhibit gap junctional intercellular communication in many types of cultured cells by various methods including metabolic co-operation, electrical coupling and dye transfer assays. Many tumour-promoting agents have subsequently been reported to inhibit gap junctional intercellular communication (Trosko et al 1990). In addition to these tumour-promoting agents, other types of tumour-promoting stimuli, such as partial hepatectomy and skin wounding, also inhibit gap junctional intercellular communication. Various oncogenes also inhibit gap junctional intercellular communication; such oncogenes include the *src*, SV-40-T antigen, *neu*, *raf*, *fps* and *ras* oncogenes. It has also been observed that chemopreventive agents enhance gap junctional intercellular communication. Retinoids, glucocorticoids and cAMP are known to antagonize the tumour-promoting ability of TPA in mouse skin, and they have been shown to up-regulate gap junctional intercellular communication. Other agents that up-regulate gap junctional intercellular communication include green tea extracts, flavones, vitamin D and components of Japanese soybean fermented food, Natto (reviewed by Trosko et al 1990, Yamasaki 1996).

## Connexin genes and tumour suppression

The first indication that normal cells may suppress growth of malignant cells in contact with them came from the work of Stoker and his colleagues (Stoker 1967). More recent evidence of a direct role of gap junctional intercellular communication in tumour suppression has come from a series of experiments in which connexin genes were transfected into gap junctional intercellular communication-deficient malignant cell lines to study the effect of connexins on cell growth *in vitro* and *in vivo*. Thus, rat glioma cells (Naus et al 1992) and chemically transformed mouse fibroblasts (Rose et al 1993) transfected with the Cx43 gene did not produce tumour growth. Similarly, human liver tumour cells transfected with the Cx32 ($\beta_1$) gene showed reduced tumour growth (Eghbali et al 1991). To examine the hypothesis that not all species of connexin genes act as tumour suppressor genes in a given cell type to the same extent, we transfected the genes encoding Cx26, Cx40 ($\alpha_5$), Cx43 and Cx32 into HeLa cells, which do not express detectable levels of any connexin genes examined. HeLa clones transfected with these connexin genes all showed an increased level of gap junctional intercellular communication. However, intriguingly, only transfectants with the gene encoding Cx26 grew more slowly *in vitro* and lost their tumorigenicity (Mesnil et al 1995). Additional studies indicated that this gene is a major connexin gene expressed in the cervix (the tissue of tumour origin from which HeLa cells were originally isolated). Thus, it appears that the establishment of gap junctional intercellular communication in tumour cells does not always lead to effects on cell growth, and our results further suggest that connexin genes exert a differential cell growth control effect, depending on the cell type in which they are expressed. The idea that there might be selective effects of specific connexins on cell growth is supported by emerging evidence that gap junctional intercellular communication mediated by specific connexins may show specific functional variations. Thus, the channels composed of these different connexins display differences in unitary conductances and voltage sensitivity. Furthermore, it is now becoming apparent that the channels encoded by the different connexin genes exhibit some selectivity in molecular transport (Elfgang et al 1995). The effects of various connexins on growth control are summarized in Table 1.

When tumour suppressor genes are mutated in germ cells, the offspring are often at high risk of cancers, as is seen with the *Rb* gene for hereditary retinoblastoma, *BRCA1* and *BRCA2* for breast cancers, and *hMSH2* for hereditary non-polyposis colon cancer (reviewed by Knudson 1996). Similarly, deletion of such genes from mice leads to a higher incidence of specific tumours. Temme et al (1997) have shown that Cx32-deficient mice are more prone to spontaneous and chemically

**TABLE 1 Examples of tumour suppression by connexin gene transfection**

| *Connexin gene* | *Host cell line* | *Growth suppression* | |
|---|---|---|---|
| | | *In vitro* | *In vivo* |
| Cx32 ($\beta_1$) | Human hepatoma | – | + |
| | Rat glioma (C6) | ± | + |
| Cx43 ($\alpha_1$) | Transformed 10T 1/2 cells | + | + |
| | Rat glioma (C6) | + | + |
| | HeLa cells | ± | ± |
| | Rat WB-F344 cells[a] | + | NT |
| | Human rhabdomyosarcoma | + | NT |
| | Rat-1 cells (antisense oligonucleotide transfection)[b] | + | NT |
| Cx40 ($\alpha_5$) | HeLa cells | – | – |
| Cx26 ($\beta_2$) | HeLa cells | + | + |

[a]Restoration of gap junctional intercellular communication by Cx43 cDNA transfection was important to suppress growth of co-cultured transformed cells.
[b]Inhibition of gap junctional intercellular communication by Cx43 antisense oligonucleotides reduced the ability of rat-1 cells to suppress foci formation of co-cultured transformed cells.
NT, not tested. Adapted from Yamasaki et al (1996).

induced hepatocarcinogenesis, supporting the idea that the Cx32 gene (which is highly expressed in the normal liver) is a liver tumour suppressor gene. Interestingly, germ line mutations of this gene have been reported in X-linked Charcot-Marie-Tooth (CMTX) syndrome (reviewed by Fischbeck et al 1996). So far, however, no increase in tumour incidence has been reported in these individuals. It may be that the number of CMTX patients is too small for an increased incidence of cancers to be detectable or that complete deletion of the Cx32 gene in mice may not exert the same effect as the mutations of it found in humans.

Martyn et al (1997) have reported the establishment of immortalized cell lines from Cx43-deficient mice and found that $Cx43^{+/+}$, but not $Cx43^{-/-}$, cells experienced crisis during immortalization. They also found that $Cx43^{-/-}$ cells grow faster and to higher saturation densities than $Cx43^{+/+}$ cells. We have also recently established $Cx43^{+/+}$, $Cx43^{+/-}$ and $Cx43^{-/-}$ fibroblast cell lines by the 3T3 method. As shown in Fig. 1, all of them went through a growth crisis during immortalization. In our preliminary studies, we saw no significant difference in growth rate or saturation density of these cell lines. It is possible that the effect of connexins on cell growth may be more clearly seen in already tumorous cells than in normal cells.

Although other tumour suppressor genes, notably the *p53* gene, are often mutated in tumours, few mutations of connexin genes have been found in rodent tumours and none has been reported for any human cancer (Fig. 2). While the results suggest that involvement of connexin gene mutations in carcinogenesis is rare, it is important to emphasize that only a few studies (all from one laboratory) on a few connexin genes (Cx32, Cx37 [$\alpha_4$] and Cx43) have been conducted in a small number of tumours. On the other hand, several polymorphisms in connexin genes in humans and rats have been discovered, although there was no apparent correlation with the polymorphism and cancer sites examined (Fig. 2).

Tumour suppressor genes are sometimes silenced in cancers by hypermethylation (Jones & Gonzalgo 1997). Recently, we studied the correlation between Cx32 gene expression and methylation of its promoter region in human colon cancer cell lines. We found that two cell lines expressed the Cx32 gene and had no methylation in its promoter region, whereas the other four cell lines did not express the gene and its promoter was hypermethylated (Zhu et al 1998). Thus, it is possible that methylation of the connexin gene promoter is a mechanism by which connexin gene expression is altered during carcinogenesis.

## Dominant negative effect of mutant connexins in tumour suppression

Six connexin molecules must be assembled to form a 'connexon', which corresponds to a hemichannel of a gap junction. Thus, dominant negative regulation of gap

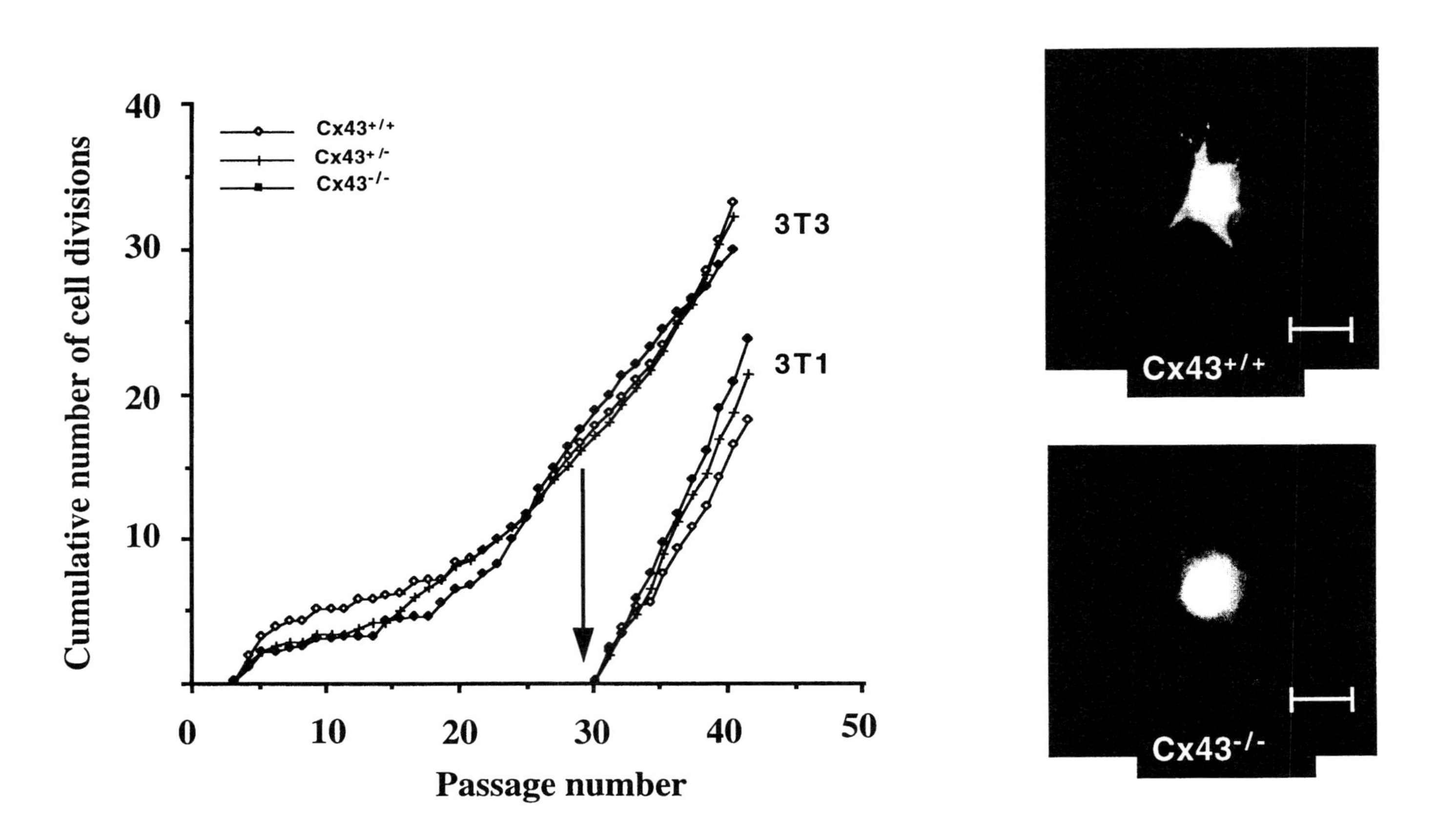

Cumulative number of cell divisions
40
30
20
10
0
10
20
30
40
50
Passage number
Cx43+/+
Cx43+/-
Cx43-/-
3T3
3T1
Cx43+/+
Cx43-/-

junctional intercellular communication may occur when non-functional mutant connexins assemble with wild-type connexins. It is likely that such mosaic connexons will be functionally defective or less effective. A mutation of a connexin gene at one allele in a given cell may, therefore, suffice to block gap junctional intercellular communication. We have proved that this is indeed the case. When HeLa cells are transfected with the Cx32 gene, their gap junctional intercellular communication is restored. However, when mutated Cx32 genes (carrying mutations found in CMTX patients) were transfected into such Cx32-expressing HeLa cells, their gap junctional intercellular communication was inhibited (Omori et al 1996). A similar dominant negative effect has been reported in the *Xenopus* oocyte model (Bruzzone et al 1994). Our results also suggested that this inhibition is associated with transfer of wild-type Cx32 from the plasma membrane into the cytoplasm, implying that hybrid connexons presumably composed of wild-type Cx32 and mutant Cx32 cannot be anchored at the plasma membrane.

In order to examine whether such a dominant negative effect on gap junctional intercellular communication leads to aberrant growth control, it was necessary to study Cx26 mutants, since only this gene, not Cx32, is tumour suppressive for HeLa cells (Duflot-Dancer et al 1997). We therefore transfected three mutant Cx26 genes (C60F, P87L and R143W) into HeLa cells in which growth was suppressed by the wild-type Cx26 gene. We found that the P87L and R143W mutants reversed the tumour suppressive effect of the wild-type Cx26 gene (Duflot-Dancer et al 1997). Similarly, we found that several Cx43 mutants also transdominantly suppressed the growth control effects of the wild-type Cx43 in rat glioma C6 cells (Omori et al 1998) and rat bladder B31 cells (Krutovskikh et al 1998). These results are summarized in Table 2.

Although various mutant connexin genes exert a dominant negative effect on cell growth, certain of them have no influence on gap junctional intercellular communication. This apparent discrepancy between the tumour suppressive effect and the gap junctional intercellular communication capacity of connexins suggests that they may exert cell growth control independently from the formation of gap junctions (Duflot-Dancer et al 1997). We have previously suggested that different connexin genes exert differential growth effects even if they apparently mediate similar levels of gap junctional intercellular

---

FIG. 1. Establishment of embryonic fibroblast 3T3 cell lines from connexin43 (Cx43; $\alpha_1$) null mice. $Cx43^{+/+}$, $Cx43^{+/-}$ and $Cx43^{-/-}$ embryonic fibroblast cells were prepared from the same litter. They were subcultured at the density of $3 \times 10^5$ cells per 60 mm petri dishes every 3 days. At passage No. 30, some culture dishes received $1 \times 10^5$ cells per dish and subculturing was continued at this cell density (3T1). The gap junctional intercellular communication ability measured by the dye (Lucifer yellow) transfer assay is shown on the right. Bars = 20 $\mu$m.

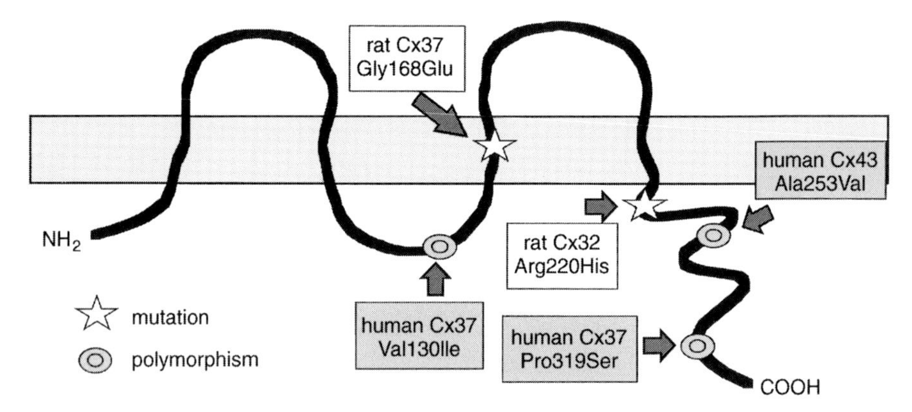


FIG. 2. Connexin (Cx) gene mutations in rodent tumours and polymorphisms in humans found at the International Agency for Research on Cancer.

communication (Mesnil et al 1995). We estimated the level of gap junctional intercellular communication based on the ability of the cells to transfer microinjected Lucifer yellow molecules, but the transfer of other molecules or ions does not necessarily parallel that of Lucifer yellow. Gap junctions formed between HeLa cells by different connexin species have qualitatively different permeabilities to different tracer dyes (Elfgang et al 1995). Thus, the gap junctional intercellular communication characteristics of each mutant may theoretically differ in such a way that some mutants would affect cell growth without altering Lucifer yellow transfer. However, it also remains possible that connexins exert cell growth control in a gap junctional intercellular communication-independent manner. This concept is comparable to the known function of $\beta$-catenin, which is a component of another cell–cell interaction system, i.e. adherens junctions. $\beta$-Catenin is a component of cadherin complexes, but it also mediates the WNT signal transduction pathways (Sanson et al 1996). Another study has suggested that $\beta$-catenin binds to the transcription factor LEF-1, and the resulting complex is considered to translocate to the nucleus and play a role in gene regulation (Behrens et al 1996). Recent studies have reported mutations of $\beta$-catenin and the genes involved in its regulation in several human tumours, suggesting the importance of such a pathway in cell growth control (Korinek et al 1997).

When dominant negative mutant genes are introduced as transgenes with tissue-specific gene promoters, tissue-specific connexin gene knockout mice can be produced. We have successfully introduced the albumin promoter-driven Cx32 V139M mutant gene into mice and found specific expression of the transgene in the liver. As expected, the level of gap junctional intercellular communication in

**TABLE 2 Dominant negative effects of mutant connexin genes on cell growth**

| | | *Transfected gene* | | *Gap junctional intercellular communication* | | *Tumorigenicity* | |
|---|---|---|---|---|---|---|---|
| *Cell type* | *Connexin* | *Wild-type* | *Mutant*[a] | *Observed* | *Dominant negative effect* | *Observed* | *Dominant negative effect* |
| HeLa | Cx26 ($\beta_2$) | – | – | – | NA | +++ | NA |
| | | + | – | ++ | NA | – | NA |
| | | + | C60F | ± | + | ± | – |
| | | + | R143W | ++ | – | +++ | + |
| | | + | P87L | ++ | – | +++ | + |
| C6 | Cx43 ($\alpha_1$) | – | – | ± | NA | ++ | NA |
| | | + | – | ++ | NA | ± | NA |
| | | + | L160M | ± | + | + | ± |
| | | + | A253V | ++ | – | + | + |
| B31 | Cx43 | – | – | + | NA | + | NA |
| | | + | – | ++ | NA | – | NA |
| | | – | Del(130–136) | – | + | +++ | + |

[a]Mutants used: C60F, cysteine to phenylalanine at codon 60 etc.; Del (130–136), deletion of amino acids 130–136 (intracytoplasmic loop). NA, not applicable.

the liver was much lower than in the liver of wild-type mice. However, we found that liver regeneration after partial hepatectomy was much lower. These mice should provide a useful model to study the role of the Cx32 gene in liver function and growth.

## Use of connexins for cancer therapy

A decade ago we reported that gap junctional intercellular communication might be used as a tool to deliver therapeutic agents among cancer cells to enhance cancer therapy (Yamasaki & Katoh 1988a). This mechanism has recently been proposed to operate in the bystander effect seen during therapy with the thymidine kinase gene from Herpes simplex virus (HSV-tk), and our results obtained with a cell culture model system have confirmed the role of connexin genes in this effect.

In gene therapy to treat cancer, typically only a fraction of the tumour cells can be successfully transfected with a gene. However, in the case of brain tumour therapy with HSV-tk, not only the cells transfected with the gene, but also neighbouring cells can be killed in the presence of ganciclovir. In order to study the role of gap junctional intercellular communication in this bystander effect, we used HeLa cells, since they show little, if any, ability to communicate through gap junctions. When HeLa cells were transfected with the HSV-tk gene and co-cultured with non-transfected cells, only HSV-tk-transfected HeLa cells ($tk^+$) were killed by ganciclovir. However, when HeLa cells transfected with a gene encoding the gap junction protein Cx43 were used, not only $tk^+$ cells but also $tk^-$ were killed, presumably due to the transfer to the $tk^-$ cells, via Cx43-mediated gap junctional intercellular communication, of toxic ganciclovir molecules phosphorylated by HSV-tk (Fig. 3). Such a bystander effect was not observed when $tk^+$ and $tk^-$ cells were co-cultured without direct cell–cell contact between these two types of cells. Thus, our results provide strong evidence that the bystander effect seen in HSV-tk gene therapy is due to connexin-mediated gap junctional intercellular communication (Mesnil et al 1996).

A similar bystander effect *in vitro* has been reported from other laboratories (e.g. Fick et al 1995). We and others now have evidence that connexin-mediated bystander killing of cancer cells also occurs *in vivo* (Dilber et al 1997, M. Mesnil, A. Duflot-Dancer & H. Yamasaki, unpublished data 1998).

We also observed similar effects when Cx26 was used instead of Cx43. These results suggest that the delivery of connexin genes themselves, together with the HSV-tk gene, will have dual effects on tumour regression, leading to tumour suppression by the connexin itself, and bystander killing via gap junctional intercellular communication (Mesnil et al 1997).

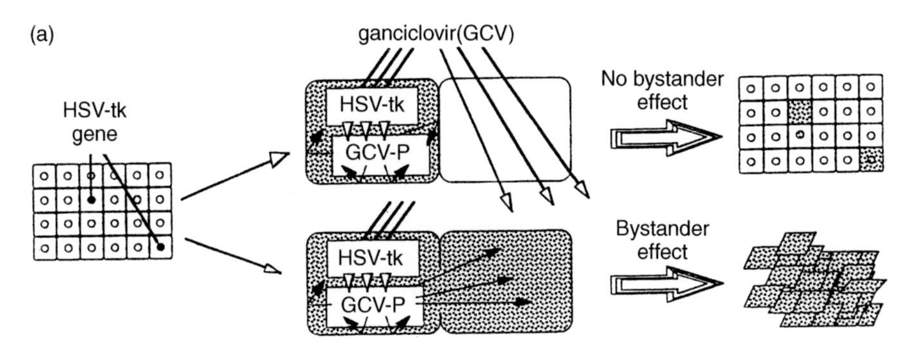


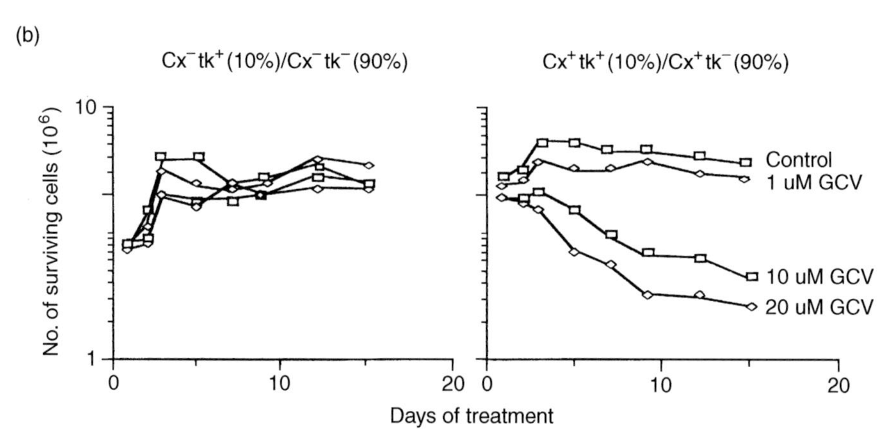


FIG. 3. Gap junctional intercellular communicated-mediated bystander effect in Herpes simplex virus thymidine kinase (HSV-tk)/ganciclovir (GCV) gene therapy. (a) Principles of the role of gap junctions in the bystander effect. The toxic phosphorylated ganciclovir (GCV-P) passes from $tk^+$ to $tk^-$ cells through the gap junctions. (b) Bystander effect of the connexin43 (Cx43; $\alpha_1$) gene. HeLa cells were transfected with Cx43 ($Cx^+$) and/or HSV-tk ($tk^+$) genes. Those cells with and without the tk gene were co-cultured in the ratio of 1:10 and exposed to ganciclovir. Note that $tk^-$ cells were also killed in such co-cultures when both $tk^-$ and $tk^+$ cells were $Cx43^+$. From Mesnil et al (1996).

## Conclusion and future perspectives

The role of connexins in tumour suppression has been rather clearly demonstrated by *in vitro* transfection studies. However, our knowledge of the extent to which they play a role in the actual process of carcinogenesis is still limited to observational evidence that altered gap junctional intercellular communication is one of the most commonly observed altered phenotypes in solid tumours. The observation of Willecke's laboratory that Cx32 null mice develop more liver tumours is so far the strongest support of *in vivo* involvement of connexin genes in tumour suppression (Temme et al 1997). Further work on this line may help to understand the exact role of connexins in growth control.

Although it is now widely believed that connexins are the major functional component of gap junctions, Finbow and his colleagues have consistently proposed that the 16 kDa protein ductin is the major component of gap junctions (Finbow et al 1995). Although most available data on connexins are consistent with the idea that they are the functional component of gap junctions, it is also fair to say that there is no convincing evidence against a role of ductin in gap junctional intercellular communication.

As described in this chapter, connexins may prove to be a useful component of cancer therapy. In the past, cancer therapy has been based on the idea of destroying cancer cells. However, application of the idea that cells form their own society may enable us to develop a new therapeutic concept. If, for example, gap junctional intercellular communication between tumours and surrounding normal cells is restored by some means, tumour growth may be controlled (Yamasaki & Katoh 1988b). These problems can now be approached more effectively since molecular probes, genetically engineered cells and animals, and various genetic manipulation tools have become available.

*Acknowledgements*

We are grateful to Chantal Déchaux for her secretarial help and to John Cheney for editing the manuscript. Our special thanks go to Colette Piccoli, who provides us with continuous technical aid in our laboratory work. Part of the work reported in this chapter has been financially supported by research grants from the National Cancer Institute (R01CA40534), the Association for International Cancer Research (No.96-27) and the Association pour la Recherche sur le Cancer.

## References

Behrens J, von-Kries JP, Kuhl M et al 1996 Functional interaction of beta-catenin with the transcription factor LEF-1. Nature 15:638–642

Bruzzone R, White TW, Scherer SS, Fischbeck KH, Paul DL 1994 Null mutations of connexin32 in patients with X-linked Charcot-Marie-Tooth disease. Neuron 13:1253–1260

Dilber MS, Abedi MR, Christensson B et al 1997 Gap junctions promote the bystander effect of herpes simplex virus thymidine kinase *in vivo*. Cancer Res 57:1523–1528

Duflot-Dancer A, Mesnil M, Yamasaki H 1997 Dominant-negative abrogations of connexin-mediated cell growth control by mutant connexin genes. Oncogene 15:2151–2158

Eghbali B, Kessler JA, Reid LM, Roy C, Spray DC 1991 Involvement of gap junctions in tumorigenesis: transfection of cells with connexin32 cDNA retards growth *in vivo*. Proc Natl Acad Sci USA 88:10701–10705

Elfgang C, Eckert R, Lichtenberg-Fraté H et al 1995 Specific permeability and selective formation of gap junction channels in connexin-transfected HeLa cells. J Cell Biol 129:805–817

Fick J, Barker FG, Dazin P et al 1995 The extent of heterocellular communication mediated by gap junctions is predictive of bystander tumor cytotoxicity *in vitro*. Proc Natl Acad Sci USA 92:11071–11075

Finbow ME, Harrison M, Jones P 1995 Ductin—a proton pump component, a gap junction channel and a neurotransmitter release channel. BioEssays 17:247–255

Fischbeck KH, Deschenes SM, Bone LJ, Scherer SS 1996 Connexin32 and X-linked Charcot-Marie-Tooth disease. Cold Spring Harbor Symp Quant Biol 61:673–677

Jones PA, Gonzalgo ML 1997 Altered DNA methylation and genome instability: a new pathway to cancer? Proc Natl Acad Sci USA 94:2103–2105

Kamibayashi Y, Oyamada Y, Mori M, Oyamada M 1995 Aberrant expression of gap junction proteins (connexins) is associated with tumor progression during multistage mouse skin carcinogenesis *in vivo*. Carcinogenesis 16:1287–1297

Kinzler KW, Vogelstein B 1997 Cancer-susceptibility genes. Gatekeepers and caretakers. Nature 386:761–763

Knudson AG 1996 Hereditary cancer: two hits revisited. J Cancer Res Clin Oncol 122:135–140

Korinek V, Barker N, Morin PJ et al 1997 Constitutive transcriptional activation by a beta-catenin-Tcf complex in $APC^{-/-}$ colon carcinoma. Science 275:1784–1787

Krutovskikh VA, Yamasaki H 1995 *Ex vivo* dye transfer assay as an approach to study gap junctional intercellular communication disorders in hepatocarcinogenesis. Prog Cell Res 4:93–97

Krutovskikh V, Yamasaki H 1997 Role of gap junctional intercellular communication disorders in experimental and human carcinogenesis. Histol Histopathol 12:761–768

Krutovskikh VA, Oyamada M, Yamasaki H 1991 Sequential changes of gap-junctional intercellular communications during multistage rat liver carcinogenesis: direct measurement of communication *in vivo*. Carcinogenesis 12:1701–1706

Krutovskikh V, Yamasaki H, Tsuda H, Asamoto M 1998 Inhibition of intrinsic gap junction intercellular communication and enhancement of tumorigenicity of rat bladder carcinoma BC31 cell line by dominant-negative Cx43 mutant. Mol Carcinog, in press

Kumar NM, Gilula NB 1996 The gap junction communication channel. Cell 84:381–388

Loewenstein WR 1987 The cell-to-cell channel of gap junctions. Cell 48:725–726

Loewenstein WR, Kanno Y 1966 Intercellular communication and the control of tissue growth: lack of communication between cancer cells. Nature 209:1248–1249

Martyn KD, Kurata WE, Warn-Cramer BJ, Burt JM, TenBroek E, Lau AF 1997 Immortalized connexin43 knock-out cell lines display a subset of biological properties associated with the transformed phenotype. Cell Growth Differ 3:1015–1027

Mesnil M, Krutovskikh V, Piccoli C et al 1995 Negative growth control of HeLa cells by connexin genes: connexin species specificity. Cancer Res 55:629–639

Mesnil M, Piccoli C, Tiraby G, Willecke K, Yamasaki H 1996 Bystander killing of cancer by HSV-tk gene is mediated by connexins. Proc Natl Acad Sci USA 93:1831–1835

Mesnil M, Piccoli C, Yamasaki H 1997 A tumour suppressor gene, CX26, also mediates the bystander effect in HeLa cells. Cancer Res 57:2929–2932

Naus CC, Elisevich K, Zhu D, Belliveau DJ, Del Maestro RF 1992 *In vivo* growth of C6 glioma cells transfected with connexin43 cDNA. Cancer Res 52:4208–4213

Omori Y, Mesnil M, Yamasaki H 1996 Connexin 32 mutations from X-linked Charcot-Marie-Tooth disease patients: functional defects and dominant-negative effects. Mol Biol Cell 7: 907–916

Omori Y, Duflot-Dancer A, Mesnil M, Yamasaki H 1998 Dominant-negative inhibition of connexin-mediated tumor suppression by mutant connexin genes. In: Werner R (ed) Gap junctions. IOS Press, Amsterdam, p 377–381
Pitts JD, Finbow ME 1986 The gap junction. Cell Sci Suppl 4:239–26
Rose B, Mehta PP, Loewenstein WR 1993 Gap-junction protein gene suppresses tumorigenicity. Carcinogenesis 14:1073–1075
Sanson B, White P, Vincent JP 1996 Uncoupling cadherin-based adhesion from wingless signalling in *Drosophila*. Nature 383:627–630
Stoker MGP 1967 Transfer of growth inhibition between normal and virus-transformed cells: autoradiographic studies using marked cells. J Cell Sci 2:293–304
Temme A, Buchmann A, Gabriel HD, Nelles E, Schwarz M, Willecke K 1997 High incidence of spontaneous and chemically induced liver tumors in mice deficient for connexin32. Curr Biol 7:713–716
Trosko JE, Chang CC, Madhukar BV, Klaunig JE 1990 Chemical, oncogene and growth factor inhibition gap junctional intercellular communication: an integrative hypothesis of carcinogenesis. Pathobiology 58:265–278
Yamasaki H 1990 Gap junctional intercellular communication and carcinogenesis. Carcinogenesis 11:1051–1058
Yamasaki H 1996 Role of disrupted gap junctional intercellular communication in detection and characterization of carcinogens. Mutat Res 365:91–105
Yamasaki H, Katoh F 1988a Novel method for selective killing of transformed rodent cells through intercellular communication with possible therapeutic applications. Cancer Res 48:3203–3207
Yamasaki H, Katoh F 1988b Further evidence for the involvement of gap-junctional intercellular communication in induction and maintenance of transformed foci in BALB/c 3T3 cells. Cancer Res 48:3490–3495
Zhu W-B, Mironov N, Yamasaki H 1998 Hypermethylation of connexin32 gene as a mechanism of disruption of cell–cell communication in tumors. Proc Am Assoc Cancer Res Annu Meet 39:199

## DISCUSSION

*Lau:* Did you observe any other differences in the connexin43 (Cx43; $\alpha_1$) knockouts, such as differences in adhesion or cell morphology?

*Yamasaki:* We could not see any clear differences, partly because these cells are non-clonal and thus the cell population was heterogeneous.

*Lau:* I was also intrigued by your suggestion that connexin gap junctions may exert a cell growth effect that may not be mediated by the transfer of molecules through the channel. Can you speculate what those mechanisms are? Are you suggesting that the connexins are involved in signalling in some way other than through the passage of molecules?

*Yamasaki:* Let me take $\beta$-catenin as a homologous example. It is a component of the cell adherence complex, but it also associates with the transcription factor LEF-1. From the two-hybrid experiments you found, on the other hand, that Cx43 associates with ZO-1. We know that ZO-1 is found in the nucleus, and $\beta$-catenin has already been found in the nucleus in several types of tumours. As $\beta$-catenin mediates signal transduction from the cellular membrane to the nucleus after binding to LEF-1, it is possible that Cx43 may be working in a similar fashion, together with associated proteins that have not yet been identified.

*Jamieson:* If your hypothesis is correct and connexins can be involved in down-regulating growth control in tumours by non-gap-junctional mechanisms, one might expect to see a specific intracellular localization of exogenous connexin in your transfectants that show reduced growth rates. Is this the case?

*Yamasaki:* Yes. Connexins are aberrantly localized in many tumour cells (Krutovskikh & Yamasaki 1997). Often they are perinuclear, and Trosko's group has claimed that Cx43 is localized inside the nucleus (de Feijter et al 1996).

*Willecke:* The hypothesis that connexins contribute to cell adhesion could be tested by determining whether cell adhesion is observed in mutants that are defective in communication but are localized at the plasma membrane. In my opinion, using immunofluoresence to demonstrate that Cx43 immunoreactivity is found in the nucleus is not sufficient to prove this hypothesis.

*Scherer:* One alternative explanation of your data is that gap junction proteins that are communication incompetent can still serve as strong cell adhesion molecules, thereby down-regulating cell proliferation. Also, you said that one of the connexin mutants was cytoplasmic when it was expressed in HeLa cells, but how did you recognize the difference between the transfected protein and the endogenous protein?

*Yamasaki:* We can't. The best control experiment to address this issue is to cross our mice with the Cx32 ($\beta_1$) null mice created by Klaus Willecke.

*Warner:* You mentioned that Lucifer yellow may not be the best probe for looking at correlations between gap junctional communication and tumour suppression. I wonder whether you should look at natural metabolites. There is no reason why Lucifer yellow should necessarily follow the same pathway or behave in the same way as a natural metabolite. This may be one of the reasons why it has been so difficult to generate reliable and reproducible correlations between gap junctional communication and tumour suppression or tumorgenicity.

*Sanderson:* I agree because only a small amount of inositol-1,4,5-trisphosphate ($InsP_3$) is required to generate a strong calcium signal in an adjacent cell. This high sensitivity results from the regenerative nature of $Ca^{2+}$ release from the $InsP_3$ receptor.

*Jamieson:* In the same vein, Ruth Sager's group did not observe Lucifer yellow dye transfer in several human breast tumour cell lines, yet with a nucleotide transfer assay we have found that some of these same cell lines can form gap junctions (S. Jamieson & M. A. Ewart, unpublished observations 1997).

*Gilula:* At this point I should say that we are all sorry that Ruth Sager passed away last year and won't be able to extend these studies. She certainly was an active contributor to the field.

*Musil:* I have a question about the expression of mutant Cx26 ($\beta_2$) in the liver of your transgenic mice. You have looked at the ratio between Cx32 mutant mRNA and the Cx32 wild-type mRNA but can you say anything about the protein levels from this?

*Yamasaki:* No.

*Musil:* I also have a question about the bystander effect. If transformed cells have a reduced ability to form functional gap junctions, then such cells would be the least likely to be therapeutically ablated by a gap junctional communication-dependent phenomenon, such as the bystander effect. Although some tumours do form functional gap junctions, I wonder if there might be a way to up-regulate gap junctional coupling in the majority of transformed cells in the hope of rendering them more susceptible to the bystander effect. Another problem is that you don't want to kill the surrounding normal cells that are capable of heterologous communication with tumour cells in which ganciclovir has been metabolized to its toxic form. Are there ways to get around this?

*Beyer:* One strategy is to target tumours, such as brain tumours, where the surrounding tissue isn't dividing. It is also important to target types of brain tumours that have lots of junctions, such as meningiomas. I'm not sure whether it is possible to up-regulate communication between tumour cells that contain few junctions.

*Yamasaki:* We have a bigenic vector that contains connexin and a suicide gene (HSV-tk), so that we can up-regulate the communication capacity of tumour cells when we introduce a cancer therapy gene.

*Musil:* This connexin protein may not be processed properly. Many tumour cells make wild-type connexin proteins but fail to assemble them into functional gap junctions, so that wouldn't help.

*Jamieson:* It's important to bear in mind that cell line systems are over-simplified and the correlation seen in many such models between reduced gap junction intercellular communication and increased tumorigenicity may not reflect what happens *in vivo*. Based predominantly on *in vitro* models, it has been proposed that Cx26 and Cx43 are tumour suppressor genes in human breast (Lee et al 1992, Hirschi et al 1996). If so, one would expect to find these connexins in the normal cell type from which human breast carcinomas derive: the ductal epithelial cell. The consensus, from several studies on rodent and human mammary tissues, is that Cx43 is not expressed by ductal epithelial cells *in vivo*, but by myoepithelial cells (Wilgenbus et al 1992, Monaghan et al 1994, 1996, Perez-Armendariz et al 1995, Pozzi et al 1995, Jamieson et al 1998). The situation for Cx26 is less certain because while some researchers find traces of this connexin in ductal epithelial cells, most don't. It may be that Cx26 is expressed only in fully differentiated (lactating) breast; if so, its absence in tumour cells would indicate only an absence of the differentiated function, and not tumorigenicity *per se*. In any case, we find Cx26 and Cx43 to be up-regulated in human breast carcinomas: the staining patterns for Cx26 are cytoplasmic, whereas punctate intercellular staining is observed for Cx43 (Jamieson et al 1998). The significance of this is not clear, but it doesn't readily fit with a simple tumour suppressor role for connexins. It may be the case that either gap junction-mediated tumour suppression does not occur *in vivo*, or that connexins have different functions during different stages of tumour formation and progression, and may even promote some stages.

*Nicholson:* It is not a simple 'down-regulation of junctions equals tumours' situation because in pre-neoplastic lesions in the liver there is a selective down-regulation of Cx32 but an up-regulation of Cx26, whereas in the full-blown tumour both connexins are down-regulated (Neveu et al 1995).

*Lo:* In collaboration with Mary Jean Sawey, we have looked at a multistage skin cancer model in mice, and found that the up-regulation of coupling may occur early on to cause increased cell proliferation, and the loss of junctions may be a late event in the final transformation process (Sawey et al 1996). The situation is complex in multistage carcinogenesis, and simple correlations probably will be hard to come by.

*Nicholson:* You used the Cx26 P87L mutant as a dominant negative to prevent the tumour suppressive ability of Cx26. You demonstrated that coupling occurs in this mutant by injecting Lucifer yellow. However, a problem with that mutant is that it has an inverted voltage sensitivity, such that it is closed when there is no voltage difference between the cells (Suchyna et al 1993). It is possible that during the Lucifer yellow injections you depolarized the cells to a point where the channels opened, so that you then observed dye transfer. If you looked at a resting tumour *in vivo* there would be no potential differences between the cells, so the channels would actually be closed.

*Yamasaki:* I agree with your comment. Although we use cultured cells because they are readily manipulatable, there are a number of reasons why cell lines may not be good models for tumours *in vivo*.

*Willecke:* I would like to add a few words of caution. We have also performed these hepatectomy studies, in which we removed two-thirds of the liver, with the Cx32-deficient mice, and we did not observe any differences in the regeneration of liver mass between the Cx32-deficient and wild-type mice. As far as I know, all the hepatectomy literature is based on the removal of two-thirds of the liver, and the data may not be relevant to experiments in which only one-third of the liver is removed.

*Gilula:* At least half the people in this room have looked at tumour cells during their gap junction careers, and we will all benefit from realizing that different changes in connexin gene expression can take place. Loewenstein & Kanno (1966) proposed the interesting and provocative concept that there is a genetic relationship between gap junction communication and growth control, but we are now at the stage where another level of complexity must be addressed. If we assume the communication pathway is important for growth control, what precise signals and molecular events are required for the correct functioning of this pathway? Is anything known about the molecule(s) involved, and what efforts are being made to address this issue?

*Nicholson:* We are taking a couple of approaches, one of which is direct and the other indirect. Based on some work by Hiroshi Yamasaki on connexins being differentially effective in regulating growth of HeLa cells (Mesnil et al 1995) and Chris Naus's work on Cx43 and Cx32 in C6 cells (Zhu et al 1991, Bond et al 1994) it's clear that efficient coupling occurs with some connexins yet they don't rescue growth, whereas others do rescue growth. We hope to use that differential screen to identify the different metabolites involved. Gary Goldberg has been trying a capture approach to identify differences in metabolites transferred by different connexins. The problem is that the metabolites get processed rapidly, so it is difficult to catch the metabolites that go through the junctions. However, our initial studies indicate that there are at least some differences between Cx43 and Cx32.

Secondly, one of the few cases in which there is a clear outcome is in the C6 cells: a secreted growth inhibitory factor seems to account for most of the inhibition of growth in those cells (Zhu et al 1992). The production of that factor is dependent on the expression of Cx43, and not Cx32. If we can identify that factor, we may be able to go backwards and find out what produced it, and with a bit of luck, if it is a known factor, we will be able to identify some candidate second messengers on which to focus. We are still a long way from answering your questions, but they are being addressed.

*Lau:* We have taken a different approach, in the sense that we have not focused on the molecules that go through the channel and induce growth suppression. Instead, we have focused on molecules that may control cell growth directly. Al Boynton's lab reported some interesting differences in cyclins and cyclin-dependent kinases (CDKs) in transformed dog epithelial cells that had been engineered to re-express Cx43. They observed that in this situation there was a down-regulation of some of the cyclins and CDKs. We have followed this line of thinking and have observed some interesting differences in the expression patterns of some of the CDK inhibitors in our Cx43 knockout cell lines that we are following up on.

*Yamasaki:* My view is similar to that of Loewenstein, i.e. gap junctional communication controls cell growth through a homeostatic mechanism in which cAMP, $Ca^{2+}$, $InsP_3$ and other signals are maintained at similar levels among gap junction-connected normal cells. We are also interested in finding out if connexin is associated with some other proteins that may eventually play a direct role in signal transduction, as has been shown for $\beta$-catenin associated with LEF-1.

*Sanderson:* Calcium is certainly important in cell regulation. We have found that trauma-inducing mechanical stimulation generates extensive calcium waves that seem to be mediated by either $InsP_3$ or calcium, depending on the number of gap junctions. We are now looking at changes in early gene expression and cell migration following traumatic events. In glial cells migration follows a large calcium wave. Because $InsP_3$ and cAMP are basic signalling molecules that pass through gap junctions these could be involved in multicellular cell regulation.

*Willecke:* Your question comes down to the point of whether there are differences in second messenger concentrations and ionic concentrations between G0 and proliferating cells. If one assumes that there are, then one can expect that cells which harbour a mutation in a growth control gene may be prevented from entering the cell cycle by small molecules passing from the surrounding cells to the mutated cell via gap junctions. One can further entertain the idea that some connexin channels may function more efficiently than others, which may explain why different connexins have different permeabilities. Another question is whether a G0 cell in the liver, for instance, has a different second messenger pattern compared to a G0 cell in the brain. These are challenges for future research.

*Lo:* I have a comment on cell cycle control. In all the different tissue culture models, generally what has been examined is doubling time or proliferation rate, rather than cell cycle regulation. The reason I bring this up is because we have generated cell lines expressing the dominant negative Cx43-lacZ fusion protein. In those cells coupling is inhibited but this had no effect on doubling time. However, we have recently been looking at cell cycle progression, and found that there are changes, i.e. a lengthening of G2/S and a shortened G1. We should therefore take note that looking at the perturbation of cell cycle controls may give us more insight than looking at the doubling time of cells.

*Gilula:* Some of these directions can be extended by making use of the available knockouts and mutants, although a lot of effort will be required because the experiments will not be simple. For example, one system in which progress has been made already is in the ovary. If the meiotic maturation of oocytes is regulated by cumulus cells, then the absence of gap junction communication will have effects on the gap junction-dependent regulatory events. When we then consider growth control and disease pathogenesis, we are forced to consider the consequences that can be defined as a result of learning more about gap junctional communication and use of the channels. Those of us who have been in the field for many years realize that there's been a lot of progress and that the issues that Klaus Willecke has addressed have to be explored. It's going to take a sizeable commitment, but there are people in this field who will make that kind of commitment.

## References

Bond SL, Bechberger JF, Khoo NK, Naus CC 1994 Transfection of C6 glioma cells with connexin32: the effects of a non-endogenous gap junction protein. Cell Growth Differ 5:179–186 de Feijter AW, Matesic DF, Ruch RJ, Guan X, Chang CC, Trosko JE 1996 Localization and function of the connexin43 gap-junction protein in normal and various oncogene-expressing rat epithelial cells. Mol Carcinog 16:203–212

Hirschi KK, Xu C, Tsukamoto T, Sager R 1996 Gap junction genes Cx26 and Cx43 individually suppress the cancer phenotype of human mammary carcinoma cells and restore differentiation potential. Cell Growth Differ 7:861–870

Jamieson S, Going JJ, D'Arcy R, George WD 1998 Expression of gap junction proteins connexin 26 and connexin 43 in normal human breast and in breast tumours. J Pathol 184:37–43

Krutovskikh V, Yamasaki H 1997 The role of gap junctional intercellular communication (GJIC) disorders in experimental and human carcinogenesis. J Histol Histopath 12:761–768

Loewenstein WR, Kanno Y 1966 Intercellular communication and the control of tissue growth: lack of communication between cancer cells. Nature 209:1248–1249

Lee SW, Tomasetto C, Paul D, Keyomarsi K, Sager R 1992 Transcriptional downregulation of gap-junction protein blocks junctional communication in human mammary tumor cell lines. J Cell Biol 118:1213–1221

Mesnil M, Krutovskikh V, Piccoli C et al 1995 Negative growth control of HeLa cells by connexin genes: connexin species specificity. Cancer Res 55:629–639

Monaghan P, Perusinghe N, Carlile G, Evans WH 1994 Rapid modulation of gap junction expression in mouse mammary gland during pregnancy, lactation and involution. J Histochem Cytochem 42:931–938

Monaghan P, Clarke C, Perusinghe NP, Moss DW, Chen XY, Evans WH 1996 Gap junction distribution and connexin expression in human breast. Exp Cell Res 223:29–38

Neveu MJ, Hully JR, Babcock KL et al 1995 Proliferation-associated differences in the spatial and temporal expression of gap junction genes in rat liver. Hepatology 22:202–212

Perez-Armendariz EM, Luna J, Aceves C, Tapia D 1995 Connexins 26, 32 and 43 are expressed in virgin, pregnant and lactating mammary glands. Develop Growth Differ 37:421–431

Pozzi A, Risek B, Kiang DT, Gilula NB, Kumar NM 1995 Analysis of multiple gap junction gene products in the rodent and human mammary gland. Exp Cell Res 220:212–219

Sawey MJ, Goldschmidt MH, Risek B, Gilula NB, Lo CW 1996 Perturbation in connexin 43 and connexin 26 gap-junction expression in mouse skin hyperplasia and neoplasia. Mol Carcinog 17:49–61

Suchyna TM, Xu LX, Gao F, Fourtner CR, Nicholson BJ 1993 Identification of a proline in M2 of Cx26 as an element involved in voltage gating of gap junctions. Nature 365:847–849

Wilgenbus KK, Kirkpatrick CJ, Knuechel R, Willecke K, Traub O 1992 Expression of Cx26, Cx32 and Cx43 gap junction proteins in normal and neoplastic human tissues. Int J Cancer 51:522–529

Zhu D, Caveney S, Kidder GM, Naus CC 1991 Transfection of C6 glioma cells with connexin43 cDNA: analysis of expression, intercellular coupling and cell proliferation. Proc Natl Acad Sci USA 88:1883–1887

Zhu D, Kidder GM, Caveney S, Naus CC 1992 Growth retardation in glioma cells cocultured with cells over expressing a gap junction protein. Proc Natl Acad Sci USA 89:10218–10221

# Summary

*Gilula:* I have two goals I would like to achieve by the end of this summary session. First, I would like us to summarize the points that have emerged in this area of gap junction-mediated signalling in health and disease. And second, I would like us to outline those areas that need to be more precisely defined experimentally. To start, I have asked Klaus Willecke to summarize the issue of molecular diversity. Why are there so many gap junction genes, and how will addressing this question influence future directions of research?

*Willecke:* There is little new information on distinct transcription control of connexin genes, which is just another way of saying that connexins are expressed in different cell types. Rudolf Werner presented some details on connexin43 (Cx43; $\alpha_1$) transcriptional control, although by and large we do not understand the transcriptional control of any of the connexin promoters in detail, and post-transcriptional control was not a key issue at this symposium. From work of Kren et al (1993), we know that following partial hepatectomy the degradation of Cx32 ($\beta_1$) mRNA is increased in rat liver. Translational control was mentioned briefly during this symposium. The levels of Cx37 ($\alpha_4$) mRNA in certain tissues, such as lung and skin, do not correspond to the levels of Cx37 protein (Traub et al 1998). Thus, translational control of connexin expression is likely to exist. Post-translational modifications, such as phosphorylation, have been mentioned, and I would like to remind you of the work of Bill Fletcher, Joerg Kistler and Bernie Gilula and their associates, as discussed during this symposium. We still don't know what the phosphorylation of connexins does. It is likely to be related to degradation (Laing & Beyer 1995, Hertlein et al 1998), but it may also have an effect on channel gating and this may differ amongst the connexins. This is still an enigma, and no-one has yet come up with a convincing answer.

To identify the functions of connexins, we will have to verify the *in vitro* results in knockout mice, e.g. we will have to express non-phosphorylated connexin proteins in the living animal and see what effect this has. The same is true for the question of degradation. It is intriguing that Gong et al (1997) have described proteolysis in Cx46 ($\alpha_3$)-deficient mice. And Joerg Kistler has discussed this in his experimental system. We have seen some progress in this area, and there seems to be a relationship between proteolysis and connexin function, but we don't know if this is an all-or-nothing effect. We also don't know whether other connexins can also be cleaved, or what this cleavage means to the cell.

Trafficking was covered by Howard Evans. At the moment it's fair to say that Cx26 ($\beta_2$) trafficking is different from all the other connexins which have been studied. We currently know at least 13 different murine connexin genes, so trafficking may be a way for the cell to distinguish one connexin from the other, and this could contribute to connexin diversity.

Assembly was covered by Nalin Kumar and Howard Evans. The question is, does the ability to form heteromeric channels have functional consequences to the cell? At the moment we still need to agree that heteromeric channels exist in living animals and are not an artefact of expressing connexins in culture. Why it would be advantageous for the cells to form heteromeric connexin channels in preference to homomeric channels is a question that has yet to be answered. Perhaps there are additional functions of heteromeric channels that can only be efficiently achieved with heteromeric channels. We don't know whether it is evolutionarily advantageous for the cell to form heteromeric channels.

A new dimension that was touched upon by Alan Lau is whether other proteins interact with connexins and if so, why? Kinases and phosphatases must interact with connexins, but if there are other interacting proteins, for example cytoskeletal proteins, what are the functional implications of these interactions *in vivo*? The approaches that Alan Lau described will help to answer this question in the future.

The compatibility of connexin channels has not been covered at length during this symposium, although it is an important issue that was originally observed by Bruzzone et al (1993). There are compatibility differences in Cx40 ($\alpha_5$) and Cx43, (cf. Haubrich et al 1996), and we now think this is also true for Cx31.1 ($\beta_4$).

There has been much discussion on the formation of rectifying junctions. Rectifying junctions have been described in invertebrates, but do they exist in vertebrates as well? Perhaps they are formed in the brain. But David Vaney and Rolf Dermietzel have pointed out that we don't know if they exist, and if they do we don't understand what kind of advantage these channels would have. This is an area that needs to be investigated in the future. I suspect that some important principles may be discovered by the clarification of these types of channels. This is a challenge for the future.

The issue of whether different gap junction channels have different permeabilities has been mentioned briefly. My supposition is that there will be differences, but that it is unlikely that connexin channels differ as much in their permeabilities as potassium channels and sodium channels. It is unlikely that one connexin channel can transport inositol-1,4,5-trisphosphate ($InsP_3$) and another one cannot. In this context, it is important to consider the short half-lives of second messenger molecules. For example, $InsP_3$ has a half-life of about 10 seconds in the cell, and the half-life of cAMP is in the same order. There may be threefold differences in the permeabilities of Cx32 and Cx26 channels to $InsP_3$, and this could be biologically relevant (H. Niessen, K. Kramer & K. Willecke, unpublished results 1998). If there are differences in permeabilities of connexin channels towards other second messenger molecules, how important are they in real life? This also needs to be addressed in the future.

Voltage-dependent gating was not a major issue during this symposium, because the focus was not on electrophysiology. However, the functional meaning of voltage-dependent gating in real life is not known. One of the dogmas is that there are few differences in membrane potential between cells of the same type, for example, in liver or in brain (although this may not be true for cells in the embryo). It is possible that new connexins which have yet to be discovered can form channels that are much more voltage dependent than those which we know already. So far, voltage dependence has been used as a landmark for different connexin channels, but there are no proven functional implications.

We have heard about effector molecule-dependent gating, and we have discussed the effects of oleamide and anandamide. David Vaney suggested during his talk that certain neurotransmitters can affect the gating of electrical synapses. I suppose we will hear more about effector molecules and their effects on gap junction coupling in the future. The interaction of chemical and electrical synapses may be an important subject to discuss in five years or so, although for now it's just an area of speculation.

When you compare our present knowledge on targeted connexin mutants that were described here by Bernie Gilula, Dan Goodenough and myself, the question arises, to what extent do connexin channels have overlapping biological functions? All the proposed functional differences should be checked in living animals. There are many advantages of cell culture experiments, but if we really want to understand the diversity of connexin channels we have to study the consequences of connexin gene deletions and mutants with single base changes in mice and humans. These experiments take a long time, but they can be done. Only then we shall be in a position to understand how gap junctions influence molecular physiology in mammals.

*Gilula:* Your summary addresses our current perspectives and helps us to identify where we have to go. All of us would agree that we are gathering pieces of information in certain areas, but you have pointed out those situations in which we don't yet have conclusive information. We have more work to do to identify the differences and understand what they mean in a functional context.

*Warner:* I have a comment on the issue of permeability. One of the points that that has come out of this symposium is a more general realization of how little we know about the permeability and selectivity properties of gap junctions. We need better and more detailed information on the properties that determine what will and what will not go through gap junctions. We also have to think carefully about how appropriate Lucifer yellow is as a probe, although we must not abandon it altogether and use only Neurobiotin instead. I suspect that the extent to which differences in permeability contribute to functional consequences of gap junctional communication will not turn out to be immediately obvious. There may be situations where a twofold difference in permeability will be extremely important.

*Nicholson:* I have a brief comment on voltage gating. There is general agreement in the field that this is not likely to have a large physiological role anywhere. Even if there are tissues that have large enough voltages so that the gap junctions respond, they are too slow. Their time courses of closure are typically in the 500–900 ms range, which is much too long to show an effect during an action potential. The only situation in which it may play an interesting role is if the closure and recovery of the channels have different time courses. If that happens, then with a repeated action potential it would be possible cumulatively to shut down gap junctions or steadily open them up. This is an interesting concept that could be used as a means of increasing or decreasing the gain in a tissue that has repetitive action potentials, but there are limited data on it.

*Warner:* Voltage gating is important when cells are forming junctions, i.e. if two cells that have different resting potentials form a channel, this channel will not open if the voltage gradient is too great.

*Kistler:* Another point related to voltage gating is the issue of hemichannels, which we have found in lens fibre cells. When these cells are depolarized the hemichannels open, and an interesting point is that such depolarization also seems to be a phenomenon of the diabetic lens.

*Gilula:* I appreciate your comment on hemichannels, although if we start to talk about hemichannels we first have to agree what a hemichannel is structurally and functionally.

*Fletcher:* I would like to add to Klaus Willecke's comments on post-translational modification. Based on work by Crow et al (1990), Musil & Goodenough (1991) and Laird et al (1991) it is clear that Cx43 is post-translationally phosphorylated. It's also clear that if purified enzymes or phosphatases are microinjected the channels can be gated. However, we do not know what level of phosphorylation is important, i.e. all the hemichannels consist of six monomers, but is phosphorylation, and therefore a conformational change, in only one of the monomers sufficient to change the 'channel open' or 'channel closed' status? This question is important because when we try to do *in vitro* phosphorylation experiments using recombinant Cx43 or a peptide corresponding to mutant Cx43, we observe different kinetics. This probably means that some phosphorylation events open and close junction channels rapidly, and others are much slower.

*Gilula:* Another important point in the context of molecular diversity is the issue of class. There are two connexin classes. Some people call them 1 and 2, or A and B. We call them $\alpha$ and $\beta$. It is possible that the members of these two classes will provide the basis for some selectivity in terms of the formation and dissociation of heteromeric channels. This also provides us with a basis for trying to understand the use of multiple connexins by individual cell types, and what that means.

I would now like to ask Anne Warner to summarize the next topic, which is the functional consequences of connexin mutations.

*Warner:* A functional map for the various connexin proteins, which links particular regions/amino acids in each connexin to a specific function, would be of immense value to all those trying to understand the physiology and functional role(s) of gap junctions. A number of laboratories are now generating observations on the consequences of mutations of individual connexin proteins. This information needs to be drawn together and then compared with the results from knockouts, peptide competition experiments and specific antibody interactions in order to build a picture of the way in which connexin sequence is linked to functional property. The increasing recognition that disease states can involve mutations in connexin proteins means that a functional map also could assist clinicians, because it might enable predictions of likely functional consequences of mutations in different regions of the protein.

However, before drawing conclusions about the relationship between mutation and functional property, the steps that lead to the formation of a functional gap junction channel must be tracked for each connexin mutant. Many studies have not paid adequate attention to these crucial issues. It is essential to know not only that the mRNA is transcribed, but also that translation to the mutant protein follows. The way in which the trafficking machinery handles the mutant protein must be determined. Does the mutant protein reach the membrane? Is the protein inserted into the membrane? Does expression of a mutant connexin reduce or improve expression of other connexins? Functional questions do not become meaningful until these issues have been resolved.

I have identified three general areas where correlation between sequence and function will advance understanding. The first is junction formation, where mutational analysis of the extracellular loops could be illuminating. Do mutated connexins in adjacent cell membranes complete the docking process to form a functional gap junction? Can a mutant connexin interact with unmodified connexin proteins? Can the ability of one connexin to form junctions with a normally incompatible partner be changed by mutation? The second concerns the functional properties of the channels formed from mutant connexins. Are there mutations that lead to permanently closed (or permanently open) channels? If functional properties change how do they change? Do these channels have different permeability or selectivity from wild-type channels; do they become more or less pH sensitive, calcium sensitive or voltage dependent? Do mutations in particular regions of the protein always lead to the same kind of functional modification? The third issue relates to the turnover and degradation of mutated connexin proteins. Do particular mutations influence the speed or efficiency of turnover and/or degradation?

Finally, the *Xenopus* oocyte system is convenient and is used widely for expression studies. A number of studies on mutant connexins have used only the oocyte system. It is essential to introduce caution. It is becoming increasingly clear that the way in which proteins are assembled and trafficked in *Xenopus* oocytes is not necessarily the same as in mammalian or insect cells. Confirmation of results obtained in *Xenopus* oocytes by assays in other cell types, preferably those in which the connexin is normally expressed, will always be necessary before firm conclusions can be drawn about the consequences of a particular mutation.

*Gilula:* This is a challenging opportunity, particularly since we will have to somehow normalize the information derived from heterologous cell systems before we can even think about constructing a map. Steve Scherer, do you have a slightly different perspective on this?

*Scherer:* I have a clinical perspective. We should pay close attention to what the knockout mice and human mutations are telling us about the issue of connexin redundancy. These individuals have phenotypes that are often less impressive than we might have anticipated based on our knowledge of which tissues express which connexins. We should carefully analyse the function of tissues that have been genetically deprived of a connexin that they would normally express.

I would also like to stress Anne Warner's point that many mutated connexins, for one reason or another, are not inserted into the membrane in mammalian cells, thus precluding a functional analysis in these cells. How mutations affect the normal trafficking of Cx32 has turned out to be complex and deserves more attention, as the aberrant trafficking of Cx32 mutations may have untoward effects within the cell related to the pathogenesis of X-linked Charcot-Marie-Tooth disease.

*Gilula:* Are there any examples of mutated proteins that are inserted into the membrane but do not form functional channels?

*Scherer:* The S26L, M34T, C60F, V39M, R215W and R220stop mutants all reach the cell membrane in mammalian cells, but only S26L and R220stop form functional gap junctions (Oh et al 1997, Omori et al 1996). The S26L mutation is interesting in that it appears to have relatively normal voltage dependence but reduced permeability.

*Gilula:* Is the S26L mutation in the transmembrane domain?

*Scherer:* Yes. It is in M1.

*Werner:* We made several mutations in the extracellular loops of Cx32. Most of these mutants exhibit severe loss of function (Dahl et al 1992). Others cause a change in pairing properties. For example, a double mutant K167T-152R is capable of pairing with Cx38 ($\alpha_2$), which wild-type Cx32 cannot (Werner et al 1993).

*Warner:* Some truncated connexins are inserted into the membrane and form channels (Fishman et al 1991).

*Gilula:* I would now like to call on Eric Beyer to summarize what has been a key subject at this symposium, and that is connexins in the heart.

*Beyer:* I am amazed by the extent of progress in this area. There is a general agreement, although there are minor discrepancies, on which connexins are present in the heart—i.e. Cx43, Cx40 and Cx45 ($\alpha_6$)—and how they are distributed in different regions. Cx43 is the major component of gap junctions in the working myocardium in the atrium and the ventricle. Cx40 is predominantly expressed in the atrium and in the conducting system. Cx45 is a minor component of gap junctions, at least in the conducting system and possibly throughout the rest of the heart. In addition, in the rest of the cardiovascular system Cx37 is a major component. There have been reports of other connexins, such as Cx46 and Cx50 ($\alpha_8$), in the literature, but there is little evidence that these represent a significant component. We know from expression systems that all of these proteins make channels that have different properties. They are also probably differentially regulated by transcriptional events and post-translational modifications. One of our challenges is to try to understand those different mechanisms of regulation.

We also have to find out how important the differential channel properties or regulation of different connexins are. Cx43, Cx40 and Cx45 each can and probably do serve as conducting channels for ions and action potential propagation. Expression systems show that these connexins can form channels with different permeability properties. How relevant are these to cardiac physiology or development?

The knockout studies, interventional studies and disease state studies suggest that these connexins have specific roles, although they probably all play a role in ion conductance and action potential propagation. Cx40 seems to be particularly important in the conducting system, but it may not have a role in the development of the heart. This contrasts with Cx43, which seems to be important for the development of the heart. How does this work? Is it mediated by additional chemical signalling or the passage of other kinds of molecules rather than just ions? How Cx45 is involved in the conducting system and/or the development of the heart is not yet known.

*Gilula:* This is a historic occasion for work in the heart because it is the first meeting where all this information has been available.

*Kistler:* In the adult working heart is there any evidence that all gap junctions in the heart could be made from Cx43 alone and that such a heart would be functional? Or are all the other connexins also required?

*Beyer:* That's a good question. I can't answer it.

*Gilula:* Nick Sperelakis (1983) has propsed that gap junctions are not required for cardiac function. There are numerous theories of how these appropriate cardiac activities are achieved in the absence of gap junction connections.

*Warner:* Noble & Winslet (1997) constructed a model of the working heart that could potentially be used to answer that question.

*Gilula:* The next subject is the issue of heterotypic connections between heterologous cells or even between homologous cells. I would like to ask Bruce Nicholson and Colin Green to summarize what we know, particularly about the relevance of model systems to the situation *in vivo*.

*Nicholson:* The model systems have helped us to lay the ground rules by which we can operate. If anything, those model systems will probably be more promiscuous, so that when we look at the situation *in vivo* we will be able to define more precisely which connexins are actually used. There are data from different labs on different systems, e.g. in HeLa cells and oocytes, and these data are pretty much in agreement (Fig. 1). There is the interesting case of Cx33 ($\alpha_7$), which may work as a dominant negative down-regulator through heteromeric interactions. There are also a couple of connexins that so far don't appear to pair with any connexins, including themselves. As far as I'm aware Cx31.1 is an example of such a connexin.

| | $\beta_2$ Cx 26 | $\beta_6$ Cx 30 | $\beta_5$ Cx 30.3 | $\beta_3$ Cx 31 | $\beta_4$ Cx 31.1 | $\beta_1$ Cx 32 | $\alpha_7$ Cx 33 | $\alpha_4$ Cx 37 | $\alpha_5$ Cx 40 | $\alpha_1$ Cx 43 | $\alpha_6$ Cx 45 | $\alpha_3$ Cx 46 | $\alpha_8$ Cx 50 |
|---|---|---|---|---|---|---|---|---|---|---|---|---|---|
| $\alpha_2$ Xe 38 | − | − | + | | − | − | | + | − | + | − | − | − |
| $\alpha_8$ Cx 50 | + | | | | | + | | | − | − | | + | + |
| $\alpha_3$ Cx 46 | + | | | | | + | | | − | + | | + | |
| $\alpha_6$ Cx 45 | − | | − | − | − | − | | + | + | + | + | | |
| $\alpha_1$ Cx 43 | − | | + | − | − | − | − | + | − | + | | | |
| $\alpha_5$ Cx 40 | − | | + | − | − | − | | + | + | | | | |
| $\alpha_4$ Cx 37 | − | | + | − | − | − | − | + | | | | | |
| $\alpha_7$ Cx 33 | | | | | | − | − | | | | | | |
| $\beta_1$ Cx 32 | + | + | − | − | − | + | | | | | | | |
| $\beta_4$ Cx 31.1 | − | | − | | − | | | | | | | | |
| $\beta_3$ Cx 31 | − | | | + | | | | | | | | | |
| $\beta_5$ Cx 30.3 | − | − | + | | | | | | | | | | |
| $\beta_6$ Cx 30 | + | + | | | | | | | | | | | |
| $\beta_2$ Cx 26 | + | | | | | | | | | | | | |

FIG. 1. This shows the allowable heterotypic pairings between connexins as determined from expression in *Xenopus* oocytes or HeLa cells. Data are an accumulation of results obtained in the laboratories of B.J. Nicholson, D. Paul and K. Willecke, and they were compiled by H. Zhu.

*Willecke:* Using dye transfer between connexin-transfected HeLa cells, we have observed that Cx31.1 can pair with Cx43, but it can hardly pair at all with itself.

*Gilula:* Where is Cx31.1 expressed *in vivo*?

*Nicholson:* It is expressed in skin, but no-one has yet performed a detailed analysis on other tissues. Apart from those connexins that don't bind anything and Cx31 ($\beta_3$), which binds only to itself, there is a tendency among the remaining connexins for the $\alpha$ connexins to pair with themselves and the $\beta$ connexins to pair with themselves. Exceptions to this rule are Cx30.3 ($\beta_5$), which is a $\beta$ connexin but has an $\alpha$ pairing preference, and Cx50, which is an $\alpha$ connexin but has a $\beta$ pairing preference. We have been looking at sequences that might allow this switch to occur. Our initial results suggest that the switch between $\alpha$ and $\beta$ pairing lies with two residues in the second extracellular loop. This provides a functional relevance to the $\alpha$ and $\beta$ categories defined by sequence alignments in Nalin Kumar's presentation here (Kumar 1999, this volume). Of course, interactions between connexins are further complicated if heteromeric interactions between connexins also occur. We have already seen that these are likely (Kumar 1999, this volume) and that they may show the same $\alpha/\beta$ preference. Colin Green is now going to list some interesting candidate tissues where these heterotypic and heteromeric interactions may play a role.

*Green:* The obvious places where there is multiple connexin expression are in the skin, liver, oocytes, granulosa cells, lens, etc.

*Gilula:* Your list is going to include everything that has been looked at. What is interesting is to find examples where there is only one connexin.

*Green:* I agree, but for all of these cases we have to determine the functional significance of particular connexins being expressed in different places at different times. In the skin or corneal epithelium, for example, Cx43 is expressed in the basal layer, and as far as I know it is the only connexin in the basal layer. As those cells are pushed upwards Cx43 expression is switched off and Cx26 is switched on. Another example is in the granulosa cell, where, as Dan Goodenough showed, processes extend across the zona pellucida, with Cx37 junctions on the ova side of the zona, but, at the same time, in the same cell, Cx43 junctions are formed with neighbouring granulosa cells. This is a case where there has to be separate trafficking of two different connexins within a single cell. Differences are also apparent between connexins in response to stimuli. For example, following neuronal loss in the Huntington's disease brain, the up-regulation of a specific connexin, Cx43, is observed, not a general up-regulation of all connexin types. In the blood vessel endothelium, Nick Severs has shown that there are three connexin types being expressed, but Dan Goodenough has suggested that it is only Cx43 that may respond to specific stimuli, such as localized turbulence. Even though other connexins may be present, therefore, there is an opportunity to up-regulate or down-regulate individual connexin types without necessarily having to postulate any likelihood of forming heteromeric or heterotypic channels.

Another question is, what is the significance when two connexin types are present in the same cell? The liver is an example where there are two connexins, Cx32 and Cx26, with a large variation in the ratio between the two in different species. In Cx32 knockout animals it doesn't seem to make any difference when you do, for example, a partial hepatectomy. The question then is whether in the liver a specific channel is necessary for at least some functions. Is it possible that in the liver any $\beta$ type connexin could function and all that is required is a relatively non-specific hole?

Another point to bear in mind is that there are huge numbers of gap junctions present in tissues. For example, an adult cardiac myocyte has over 1000 gap junctions. This is also true for the cochlea, which has high levels of Cx26, and for Cx50 and Cx46 in the lens. Therefore, the gap junction channel may not be as efficient as we might think, and we have to question whether it is functionally significant when small amounts of another connexin type are present. Finally, as Howard Evans has shown, there are different trafficking routes. Trafficking at different times, to different places or by different routes allows connexin separation. Therefore, although we have a long list of tissues expressing multiple connexin types, just what is the requirement, or indeed evidence *in vivo*, for junctions made up of channels that are predominantly heteromeric or heterotypic?

*Gilula:* When two cells are heterologous do we have evidence for the utilization of different connexins to form a gap junction? The oocyte system provides us with an opportunity to demonstrate this, and Dan Goodenough has presented evidence that the same connexin, in this case Cx37, can be used in the interacting cumulus cell. Also, in the lens the epithelial cells make a connexin that is not detected in the fibre cells immediately below, but do we know whether heterotypic junctions are formed?

*Goodenough:* This is an open question. The lens epithelium also expresses the two connexins that are found in fibre cells in the newly hatched chick (Jiang et al 1995).

*Nicholson:* There is some evidence for heterotypic channels in the liver based on Vyto Verselis' data on isolated hepatocyte pairs (Verselis et al 1993). They have seen occasional examples of rectification that are the same as that reported in Barrio et al (1991) for Cx32/Cx26 heterotypic channels. The problem is that Cx32 and Cx26 are usually both expressed in most hepatocytes, so 99% of the time you get a symmetrical response because rectifying responses in apposed directions will cancel out.

*Vaney:* The best example is in neuroglial cells, where astrocytes and oligodendrocytes are almost certainly coupled by heterotypic junctions.

*Dermietzel:* Yes. We showed that oligodendrocytes and astrocytes have different connexin types (Dermietzel et al 1989) and we know they are coupled, so it is probable that heterotypic channels are present. We injected Lucifer yellow and Neurobiotin into oligodendrocytes in brain slices, and we did not observe transfer of these dyes into astrocytes (Pastor et al 1998). We haven't done the opposite experiment, i.e. inject astrocytes and look at transfer to oligodendrocytes.

*Vaney:* Last year there were two papers from Nagy's laboratory on the electron microscopic localization of Cx43 and Cx30 in astrocytes (Ochalski et al 1997, Nagy et al 1997). The immunolabelling was confined to the astrocytic side of the gap junctions between astrocytes and oligodendrocytes; the unlabelled oligodendrocytic side of these heterologous gap junctions may be formed by Cx32.

*Dermietzel:* The next step is to demonstrate asymmetrical localization of both connexins in double-labelled preparations.

*Sanderson:* Calcium waves seem to go in both directions in connexin channels formed between astrocytes and oligodendrocytes, so while there may be rectification of the dye coupling, there may be less rectification of the signalling molecules if these molecules are less charged or polarized. We have also been looking at the idea of different types of junctions across biological interfaces. For example, airway epithelial cells and smooth muscle cells seem to be able to communicate in both directions. They both make Cx43, Cx32 and Cx26, so they could use homotypic or heterotypic channels. Similarly, at the bloodbrain barrier endothelial cells, smooth muscle cells and glial cells are in close proximity to each other and propagate $Ca^{2+}$ waves perhaps through heterotypic junctions.

*Gilula:* These summaries have provided us with perspectives on issues that will be a legacy for those who continue to work in the field and for those who are new to the field. The published book will demonstrate to those people who weren't able to attend this symposium the positive and constructive attitudes extended by everybody here towards progress in this field. We can only benefit from co-operation and identifying the important progress that will take us forward, as opposed to counterproductive information. A consensus could result from this symposium that would make it possible for people outside the gap junction field to be better informed and to better appreciate the progress that is being made. I hope that the book will be a signpost to this end. I also hope that people in the field will think carefully about the nomenclature decision that was made about 10 years ago to use the sizes deduced from sequences of the gene products as a way to identify connexins. This approach has a serious limitation — people in related fields have difficulties understanding the progress that is being made in this area of research — and we now know enough about the connexin multigene family to adopt a better classification system.

## References

Barrio LC, Suchyna T, Bargiello T et al 1991 Gap junctions formed by connexin 26 and 32 along and in combination are differently affected by applied voltage. Proc Natl Acad Sci USA 88:8410–8414 (erratum: 1992 Proc Natl Acad Sci USA 89:4220)

Bruzzone R, Haefliger JA, Gimlich RL, Paul DL 1993 Connexin40, a component of gap junctions in vascular endothelium is restricted in its ability to interact with other connexins. Mol Biol Cell 4:7–20

Crow DS, Beyer EC, Paul DL, Kobe SS, Lau AF 1990 Phosphorylation of connexin 43 gap junction protein in uninfected and Rous sarcoma virus transformed mammalian fibroblasts. Mol Cell Biol 10:1754–1763

Dahl G, Werner R, Levine E, Rabadan-Diehl C 1992 Mutational analysis of gap junction formation. Biophys J 62:172–182

Dermietzel R, Traub O, Hwang TK et al 1989 Differential expression of three gap junction proteins in developing and mature brain tissues. Proc Natl Acad Sci USA 86:10148–10152

Fishman GI, Moreno AP, Spray DC, Leinwand LA 1991 Functional analysis of human cardiac gap junction channel mutants. Proc Natl Acad Sci USA 88:3525–3529

Gong X, Li E, Klier G et al 1997 Disruption of the α3 connexin gene leads to proteolysis and cataractogenesis in mice. Cell 91:833–843

Haubrich S, Schwarz HJ, Bukauskas F et al 1996 Incompatibility of connexin 40 and 43 hemichannels in gap junctions between mammalian cells is determined by intercellular domains. Mol Biol Cell 7:1995–2006

Hertlein B, Butterweck A, Haubrich S, Willecke K, Traub O 1998 Phosphorylated carboxy terminal serine residues stabilize the mouse gap junction protein connexin45 against degradation. J Membr Biol 162:247–257

Jiang JX, White TW, Goodenough DA 1995 Changes in connexin expression and distribution during chick lens development. Dev Biol 168:649–661

Kren BT, Kumar NM, Wang S, Gilula NB, Steen CJ 1993 Differential regulation of multiple gap junction transcripts and proteins during rat liver regeneration. J Cell Biol 123:707–718

Kumar NM 1999 Molecular biology of the interactions between connexins. In: Gap junction-mediated intercellular signalling in health and disease. Wiley, Chichester (Novartis Found Symp 219) p 6–21

Laing JG, Beyer EC 1995 The gap junction protein connexin43 is degraded via the ubiquitin proteasome pathway. J Biol Chem 270:26399–26403

Laird DW, Puranam KL, Revel JP 1991 Turnover and phosphorylation dynamics of connexin43 gap junction protein in cultured cardiac myocytes. Biochem J 273:67–72

Musil LS, Goodenough DA 1991 Biochemical analysis of connexin 43 intracellular transport, phosphorylation, and assembly into gap junctional plaques. J Cell Biol 115:1357–1374

Nagy JI, Ochalski PA, Li J, Hertzberg EL 1997 Evidence for the co-localization of another connexin with connexin43 at astrocytic gap junctions in the brain. Neuroscience 78:533–548

Noble D, Winslet R 1997 Reconstructing the heart: network models of SA node-atrial interactions. In: Panifiliv AV, Holden AV (eds) Computational biology of the heart. Wiley, Chichester, p 49–64

Ochalski PA, Frankenstein UN, Hertzberg EL, Nagy JI 1997 Connexin-43 in rat spinal cord: localization in astrocytes and identification of heterotypic astro-oligodendrocytic gap junctions. Neuroscience 76:931–945

Oh S, Ri Y, Bennett MVL, Trexler EB, Verselis VK, Bargiello TA 1997 Changes in permeability caused by connexin 32 mutations underlie X-linked Charcot-Marie-Tooth disease. Neuron 19:927–938

Omori Y, Mesnil M, Yamasaki H 1996 Connexin 32 mutations from X-linked Charcot-Marie-Tooth disease patients: functional defects and dominant negative effects. Mol Biol Cell 7:907–916

Pastor A, Kremer M, Möller T, Kettenmann H, Dermietzel R 1998 Dye-coupling between spinal cord oligodendrocytes: differences in coupling efficiency between gray and white matter. Glia, in press

Traub O, Hertlein B, Kasper M et al 1998 Characterization of the gap junction protein connexin37 in murine endothelium, respiratory epithelium and after transfection in human HeLa cells. Eur J Cell Biol, in press

Verselis VK, Bargiello T, Rubin JB, Bennett MVL 1993 Comparison of voltage-dependent properties of gap junctions in hepatocytes and *Xenopus* oocytes expressing Cx32 and Cx26. In: Hall JE, Zampighi GA, Davis RM (eds) Gap junctions: progress in cell research, vol 3. Elsevier, New York, p 105–112

Werner R, Rabadan-Diehl C, Levine E, Dahl G 1993 Affinities between connexins. In: Hall JE, Zampighi GA, Davis RM (eds) Progress in cell research, vol 3. Elsevier Science, New York, p 21–24

# Index of contributors

*Non-particpating co-authors are indicated by asterisks. Entries in bold type indicate papers; other entries refer to discussion contributions.*


**A**

*Abel, A. **175**
*Ahmed, S. **44**

**B**

*Balice-Gordon, R. J. **175**
Becker, D. 20, 73, 74, 75, 127, 128, 132, **134**, 237
Beyer, E. C. 17, 20, 39, 58, 59, 88, 89, 90, 91, 92, 94, 108, 110, 111, 151, 206, 210, 221, 224, 225, 237, 238, 239, 256, 267
*Bond, J. **97**
*Bone, L. J. **175**

**C**

*Casalotti, S. **134**

**D**

*Dasgupta, C. **212**
Dermietzel, R. 32, 33, 91, 128, 152, 170, 172, 186, 270, 271
*Deschênes, S. M. **175**
*Diez, J. **44**
*Donaldson, P. **97**
*Duncan, J. **212**

**E**

*Eckert, R. **97**
*Edwards, J. **134**
*Escobar-Poni, B. **212**
Evans, W. H. 17, **44**, 54, 55, 56, 57, 58, 59, **134**, 187

**F**

*Fischbeck, K. H. **175**
Fletcher, W. H. 18, 19, 39, 74, 109, 132, **212**, 221, 222, 223, 224, 225, 238, 264
Forge, A. 93, **134**, 151, 152, 153, 154, 155

**G**

*George, C. H. **44**
Giaume, C. 31, 40, 129, 131, 132, 172
Gilula, N. B. **1**, 16, 18, 19, 20, **22**, 31, 32, 37, 39, 40, 41, 42, 43, 55, 56, 57, 59, 73, 75, 88, 90, 91, 93, 94, 95, 108, 110, 111, 112, 125, 126, 132, 151, 152, 153, 154, 155, 171, 172, 173, 185, 206, 207, 209, 211, 222, 224, 225, 235, 236, 237, 239, 255, 257, 259, 261, 263, 264, 266, 267, 268, 269, 270, 271
Goodenough, D. A. 20, 34, 38, 41, 42, 56, 57, 58, 74, 75, 88, 89, 90, 91, 94, 95, 110, 111, 112, 125, 126, 127, 128, 131, 152, 153, 209, 210, 222, **227**, 235, 236, 237, 238, 270
Green, C. R. 19, 20, 56, 90, **97**, 173, 207, 269

**J**

Jamieson, S. 255, 256

**K**

*Kendall, J. M. **44**
*Kirchhoff, S. **76**
Kistler, J. 18, 31, 32, 34, 41, 57, 73, 94, **97**, 108, 109, 110, 111, 112, 155, 171, 186, 237, 240, 264, 267

*Krutovskikh, V. **241**
Kumar, N. M. **6**, 16, 17, 18, 19, **22**, 41, 42, 108, 238

L

Lau, A. F. 42, 75, 109, 222, 223, 224, 254, 258
Lench, N. **134**, 151, 152, 154, 187
*Lin, J. S. **97**
Lo, C. W. 43, 72, 75, 89, 90, 92, 93, 94, 153, 207, 210, 222, 224, 257, 259

M

*Martin, P. E. M. **44**
*Merriman, R. **97**
*Mesnil, M. **241**
*Mironov, N. **241**
Musil, L. 16, 36, 41, 42, 54, 55, 58, 72, 90, 108, 173, 209, 223, 224, 256

N

*Nadarajah, B. **157**
Nicholson, B. J. 16, 18, 31, 33, 34, 36, 37, 39, 55, 57, 93, 109, 110, 111, 127, 151, 154, 155, 173, 240, 257, 258, 264, 268, 269, 270

O

*Omori, Y. **241**
*Ott, T. **76**

P

Parnavelas, J. G. **157**, 170, 171, 172, 173
*Paul, D. L. **227**
*Plum, A. **76**

S

Sanderson, M. J. 33, 34, 40, 89, 90, 93, 131, 154, 155, 207, 255, 258, 271
Scherer, S. S. 16, 152, 154, 171, **175**, 186, 187, 255, 266
Severs, N. J. 38, 39, 89, 93, **188**, 207, 209, 210, 211, 222
*Shah, M. **212**
*Simon, A. M. **227**
*Souter, M. **134**

T

*Temme, A. **76**
*Thönnissen, E. **76**
*Tunstall, M. **97**

U

*Unger, V. M. **22**

V

Vaney, D. I. 32, 42, **113**, 125, 126, 128, 129, 152, 171, 172, 270, 271

W

Warner, A. 19, 33, 34, 38, 40, **60**, 72, 73, 75, 95, 127, 171, 222, 236, 237, 255, 263, 264, 265, 267, 268
Werner, R. 17, 31, 38, 91, 92, 186, 208, 209, 236, 266
Willecke, K. 16, 74, 75, **76**, 89, 90, 91, 93, 94, 95, 129, 153, 155, 235, 238, 255, 257, 259, 261, 269

Y

*Yamakage, K. **241**
Yamasaki, H. 19, 93, 94, 186, 187, 223, **241**, 254, 255, 256, 257, 258
Yeager, M. **22**, 31, 32, 33, 34, 36, 41, 89, 90, 109, 152, 206

Z

*Zhu, W. **241**

# Subject index


**A**

$\alpha_1$ *see* connexin 43 ($\alpha_1$)
$\alpha_2$ *see* connexin 38 ($\alpha_2$)
$\alpha_3$ *see* connexin 46 ($\alpha_3$); connexin 50 ($\alpha_3$)
$\alpha_4$ *see* connexin 37 ($\alpha_4$)
$\alpha_5$ *see* connexin 40 ($\alpha_5$)
$\alpha_6$ *see* connexin 45 ($\alpha_6$)
$\alpha_7$ *see* connexin 33 ($\alpha_7$)
$\alpha_8$ *see* connexin 50 ($\alpha_8$)
aequorin (Aeq) 48–49
altricial mammals 147–48
anandamide 4, 129–132, 263
 *see also* bioactive lipids; oleamide
antibody studies 8, 189
 connexin 26 ($\beta_2$) 54–55, 171
 connexin 43 ($\alpha_1$) 25
 connexin 45 ($\alpha_6$) 196
 heart connexins 211
 inner ear 136
 mouse ovarian follicle 228–230
aorta 209
apical ectodermal ridge (AER) 62–69
 in connexin 43 ($\alpha_1$) knockout mouse 74
 fibroblast growth factor 4 (FGF4) and 65
apoptosis 74–75
arachidonic acid 129
arrhythmias 23, 88–89, 188, 196–197, 206–207
 connexin 43 ($\alpha_1$) and 197
 connexin 40 ($\alpha_5$) knockout mouse 83–84
 re-entry 39, 196–197
 therapies 206–207
arterial wall 199
 diseased 201–203, 207
 *see also* blood vessel endothelium
arthropods 7, 20–21
asthma 207
astrocytes 39–40, 131, 164
atherosclerosis 188–189, 201–203, 209
atrioventricular conduction 38
atrioventricular node 193
aurothioglucose 34

**B**

$\beta_1$ *see* connexin 32 ($\beta_1$)
$\beta_2$ *see* connexin 26 ($\beta_2$)
$\beta_3$ *see* connexin 31 ($\beta_3$)
$\beta_4$ *see* connexin 31.1 ($\beta_4$)
$\beta$ catenin 258–259
 cell growth control and 248, 254–255
$\beta$ connexin mutations 19
balloon angioplasty 201–203
BHK hamster kidney cells 12, 23, 26, 27–28, 41–42
bioactive lipids 3–4, 129, 131
 *see also* anandamide; oleamide
blood vessel endothelium 83, 230
 *see also* arterial wall
breast cancer 244, 256–257
brefeldin A 48–49, 56, 57
bystander effect 250, 256

**C**

C6 cells 258
cadherin 209–210
calcium
 calpain and 108
 intercellular signalling 40–41
 intracellular stores 56, 131–132
 waves 258–259, 271
calmodulin 53
calpain 101, 108, 110–111
calpastatin 108
cAMP 121, 132, 258–259
cancer therapy 241–260
5,6-carboxyfluoresein 182–184
carcinogenesis 241–243
cardiovascular disease 188–211
 connexin distribution 196–197
 hibernating myocardium 197
 incidence 188
  *see also* cardiovascular system; heart
cardiovascular system 189
 *see also* cardiovascular disease

cataractogenesis 97–99, 110–112, 239
  diabetic 97–112
  *see also* cataracts
cataracts 84
  *see also* cataractogenesis
cell adhesion 255
cell coupling 127, 129–130
cell culture 153–154
cell cycle regulation 259
cell growth control 247–248, 254–255, 257–259
central nervous system (CNS) 4, 113, 114, 158–159
  *see also* cerebral cortex
cerebral cortex 157–174
  adult 162–166
  connexin expression 162–164
  development 157–174
  *see also* central nervous system (CNS); corticogenesis
cGMP 128, 132
channel compatibility 262
channel rectification 38–41, 121, 129, 262
Charcot-Marie-Tooth (CMT) disease 19, 56, 151, 175–176
  *see also* X-linked Charcot-Marie-Tooth (CMTX) disease
chemical–electrical synapse interaction 263
chemopreventive agents 243
Ciba Foundation symposium 125 1
CMT *see* Charcot-Marie-Tooth (CMT) disease
CMTX *see* X-linked Charcot-Marie-Tooth (CMTX) disease
CNS *see* central nervous system (CNS)
cochlea 78, 148
  connective tissue 143–148
  maturation 147
    *see also* inner ear
cone bipolar cells 116–126
  carbenoxolone 123
  cat retina 118
  glycine and 123, 125–126
  rod amacrine cells and 120, 123, 125–126, 128
congenital heart defects 212–225
  *see also* visceroatrial heterotaxia (VAH)
connexin 35 8
connexin 36 8
connexin 26 ($\beta_2$) 39, 55–57, 63
  antibody studies 54–55, 171
  in breast cancer 256–257
  C-terminal domain 155
  in cancer 242
  in central nervous system (CNS) 114, 158–159
  in cerebral cortex 162–164
  co-translational insertion 55
  in cochlea 78, 148
  in connexin 32 ($\beta_1$) knockout mouse 81
  connexin 26 ($\beta_2$)–aequorin chimeras 48
  deafness 78, 83
  in Deiters' cells 141
  dominant negative effects 79, 257
  endocochlear potential and 149
  Golgi apparatus 50, 59
  in Hensen's cells 141
  Herpes simplex virus thymidine kinase gene 250
  Huntington's chorea 173
  in inner ear 136
  intracellular stores 54–55, 171
  intracellular trafficking 46–49, 50, 54–55, 57, 262
  knockout mouse 82–83, 90, 153, 171
  in liver 11, 16, 45, 46, 257
  microtubules and 49
  mutants 256
  mutations 148–149, 151–153
  in neurogenesis 161–162
  neuronal coupling and 163–164, 170
  in neurons 170, 173
  in organ of Corti 139–141
  in placenta 155
  potassium and 154–155
  recycling 57, 58
  in retina 127–128
  in skin 269
  specific permeability 62
  in tumour suppression 243
  voltage sensitivity 155
connexin 31 ($\beta_3$)
  in brain 85, 90–91
  knockout mouse 84–85
  *lacZ* reporter gene and 84, 91
connexin 31.1 ($\beta_4$) 262, 268–269
connexin 32 ($\beta_1$) 17, 39, 56–57, 63
  in C6 cells 258
  in central nervous system (CNS) 114, 158–159
  in cerebral cortex 162–164
  Charcot-Marie-Tooth (CMT) disease 151

connexin 32 ($\beta_1$) (*cont.*)
Charcot-Marie-Tooth (CMT) disease and 19, 56
connexin 43 ($\alpha_1$) and 25
connexin 32 ($\beta_1$)–Aeq chimeras 48–49
dominant negative effects 178–180, 181–182
fibroblast growth factor 4 (FGF4) and 65–66
in Huntington's chorea 173
in inner ear 136, 139
intracellular concentrations 46
intracellular trafficking 49, 55, 266
knockout mouse 80–82, 171, 180
connexin 26 ($\beta_2$) and 81
in connexin 43 ($\alpha_1$)–connexin 32 ($\beta_1$) double knockout mouse 74
glucose mobilization in 80–81
liver in 81, 93–94, 244–245, 250, 257
tumours in 81, 82, 244–245, 250
X-linked Charcot-Marie-Tooth (CMTX) disease in 81
in limb bud 95
in liver 45, 46, 257, 261
mutations 176–180, 186–187, 247, 266–267
trafficking 178–182, 187
in organ of Corti 142
in retina 127–128
specific permeability 62
in strial basal cells 148
in teeth 94–95
in thyroid gland 12, 94–95
tumour suppression 243
tumours 257
X-linked Charcot-Marie-Tooth (CMTX) disease and 51–53, 78, 175–187, 245, 266
connexin 33 ($\alpha_7$) 79, 164, 268
connexin 37 ($\alpha_4$)
in aorta 209
in arterioles 236
in artery 199
in cardiovascular system 189
channel activity 235–236
dominant negative effects 79
in granulosa cell 235, 269
in heart 267
infertility 233
knockout mouse 83, 230–231
connexin 37 ($\alpha_4$)–connexin 40 ($\alpha_5$) double knockout mouse 237
in lung 236, 261
in mouse ovarian follicle 227, 228–230
mutations 230, 232–233
in retina 127–128
in skin 261
in testes 236
connexin 38 ($\alpha_2$) 74
connexin 40 ($\alpha_5$) 18
in aorta 209
in artery 199
in central nervous system (CNS) 159
in connexin 43 ($\alpha_1$) knockout mouse 80
connexin 43 ($\alpha_1$) and 262
in heart 38, 89, 189, 194, 267
knockout mouse 83–84
connexin 43 ($\alpha_1$) in 90
connexin 37 ($\alpha_4$)–connexin 40 ($\alpha_5$) double knockout mouse 237
connexin 43 ($\alpha_1$)–connexin 40 ($\alpha_5$) double knockout mouse 85
heart in 88, 238
in mouse ovarian follicle 228–230
in neuroblasts 164, 165
in retina 127–128
in tumour suppression 243
connexin 43 ($\alpha_1$) 9, 17, 34, 56–57, 63, 213–215, 219
antibody studies 25
in aorta 209
apoptosis and 75
in astrocytes 164
asymmetry 222–223
in atherosclerosis 209
in breast cancer 256–257
in C6 cells 258
calpain 103
in cancers 242
in cardiac gap junctions 23
in cardiovascular system 189
CDK expression 258
in central nervous system (CNS) 114, 158–159
in cerebral cortex 162–164
channel gating 218–219
compatibility with connexin 40 ($\alpha_5$) 262
congenital heart defects and 212–225
connexin 32 ($\beta_1$) and 25
in connexin 50 ($\alpha_8$) knockout mouse 237
connexin 31.1 ($\beta_4$) and 269

connexin 43 ($\alpha_1$)–lacZ fusion protein 207
cytoplasmic sequences 25, 216–217, 218–219
dominant negative mutation 18, 215, 222
in embryos 207, 213, 216, 222–223
fibroblast growth factor 2 (FGF2) and 167
fibroblast growth factor 4 (FGF4) and 65–66
in glial cells 32
glial fibrillary acidic protein (GFAP) and 173
in granulosa cell 269
in heart 38, 194, 210–211, 213, 216, 267
development 80, 207–208, 215, 267
Herpes simplex virus thymidine kinase gene and 250
Huntington's chorea and 173, 269
in inner ear 136
intracellular trafficking 18, 55
knockout mouse 80, 89, 94, 254
apical ectodermal ridge (AER) in 74
calpain 110–111
cell growth in 245
connexin 45 ($\alpha_6$) 210
connexin 43 ($\alpha_1$)–connexin 32 ($\beta_1$) double knockout mouse 74
connexin 43 ($\alpha_1$)–connexin 40 ($\alpha_5$) double knockout mouse 85
fibroblast growth factor 4 (FGF4) in 74
impulse conduction in 210–211
limb bud in 73–74
pulmonary outflow obstruction in 207
*Sonic hedgehog* (*Shh*) in 73
thymus function in 224–225
visceroatrial heterotaxia (VAH) in 213
*lacZ* reporter gene and 92–93
in lens 101
in limb bud 95
in lung 209
membrane topology 25
in mouse ovarian follicle 227, 228–230
mutations 19, 213, 219–220
S364P mutation 213–215, 223
in neural crest cells 207–208
in neurogenesis 161–162, 164
oestrogen and 91–92
in organ of Corti 142
$^{32}$P-orthophosphate and 217
phosphorylation 216, 217, 218, 224–225, 264
plaque formation 208–209
progesterone and 91–92
in retina 127–128
in skin 269
in smooth muscle cells 199, 201
in teeth 94–95
in testes 209
thrombin and 203
in thyroid gland 12, 94–95
transcriptional control 91–92, 261
transmembrane domains 25–27
truncation mutation 18, 23, 26, 27–28
in tumour suppression 243
tyrosine kinase and 167
in viseroatrial heterotaxia (VAH) 78
ZO-1 and 42, 254–255
connexin 45 ($\alpha_6$)
antibody studies 196
atrioventricular block and 196
in cardiovascular system 189
in central nervous system (CNS) 159
in connexin 40 ($\alpha_5$) knockout mouse 89
in connexin 43 ($\alpha_1$) knockout mouse 80
in heart 38, 196, 267
in mouse ovarian follicle 227
specific permeability of 62
connexin 46 ($\alpha_3$)
in BHK hamster kidney cells 41–42
cataracts and 84
cleavage 101–103, 110
in connexin 50 ($\alpha_8$) knockout mouse 237
in heart 267
knockout mouse 3, 84
cataractogenesis and 239
connexin 50 ($\alpha_8$) in 84, 94, 239–240
$\gamma$ crystallin 84
in lens 84, 94, 110
in lens 3, 99–101, 110
in Schwann cells 186
in *Xenopus* oocyte system 101
*see also* connexin 56 ($\alpha_3$)
connexin 50 ($\alpha_8$) 239–240
cataractogenesis 110
cleavage 101–103, 110
in connexin 46 ($\alpha_3$) knockout mouse 84, 94, 239–240
eye size and 84
in heart 267
knockout mouse 84, 110, 237–238
in lens 84, 99–101, 110, 237–238
in *Xenopus* oocyte system 101

connexin 56 ($\alpha_3$) 108
  *see also* connexin 46 ($\alpha_3$)
connexin–protein interactions 262
connexins 6–21
  aequorin and 48
  animal models 3
  antibody studies 8
  in arrhythmias 88–89
  in arterial wall 199
  assembly 58
  bioactive lipids and 3
  in breast cancer 256–257
  cancer therapy and 241–260
  in cardiovascular disease 196–197
  cell adhesion and 255
  in cell cycle 20
  cell growth control and 254
  cell specialization and 94
  cellular expression 6–21, 76–77
  in cerebral cortex 159, 162–164, 171–172, 172
  chimeras 13, 17
  classification 2, 8, 271
  cleavage 101–103, 108, 111, 261
  coexpression 13, 17, 270
  compensatory expression 85, 263
  degradation 57–58
  in diabetic lens 108
  docking 34–37
  domains 158, 212
  domains, extracellular 9, 10, 11, 17, 18
    cysteine residue mutagenesis 34–37
  domains, transmembrane 9–10, 23, 31–32
  dominant negative effects 12–13, 79, 247–248
  expression *in vivo* vs *in vitro* 92
  fibroblast growth factor 4 (FGF4) and 72
  following trauma 93
  genes 76–96
  growth factors and 203
  in heart 38, 39, 194–196, 267
  heteromeric 17
  human genetic diseases and 77–78
  in inner ear 93, 134–156, 136
  interactions 11–12
  intracellular concentrations 46
  intracellular trafficking 45, 46, 51–53, 55, 56–57, 262, 266, 270
    site-directed mutations 12
  knock-in experiments 86
  knockouts 3, 43, 261
  in lens 97–112
  in limb bud 63, 69
  membrane insertion 50
  metabolite transfer 258
  multigene family 2, 6, 8, 41, 76–77, 176
  oligomerization 11, 13, 50
  packing 41–42
  post-translational modification 13, 166–167
  recombinant 26–27
  redundancy 266
  regulation 3, 166–167
  in retina 42–43
  in rodents 45
  specific permeability 62
  structure 910, 41–43, 77, 212–213
  temporal pattern 20
  transcription 166–167, 261
  translational control 91
  tumour suppression and 241–260, 243
  in tumours 245, 255, 257
  turnover within cell 53, 265–266
  *see also* individual connexins
connexons
  assembly 50–51, 262
  classification 158
  connexin 43 ($\alpha_1$) and connexin 32 ($\beta_1$) 17
  connexin 32 ($\beta_1$) and connexin 43 ($\alpha_1$) 16
  connexin 32 ($\beta_1$) and connexin 26 ($\beta_2$) 16, 59
  dominant negative effect 247
  heteromeric 8, 11, 13–14, 16, 20, 50, 268
  heteromeric vs homomeric 58–59
  homomeric 11, 17, 50
  in lens fibre cells 264
  structure 176–178
corticogenesis 159–162
  *see also* cerebral cortex
COS African green monkey kidney cells 48–49
crystallin 110–111
cysteine mutants 35–37
cystic fibrosis channel 50


**D**


deafness 134–135, 148–149, 154
  hereditary non-syndromic sensorineural 78, 135, 152
  Jervel and Lange-Nielsen syndrome 154
Deiters' cells 142
  *see also* inner ear

demyelinating neuropathy *see* X-linked Charcot-Marie-Tooth (CMTX) disease
disulfide bonds 35–37
dominant negative effects 79
  on intracellular trafficking 13
  mutations 17, 18, 215, 245–250
dopamine 119, 128–129
*Drosophila* 20–21
ductin 250
dye transfer 158, 182–184, 235

**E**

embryo 60–75
  amphibian 61–62
  brain 90–91
  chick 222–223
  connexin 43 ($\alpha_1$) 213
endocochlear potential 146–148
  connexin 26 ($\beta_2$) and 149
  potassium circulation 148
endoplasmic recticulum 17, 46
epidermal growth factor (EGF) 75
extracellular domains 34–37
  $\beta$ zip 35–37

**F**

fibroblast growth factor 2 (FGF2) 167
fibroblast growth factor 4 (FGF4) 62–69, 70
  anterior mesenchyme and 65–66, 67–68
  apical ectodermal ridge (AER) and 65
  cell polarization and 75
  connexin 32 ($\beta_1$) and 65–66
  connexin 43 ($\alpha_1$) and 65–66
  mesenchyme cell communication 66–68
  posterior mesenchyme and 65–66, 68–69
  *Sonic hedgehog* (*Shh*) and 65, 69, 70, 73
  in undifferentiated cells 66
folliculogenesis 227
  *see also* ovarian follicle, mouse

**G**

$\gamma$ crystallin 84, 237, 239
ganciclovir 250
ganglion cell dendrites 116–117
gap junction proteins *see* connexins
gap junctions 7
  amino acid transmitters 123
  in arthropods 7, 20–21
  assembly pathways 46–49
  asymmetry 121, 129, 165
  in atrial cells 38–39
  in cerebral cortex 157–174
  in cochlear connective tissue 143–148
  in cone bipolar cells 118
  in crayfish 6
  in development 60–75
  in diabetic cataractogenesis 106
  dopamine and 128–129
  electron cryo-crystallography of 22–38
  exocrine secretion and 82
  fibroblast growth factor 4 (FGF4) and 65–66
  formation 44–59
  function 176–178
  in ganglion cell dendrites 116–117
  growth factors and 60–75, 167
  in heart 22–38, 23
  heteromeric vs homomeric 262
  heterotypic channels 270–271
  in inner ear 93, 134–156
  in lens 98–103
  in limb bud 69
  molecular heterogeneity of 25–26
  neuronal firing and 166
  neurotransmitter coupling 121–123
  non-connexin proteins in 20–21, 53
  oleamide and 4, 27–28, 131
  in organ of Corti 139–141
  other signalling pathways and 74–75
  permeability 262–263
  protein kinase A and 128–129
  in retina 113–133
  in rod signal circuit 118–119
  second messengers 121–123
  selectivity 263–264
  in squid 60
  in stria vascularis 143–148
  structure 1–2, 6–7, 23, 44–45, 158, 176–178, 227
glial cells 32, 129, 130
glial fibrillary acidic protein (GFAP) 164–165
glycine 125–126
Golgi apparatus 17, 46, 50–52, 56–57, 187
  brefeldin A 48–49
graafian vesicle breakdown (GVB) 231–232
granulosa cells 236, 269
  *see also* ovarian follicle, mouse
growth factors 203

**H**


hair cells 137, 139, 142, 143
 *see also* inner ear
heart 88, 193–194, 197, 267–268
 cAMP-dependent protein kinase 216
 connexin 37 ($\alpha_4$) and 267
 connexin 40 ($\alpha_5$) and 38, 267
 connexin 43 ($\alpha_1$) and 38, 80, 213, 216, 267
 connexin 45 ($\alpha_6$) and 38, 267
 connexin 46 ($\alpha_3$) and 267
 connexin 50 ($\alpha_8$) and 267
 in connexin 40 ($\alpha_5$) knockout mouse 88
 development 80, 215, 267
 impulse conduction 194–196
 infarct border zone 210
 intercalated disks 193
 protein kinase C activity 216, 224
 *see also* cardiovascular disease; visceroatrial heterotaxia (VAH)
hemichannels *see* connexons
Hensen's cells 141, 142
 *see also* inner ear
Herpes simplex virus thymidine kinase gene 250
heterotaxia *see* visceroatrial heterotaxia (VAH)
Huntington's chorea 173, 269


**I**


infertility 230, 237
inner ear 134–156
 anatomy 136–139
 antibody studies 136
 cochlear amplifier 137
 connexin 43 ($\alpha_1$) and 136
 connexin 26 ($\beta_2$) and 136, 148–149
 connexin 32 ($\beta_1$) and 136, 139
 Deiters' cells 141
 endolymph 136, 143, 152–153
 hair cells 137, 139, 142, 143
 Hensen's cells 141
 ion transfer 93, 137, 143
 $Na^+/K^+$ ATPase 143, 145
 $Na^+/K^+/Cl^-$ co-transporter 143
 organ of Corti 137
 perilymph 136
 physiology 136–139
 sensory epithelia 139–143
 stria vascularis 137–139
 *see also* cochlea; hair cells; stria vascularis
inositol-1,4,5-trisphosphate ($InsP_3$) 40, 166, 255, 258–259
intracellular trafficking 18, 45–50, 55–57, 178–182, 262, 266, 270
ion transfer 93, 137, 143
*Isk* knockout mouse 154


**K**


knockout mouse 79–85
 connexin redundancy 266
 embryonic lethality 79
 potassium channel 154
 tissue specific 248–250
 tumours 90
 *see also* individual connexins


**L**


lactacystin 57
*lacZ* reporter gene 91–93, 208
lens 97–112, 270
 calpain 108
 in chick embryo 91, 108
 connexin 43 ($\alpha_1$) and 101
 connexin 46 ($\alpha_3$) and 99–101
 connexin 50 ($\alpha_8$) and 99–101
 in connexin 50 ($\alpha_8$) knockout mouse 84, 237–239
 in connexin 46 ($\alpha_3$) knockout mouse 84, 94, 238–239
 crystallin 238
 diabetic 103–106, 108
 fibre cells 3, 97, 99, 101, 264
 $\gamma$ crystallin 84, 237, 239
 $Na^+/K^+$ pumps 99
 pH and 103
 protease activity 111
 in rats 97–98
 solute and water flow 99
 tissue liquefaction 98, 103–106
leupeptin 57
limb bud 62–69
 anterior mesenchyme 65–66
 anteroposterior patterning 62
 connexin 32 ($\beta_1$) and 95
 connexin 43 ($\alpha_1$) and 95
 in connexin 43 ($\alpha_1$) knockout mouse 73–74
 in connexin 43 ($\alpha_1$)–connexin 32 ($\beta_1$) double knockout mouse 94–95

fibroblast growth factor 4 (FGF4) and 67–68
posterior mesenchyme cells 62
signalling systems in 69–70
liver 11, 16–17, 46–49
connexin 26 ($\beta_2$) and 16, 256
connexin 32 ($\beta_1$) and 16, 261
in connexin 32 ($\beta_1$) knockout mouse 93–94
heterotypic channels 270
tumours 81, 82, 257
Lucifer yellow 31–32, 40, 126–127, 263–264
lung 261

**M**

model systems 268–270
mouse, knockouts 76–96
*see also* individual connexins
multistage skin cancer model 257
mutant channels, properties 265
myelin sheath 182–184
*see also* Schwann cells

**N**

N2A cells 17
neocortical circuit formation 162–166
*see also* central nervous system (CNS)
neural crest cells 94, 208
Neurobiotin 118–119, 126–127, 230–231
neuroblasts 164–165
neurogenesis 161–162
nitric oxide 128
nocodazole 48–49, 57
non-polyposis colon cancer 244
*see also* cancer therapy

**O**

oestrogen 91–92
oleamide 3–4, 27–28, 32–34, 131, 263
*see also* anandamide; bioactive lipids
oocyte 228, 270
meiotic competence 228, 231–232, 259
*see also* ovarian follicle, mouse
Opus protein family 20–21
organ of Corti 137, 139–142, 152
cell culture 154
connexin 26 ($\beta_2$) and 139–141
ovarian follicle, mouse 226–240
antibody studies 228–230
connexin 37 ($\alpha_4$) and 227, 228–230
connexin 40 ($\alpha_5$) and 228–230
connexin 43 ($\alpha_1$) and 227, 228–230
connexin 45 ($\alpha_6$) and 227
development of 226–227
graafian vesicle breakdown (GVB) 228
at puberty 227
*see also* folliculogenesis; granulosa cells; oocyte
ovulation 83

**P**

pertussis toxin (PTX) 129
pH
gating 109, 110
lens fibre cells and 101
placenta 90, 155
polysplenia syndrome 238
post-translational modification 42, 119, 166–167, 261, 264
posterior mesenchyme, limb bud 64–65, 72–73
fibroblast growth factor 4 (FGF4) and 65–66, 68–69
progesterone 91–92
protein kinase A 128–129
protein kinase C 223
proteosomes 57–58
Purkinje system 89

**R**

retina 113–133
cat 117, 118
cellular coupling 42–43
channel gating 132
connexin 37 ($\alpha_4$) and 127–128
connexin 40 ($\alpha_5$) and 127–128
connexin 43 ($\alpha_1$) and 127–128
connexin 26 ($\beta_2$) and 127–128, 152
connexin 32 ($\beta_1$) and 127–128
dye transfer 126–127
fish 117–118, 121, 128
a ganglion cells 127
gap junction modulation 118
neuronal architecture 114
neuronal coupling 115–118
rabbit 119
salamander 117
retinoblastoma 244
rod amacrine cells 116, 118–121

rod amacrine cells (*cont.*)
  ambient illumination and coupling activity 119–120
  cat retina 118
  cone bipolar cells and 120, 123, 125–126, 128
  distribution 119
  Neurobiotin 118–119

**S**

Schwann cells 77–78, 178, 182–184, 186
  intracellular trafficking 56
  radial pathway 184
  *see also* myelin sheath
Shaking B protein 20–21
sinoatrial node 38–39
skate 8
skin 261, 269
smooth muscle cells 199, 201, 209
*Sonic hedgehog* (*Shh*) 65, 69, 70, 73
squid 60
stria vascularis 143–148
  strial astrocytes 130
  strial basal cells 148
  vascular structure 153
  *see also* inner ear

**T**

teeth 74, 94–95
thrombin 203
thyroid gland 12, 42, 74, 94–95
TPA (12-*O*-tetradecanoylphorbol 13-acetate) 223, 243
trafficking pathways 56
transcriptional regulation 91–92
tumour suppression 241–260
  connexin 40 ($\alpha_5$) gene and 243
  connexin 43 ($\alpha_1$) gene and 243
  connexin 26 ($\beta_2$) gene and 243, 247
  connexin 32 ($\beta_1$) gene and 243
  dominant negative mutations 245–250
  germ cell mutations 244–245
  hypermethylation 245
  *p53* 245
tumours 256–257
  connexin 32 ($\beta_1$) and 19
  liver 81, 82, 257
  pancreatic 82

**U**

uncoupling agents 32

**V**

VAH *see* visceroatrial heterotaxia (VAH)
ventricular conduction 83–84, 89
ventricular myocardium 193–194
*viable dominant spotting* mutant mouse strain 145–146, 147, 153
visceroatrial heterotaxia (VAH) 19
  connexin 43 ($\alpha_1$) and 78, 213, 219–220
  definition 219–220, 222
  Linda Loma subtype 220, 223
  *see also* congenital heart defects; heart
voltage-dependent gating 263, 264

**W**

Wolff-Parkinson White syndrome 196–197

**X**

X chromosome inactivation 180
X-linked Charcot-Marie-Tooth (CMTX) disease 51–53, 171, 175–187
  connexin 32 ($\beta_1$) and 77–78, 81, 186–187, 245, 266
  liver tumours and 82
  phenotypic variation 187
  sural nerve biopsy 186–187
  *see also* Charcot-Marie-Tooth (CMT) disease
*Xenopus* oocyte system 62, 74, 101, 266
  connexin 32 ($\beta_1$) mutation studies 178
  Shaking B protein 20–21

**Z**

ZO-1 42, 254–255

*Other Novartis Foundation Symposia:*

No. 209 **Oligonucleotides as therapeutic agents**
Chair: M. Caruthers
1997 ISBN 0-471-97279-7

No. 214 **Epigenetics**
Chair: A. Wolffe
1998 ISBN 0-471-97771-3

No. 225 **Gramicidin and related ion channel-forming peptides**
Chair: B. Wallace
1999 ISBN 0-471-98846-4